STUDENT STUDY GUIDE
to accompany

CHEMISTRY

The Molecular Nature
of Matter
and Change

by MARTIN SILBERBERG

Elizabeth Bent Weberg

 Mosby

St. Louis Baltimore Boston Carlsbad Chicago Naples New York Philadelphia Portland
London Madrid Mexico City Singapore Sydney Tokyo Toronto Wiesbaden

Mosby
Dedicated to Publishing Excellence

A Times Mirror Company

Printed in the United States of America

Mosby–Year Book, Inc.
11830 Westline Industrial Drive
St. Louis, MO 63146

ISBN 0-8151-8018-7

A problem
that's not a problem
isn't a problem

Henry A. Bent

Acknowledgments

I thank Professor Arlan D. Norman, University of Colorado, for getting me involved in this project and for his helpful advice during the project. I also thank Paul Hunter and Joe Branch for providing practice problems for the ends of the study guide chapters.

I gratefully acknowledge the helpful comments of the following reviewers: Rolf T. Weberg, Henry A. Bent, Anne M. Bent, Stacey F. Bent, and Brian E. Bent. I thank my father, Henry Bent, for his suggestions and comments, many of which have been incorporated throughout the study guide (for example, the treatment of temperature scales in chapter 1, relating electron configurations to a periodic table in chapter 8, bonding models in chapter 9, and practice problems at the end of chapter 13). Finally, I thank my husband, Rolf Weberg, for the many hours he spent formatting equations and reviewing this document.

Contents

CHAPTER 1

Keys to the Study of Chemistry

TOPICS:

- Matter and Energy
- Conversion factors
- Measurements
- Significant digits
- Precision and Accuracy

SECTION 1.1: Chemistry, Matter, and Energy

Chemistry is the science that deals with the specific properties, especially the composition and structure, of matter and its transformations. It isn't a chem-mystery! We are all familiar with matter – the stuff that is all around us. What you may not be as familiar with are the terms that chemists use to describe matter and the way in which chemists view matter as composed of tiny particles in random motion. Chemistry is a <u>physical</u> science because it deals with "physical quantities" – things that can be expressed by a number followed by a unit.

A Chemist's Concern:

Which of the following would likely concern a chemist?

A) Water, like all matter, falls a distance of $1/2gt^2$ during a given time, t.
B) Water has a density of 1.0 g/mL at 25°C and 1 atm.
C) Water can be electrolytically decomposed into the elements hydrogen and oxygen.
D) The formula for water is H_2O
E) The structure of a molecule of water is:

$$H^{-O}{}^{-}H$$

F) Water reacts violently with sodium metal to produce aqueous sodium hydroxide and hydrogen gas.
G) Water is one of the products when natural gas burns.
H) Water is the only product of combustion of hydrogen.
I) Water boils at about 90°C at an elevation of 10,000 ft.
J) Salt water boils at a higher temperature than pure water.
K) A cupful of water molecules as a liquid occupies a volume of approximately 100 gallons at 100°C and atmospheric pressure, as a gas.
L) Water in rivers can erode land to form canyons.
M) Hard water contains calcium and magnesium ions.
N) Water is a good solvent for table salt and for table sugar.
O) Water and oil do not mix.
P) Water expands when it freezes.
Q) Water is the chief constituent of our bodies.

All of the above might concern a chemist except for "A", "L", and "Q" which would be of more concern to a physicist, geologist, and biologist respectively.

Properties of matter:

Matter is anything that has mass and occupies a volume. We describe matter by its **physical** and **chemical properties**. The first time you filled a car with gasoline, you most likely made observations about the properties of gasoline. First, as you began to pump the gasoline into the car, you may have noticed that gasoline has a distinct odor. When you overfilled the tank, and the gas spilled down the car, you probably noticed it was a clear liquid, but it (fortunately) did not dissolve the paint off your parent's car. As the gas landed in a puddle of rainwater, you might have noticed that it floated on top of the water. You noticed the sign that asked you to stop your engine and refrain from smoking, so you speculated that gasoline is flammable.

Gasoline's odor, color, and density are some of its **physical properties** because they do not involve interaction with other substances. The flammability of gasoline is a **chemical property** because it involves interactions with other substances. To react chemically, and rearrange their atoms, molecules must collide physically. When gasoline ignites, it reacts with oxygen to form new substances including carbon dioxide and water. This is an example of a **chemical change** or **chemical reaction**.

A **physical change** occurs when a substance changes state but not composition. When water boils or freezes, it undergoes a physical change because its composition remains H_2O whether it is ice, liquid water, or steam. A **chemical change** or **chemical reaction** occurs when a substance is converted into a different substance. When iron rusts, it reacts with oxygen to form iron oxide. A chemical change occurs; iron oxide is a substance with a set of properties different from iron.

The states of matter:

Broadly speaking, there are three ways molecules may be arranged:

Touching & ordered

Touching & disordered

Not touching

States of matter exercise:

A) The above arrangements of molecules represent the three phases of matter called:

_____   _____   _____.

B) For the most common compound, water, these three phases of matter have the special names:

_____   _____   _____.

C) The Greeks gave these three states of matter the names:

_____   _____   _____.

answers:

A) solid, liquid, gas; B) ice, water, steam; C) earth, water, air

(Note that the fourth type of matter that the Greeks classified, fire, is the thermal energy (heat) absorbed or liberated in passing from one state of matter to another).

Energy:

What _is_ energy? When we say we feel energetic, or we want a candy bar for quick energy, we have an intuitive idea that energy is something that makes us move and get things done. In fact, we define mechanical energy as the capacity to do work. We define work as the distance an object moves times the force acting on the object (W=f x d).

Energy of mechanical systems (systems in which there are no changes in temperature or chemical composition) is divided into potential energy (energy of position) and kinetic energy (energy of motion). When we hold a ball above the ground, it contains potential energy which is converted to kinetic energy as the ball is dropped. The sum of the potential energy and kinetic energy for the ball as it falls is constant:

$$E_{kinetic} + E_{potential} = constant.$$

As the ball comes to rest, the energy is not lost, it is converted to other forms through **work** and **heat**. Thermochemistry, the study of heat transferred in chemical reactions, will be discussed in Chapter 6. We know from experience that the energy of the universe is constant; energy cannot be created or destroyed. Mathematically we write:

$$E_{univ.} = constant$$

Therefore:

$$(E_{univ.})_{final} = (E_{univ.})_{initial}$$
$$(E_{univ.})_{final} - (E_{univ.})_{initial} = 0$$
$$\Delta E_{univ.} = 0$$

where ΔE_{univ} means the difference between E_{final} and $E_{initial}$.

SECTION 1.2: Chemical Problem Solving

The Scientific Approach:

The scientific method is a series of steps often used to solve problems. You have probably used the scientific method in daily life without analyzing the steps. Let's consider a problem you might face, and then notice the steps you go through as you solve the problem.

You commute to class on a bicycle and get a flat rear tire. An unobservant person might never notice the tire is flat and simply wonder why the hills seem so tough on the way home. However, you **observe** that the tire is flat and you remember seeing glass on the road. You **hypothesize** that the tire is flat because glass punctured the inner tube.

You place the inner tube in water (**experiment**) and observe bubbles which supports your hypothesis that glass punctured the inner tube. You patch the inner tube, and ride your bike to class the next day. Again, you obtain a flat rear tire. You hypothesize that your patch leaked so you check the inner tube again in water (experiment). Bubbles escape from the patch, so you hypothesize that perhaps the glue to your patch kit was old, and to avoid patching problems, you replace the inner tube with a new one.

After riding your bike, <u>again</u> the rear tire is flat. This time you notice that the tube always punctures in approximately the same place. A **model** begins to emerge that something on the bike must be causing the flat tires. You hypothesize that there is a sharp object inside the tire, so you remove the tire and check it carefully for a foreign object. Finding nothing, you continue to check the bike and notice that there is no rim tape, and a spoke pokes through the rim in the same place you found the punctures in the inner tubes. You apply rim tape and ride to class the next day. After class, you return to your bike to find that the FRONT tire is flat. The next day you walk to class and order Kevlar® inner tubes.

You put the scientific method to work: you made **observations**, formed **hypotheses** via imagination, knowledge, and logic to explain the observations, ran **experiments** to support or disprove the hypotheses, and eventually formed a **model** based on experiments. Even though you solved the immediate problem, you determined that new technology was needed to avoid future problems.

Units and Conversion factors:

When we measure a physical quantity, we must always express a number <u>and</u> a unit for it to be meaningful. "I was doing 60" may be understood to mean 60 <u>miles per hour</u> in America, but in Germany, "doing 60" would be poking along at about 37 mph, since Germans measure speed in kilometers per hour. A physical quantity, is therefore, a quantity that can be expressed by the product of a <u>pure number</u> and a <u>physical unit</u>:

$$\text{physical quantity} = \text{pure number} \times \text{physical unit}$$

For example, if the mass of an average college student is 140 lbs, we write:

$$m_{avg. college student} = 140 \times lb$$

We usually leave out the multiplication sign and write:

$$m_{avg. college student} = 140 \ lb$$

Other examples are:

$$m_{avg. water molecule} = 18.0 \ amu \ \text{(abbreviation for atomic mass units)}$$

$$m_{avg. O atom} = 16.0 \ amu$$

$$m_{avg. H atom} = 1.008 \ amu$$

$$m_{avg. 2H atoms} = 2.016 \ amu$$

Physical Quantities:

Which of the following concepts are "physical quantities" (expressible by a pure number and a physical unit)?

a) distance, area, volume

b) speed, acceleration, momentum

c) force, energy, power

d) density, viscosity, compressibility

e) temperature, pressure, specific heat

f) beauty, love, virtue

<u>answer</u>: all but "f".

Conversion factors convert one unit to another but do not change the physical quantity because a conversion factor equals 1. If we take an equation such as:

$$1 \text{ lb} = 454 \text{ grams}$$

we can obtain two conversion factors by:

A) dividing both sides of the above equation by 454 g:

$$\frac{1 \text{ lb}}{454 \text{ g}} = \frac{454 \text{ g}}{454 \text{ g}}$$

The conversion factor is:

$$\frac{1 \text{ lb}}{454 \text{ g}} \quad \text{which equals 1.}$$

B) dividing both sides of the first equation by 1 lb:

$$\frac{1 \text{ lb}}{1 \text{ lb}} = \frac{454 \text{ g}}{1 \text{ lb}}$$

The conversion factor is:

$$\frac{454 \text{ g}}{1 \text{ lb}} \quad \text{which equals 1.}$$

If we multiply a physical quantity by a "conversion factor", the pure number and physical unit will change, but the physical quantity will not change since we multiply by 1. The use of conversion factors in calculations is often referred to as the **unit-factor** method or **dimensional analysis**.

How to solve chemistry problems:

If the average college student weighs 140 lb, how much does the average college student weigh in kilograms?

Often you will not know how to solve a problem at the outset – in fact, for real problems, we never do. Don't panic! Instead of concentrating on what you don't know, <u>concentrate on what you do know</u>. The first step to solving a problem is to:

1) Write down what you know as mathematical equations

This exercise will serve a couple of purposes. It gets you started (often the hardest part), and it will give you a chance to think about tools you may need to solve the problem.

We know that "the average college student weighs 140 lb", or:

$$m_{avg.\ college\ student} = 140\ lb$$

2) Write down what you want to know

This will remind you of where the problem eventually needs to end up.

$$m_{avg.\ college\ student}\ in\ kg = ?$$

3) Write down equations you know that relate to the given information

Even if you do not have a specific road map in mind, writing down conversions or equations you might use will eventually show you how to link the known to the unknown.

Equalities we could use to convert from lb to kg are:

$$1\ lb = 454\ g$$
$$1\ kg = 1000\ g$$

4) Perform calculations

We convert 140 lb first to grams and then to kilograms. The conversion factors must contain the unit <u>you wish to convert to</u> in the numerator so that the unit <u>you wish to convert from</u> will be in the denominator and cancel out:

$$140\ lb \times \frac{454\ g}{1\ lb} \times \frac{1\ kg}{1000\ g} = 63.6\ kg$$

5) Check the answer to see if it makes sense

Are the units correct? Is the answer approximately the right size? A kilogram is larger than a pound, so our average college student should weigh fewer kilograms than pounds: 63.3 is smaller than 140, so the answer appears to be correct.

A few hints:

• Include units when working problems. Take the time to write them legibly so you won't make careless errors when canceling units that appear in both the numerator and denominator. Sometimes all this unit writing may seem cumbersome, but it is a good way to avoid mistakes.

• If you get through the first two steps and cannot see a route to the answer, relax! Play around with the information given, and it will often become clear how to obtain the answer.

• Teach the material to someone else (or explain it to yourself out loud). This chemist found that she didn't really understand general chemistry until she was a teaching assistant for general chemistry.

• Don't work problems as quickly as you can, but try working fewer problems more slowly. Work problems more than one way. Don't make the goal of a problem set to "get all the right answers", but to understand what the answers mean.

• Remember that mathematical equations, which may look "foreign" are just a short-hand way to say something that most likely takes up lots of space if written with words. Try to get into the habit of translating short-hand (math) into long-hand (English) in your mind. Eventually you may reach the point where you no longer need to translate into English before the meaning is clear.

• Review! If you can work problems again the next day with no notes, chances are you understand the material and will be able to apply it for an exam.

How to solve chemistry problems (continued):

What other information could we obtain from the problem worked above?

We could determine the number of lb in a kg by using our relationship that:

$$140 \text{ lb} = 63.6 \text{ kg}$$

Dividing each side of the above equation by 63.3 kg gives:

$$\frac{140 \text{ lb}}{63.6 \text{ kg}} = \frac{63.6 \text{ kg}}{63.6 \text{ kg}} = 2.2 \frac{\text{lb}}{\text{kg}}$$

Alternatively, we could start with a kilogram, and use the knowledge that there are 1000g in a kg, and 454 g in a pound to obtain:

$$1 \text{ kg} \times \frac{1000 \text{ g}}{1 \text{ kg}} \times \frac{1 \text{ lb}}{454 \text{ g}} = 2.2 \text{ lb}$$

Conversion problem:

1 gram of water at 25°C and atmospheric pressure occupies a volume of 1 cm^3 (about a thimble full). How much volume does it occupy in cubic inches? (Use the conversion factor 1 inch = 2.54 cm)

solution:

$$1 \text{ cm}^3 \times \left(\frac{1 \text{ in}}{2.54 \text{ cm}}\right)^3 = 0.0610 \text{ in}^3$$

Note: be sure to cube the entire conversion factor.

SECTION 1.3: Measurement in Scientific Study

SI Units:

The seven fundamental units of the SI system are based on the metric system. We use decimal prefixes and scientific notation to write quantities that are much larger or smaller than the base units. We will discuss the five fundamental SI units that will be used in this study guide (mass, length, time, temperature, and amount). We will also discuss density and volume whose units are derived from the fundamental units of length and mass.

Mass

The SI unit for mass is the kg. A kilogram is a little more than twice the size of a pound.

$$1 \text{ kg} = 2.2 \text{ lb}$$

Conversion between grams and pounds:

How many grams are in 1 pound?

solution:

$$1 \text{ lb} \times \frac{1 \text{ kg}}{2.2 \text{ lb}} \times \frac{1000 \text{ g}}{1 \text{ kg}} = 454 \text{ g}$$

Mass and weight are not synonyms. Mass refers to a quantity of matter; it does not change depending on where it is measured. Weight is dependent on the quantity of matter and the gravitational field pulling on it. Your weight is different on the moon than on earth. Your mass is the same on any planetary body.

Length

The SI unit for length is the meter. A meter is a little longer than a yard. A centimeter (1/100th of a meter) is a little smaller than half an inch.

$$1 \text{ inch} = 2.54 \text{ cm}$$

Conversion between yards and meters:

How many yards are in 1 meter?

<u>solution</u>:

$$1 \text{ m} \times \frac{100 \text{ cm}}{1 \text{ m}} \times \frac{1 \text{ in}}{2.54 \text{ cm}} \times \frac{1 \text{ yd}}{36 \text{ in}} = 1.09 \text{ yd}$$

A meter is slightly longer than a yard.

Conversion between angstroms and nanometers:

Two units frequently used in chemistry to express dimensions useful on an atomic scale are the angstrom ($1 \text{ Å} = 10^{-8}$ cm or $1 \text{ Å} = 10^{-10}$ m) and the nanometer ($1 \text{ nm} = 10^{-7}$ cm or $1 \text{ nm} = 10^{-9}$ m). How many angstroms are in one nanometer?

<u>solution</u>:

$$1 \text{ nm} \times \frac{10^{-7} \text{ cm}}{1 \text{ nm}} \times \frac{1 \text{ Å}}{10^{-8} \text{ cm}} = 10 \text{ Å}$$

An angstrom is 1/10th the size of a nanometer.

Temperature

Temperature is an important physical quantity in chemistry because it affects chemical reaction rates and chemical stability. The three temperature scales you will encounter and their relationships to one another are sketched on the next page.

Comments about the temperature scales:

- The Kelvin scale is also known as the **absolute scale** because 0 degrees K is based on absolute zero, the hypothetical temperature at which all molecular motion would cease.

- A degree Kelvin is the same size as a degree Celsius; a Celsius or Kelvin degree is 1.8 (or $180/100 = 9/5$) times larger than a Fahrenheit degree.

- Absolute temperature is referred to with a capital "T". It does not require the degree sign: the freezing point of water on the absolute scale is $T = 273.15\,K$.

Warm-up problem:

What is the temperature on the Celcius scale of a thermal state halfway between the normal boiling point and the normal freezing point of water ($t_{1/2}$)? On the Fahrenheit scale? On the Kelvin scale?

solutions:

Celsius: $t_{1/2} = 50°C$

Fahrenheit: $t_{1/2} = 122°F$

Kelvin: $t_{1/2} = 323.15\,K$

How did you get 122°F? Probably you took the difference between the freezing and boiling points (212-32=180) and divided by 2 (180/2 = 90), and added it to 32 (90+32 = 122). Or, you might have taken the average of 212 and 32: $\frac{(212+32)}{2}$ = 122.

How can we show algebraically that $t_{1/2}$ is half-way up the temperature scales? We take the ratio of the difference in degrees between $t_{1/2}$ and t_{nfp} to the difference in degrees between t_{nbp} and t_{nfp}:

Fahrenheit: $\dfrac{t_{1/2}°F - 32°F}{212°F - 32°F} = \dfrac{122°F - 32°F}{180°F} = \dfrac{1}{2}$

Celsius: $\dfrac{t_{1/2}°C - 0°C}{100°C - 0°C} = \dfrac{50°C - 0°C}{100°C} = \dfrac{1}{2}$

Kelvin: $\dfrac{t_{1/2}K - 273.15K}{373.15K - 273.15K} = \dfrac{323.15K - 273.15K}{100K} = \dfrac{1}{2}$

We can generalize this formula for any temperature:

$$\dfrac{t_F - 32°F}{180°F} = \dfrac{t_C}{100°C} = \dfrac{T - 273.15K}{100K}$$

Temperature conversion practice:

Rearrange the equations in bold above to obtain equations which will a) convert from Fahrenheit to Celsius, b) convert from Celsius to Fahrenheit, c) convert from Celsius to Kelvin.

solutions:

a) We will use the equations containing t_C and t_F, and solve for t_C:

$$\dfrac{t_F - 32°F}{180°F} = \dfrac{t_C}{100°C}$$

$$t_C = \dfrac{100°C}{180°F} \times \left(t_F - 32°F\right) = \dfrac{5°C}{9°F} \times \left(t_F - 32°F\right)$$

Check: If t_F = 122°F, what is t_C?

$$t_C = \dfrac{5°C}{9°F} \times \left(122°F - 32°F\right) = 50°C$$

b) We will use the equations same equations as part "a", and solve for t_F:

$$t_F = \left(\frac{180°F}{100°C} \times t_C\right) + 32°F = \left(\frac{9°F}{5°C} \times t_C\right) + 32°F$$

Check: If $t_C = 50°C$, what is t_F?

$$t_F = \left(\frac{9°F}{5°C} \times 50°C\right) + 32°F = 122°F$$

c) We will use the equations containing t_C and T, and solve for T:

$$\frac{t_C}{100°C} = \frac{T - 273.15K}{100K}$$

$$T = (t_C/°C + 273.15)K$$

Check: If $t_C = 50°C$, what is T?

$$T = (50°C/°C + 273.15)K = 323.15K$$

Additional checks for temperature conversions:

A) If $t_C = 0°C$ (the freezing point of water), what is t_F?

$$t_F = \left(\frac{9°F}{5°C} \times 0°C\right) + 32°F = 32°F$$

B) If $t_C = 100°C$ (the boiling point of water), what is t_F?

$$t_F = \left(\frac{9°F}{5°C} \times 100°C\right) + 32°F = 212°F$$

C) If $t_F = 32°F$, what is t_C?

$$t_C = \frac{5°C}{9°F} \times \left(32°F - 32°F\right) = 0°C$$

D) If $t_F = 212°F$, what is t_C?

$$t_C = \frac{5°C}{9°F} \times \left(212°F - 32°F\right) = \frac{5°C}{9°F} \times 180°F = 100°C$$

E) If $t_C = 0°C$, what is T?

$$T = (0°C/°C + 273.15)K = 273.15 \ K$$

F) If t_C = 100°C, what is T?

$$T = (100°C/°C + 273.15)K = 373.15 \text{ K}$$

Temperature conversion problems:

A) What is 98.6°F (body temperature) on the Celsius scale?

$$t_C = \frac{5\,°C}{9\,°F} \times \left(t_F - 32°F\right) = \frac{5\,°C}{9\,°F} \times \left(98.6°F - 32°F\right) = 37.0°C$$

B) What is 25°C (room temperature) on the Fahrenheit scale? on the absolute scale?

$$t_F = \left(\frac{9\,°F}{5\,°C} \times t_C\right) + 32°F = \left(\frac{9\,°F}{5\,°C} \times 25°C\right) + 32°F = 77°F$$

$$T = \left(\frac{t_C}{°C} + 273.15\right) K = \left(25 + 273.15\right) K = 298.15 K$$

C) When is $t_F = t_C$?

Let $t_F = t_C = t$ and substitute into the equation for t_F:

$$t = \left(\frac{9}{5} \times t\right) + 32$$

Solving for t gives:

$$t = \left(-\frac{5}{4}\right) \times 32 = -40 = t_F = t_C$$

So −40°F = −40°C.

D) What is absolute zero (T=0) on the Celsius and Fahrenheit scales?

$T = (t_C/°C + 273.15)K$, therefore, $t_C = (T/K - 273.15)°C$

$t_C = (0/K - 273.15)°C = -273.15°C$

$$t_F = \left(\frac{9\,°F}{5\,°C} \times t_C\right) + 32°F = \left(\frac{9\,°F}{5\,°C}\right) \times \left(-273.15°C\right) + 32°F = -459.67°F$$

Amount:

The SI unit for amount is the mole. "Amount" can be a confusing term, because we normally also think of mass and volume as amounts. You can think of the SI definition of amount as a number of objects, or a populationofentities. A mole = 6.02×10^{23}, a very large number! We work with moles of atoms or molecules because we need a huge number of these minuscule particles in order to measure them conveniently in the laboratory. Chapter three will discuss the origin of the mole and show why it is an extremely useful number.

Conversion between grams and atomic mass units:

Given that 6.02×10^{23} amu = 1 g, and the average mass of a hydrogen atom is 1.008 amu ($m_{avg.\ H\ atom}$ = 1.008 amu), what is the mass in grams of 6.02×10^{23} H atoms?

solution:

$$\frac{1.008\ amu}{H\ atom} \times \frac{1\ g}{6.02 \times 10^{23}} \times \frac{6.02 \times 10^{23}\ H\ atoms}{mole\ H\ atoms} = \frac{1.008\ g}{mole\ H\ atoms}$$

Notice that the mass of one mole of hydrogen atoms (in grams) is the same as the average mass of one hydrogen atom (in atomic mass units).

Common Prefixes used in the SI System:

Prefix	Symbol		Meaning		Exponential notation
mega	m	=	1,000,000	=	10^6
kilo	k	=	1,000	=	10^3
hecto	h	=	100	=	10^2
deka	da	=	10	=	10^1
–	–	=	1	=	10^0
deci	d	=	0.1	=	10^{-1}
centi	c	=	0.01	=	10^{-2}
milli	m	=	0.001	=	10^{-3}
micro	μ	=	0.000001	=	10^{-6}
nano	n	=	0.000000001	=	10^{-9}

Summary of SI units:

Physical quantity	Quantity's symbol	Name of SI unit	Abbreviation	Other common units
Mass	m	kilogram	kg	gram, pound, ton
Length	l	meter	m	foot, yard, mile, km
Time	t	second	s	minute, hour
Temperature	T	Kelvin	K	mK (milliKelvin)
Amount	n	mole	mol	dozen, gross

Volume:

Volume is <u>derived</u> from the SI unit for length, the meter. A cube that measures one meter on each edge has a volume of $(1m)^3 = 1\ m^3$. A cubic meter is pretty large. In chemistry we generally base volume on a cubic decimeter which is also called a liter (L). A liter is slightly larger than a quart. We often do not use entire liters of solutions in the laboratory, so in these cases we measure volume in milliliters (mL). A mL is one thousandth the size of a liter. One milliliter equals one cubic centimeter (cm^3).

The equivalence of a mL and a cm³:

$1\ cm^3$ means: $1\ (cm)^3$

$1\ (cm)^3 = (10^{-2}m)^3 = (10^{-2})^3 m^3 = 10^{-6} m^3$

$1\ mL = 10^{-3}(dm)^3 = 10^{-3}(10^{-1}m)^3 = 10^{-3}(10^{-1})^3 m^3 = 10^{-3} \times 10^{-3} m^3 = 10^{-6} m^3$

Volume determination by water displacement:

The volume of irregularly shaped solids are often determined from the volume of water they displace. A graduated cylinder contains 15.50 mL of water. When a nickel is added, the volume increases to 16.05 mL. What is the volume of the nickel in cm^3 and in^3?

<u>solution:</u>

The change in volume of the water in the graduated cylinder must be due to the nickel's volume. The volume change is:

$$\Delta V = \text{volume after} - \text{volume before}$$
$$= 16.05 \text{ mL} - 15.50 \text{ mL}$$
$$= 0.55 \text{ mL}$$

Since a mL = cm^3, the volume of the nickel in cubic centimeters is 0.55 cm^3. We will use the relationship 1 inch = 2.54 cm to find the volume of the nickel in inches:

$$0.55 \text{ cm}^3 \times \left(\frac{1 \text{ inch}}{2.54 \text{ cm}}\right)^3 = 0.034 \text{ in}^3$$

You measure the diameter of a nickel to be 2.10 cm and you measure its thickness to be 0.18 cm. Calculate the volume of the nickel from these measurements.

$$V = \text{area} \times \text{thickness}$$

The area of a circle is πr^2, so we can write:

$$V = \pi r^2 \times \text{thickness}$$

$$V = \pi \times (2.10 \text{ cm}/2)^2 \times 0.18 \text{ cm}$$

$$V = 0.62 \text{ cm}^3$$

Compare this volume to the volume obtained by the water displacement method (0.55 cm^3). Which is likely to be more accurate? (The accuracy of each measurement depends upon the accuracy of your measuring devices, but water displacement is the potentially more accurate method since the nickel could have rounded edges and other deviations from a perfect cylinder).

Density

Density, _derived_ from the SI units for mass and volume, is the mass per unit volume of a substance.

$$d = \frac{m}{V}$$

The SI unit for density is therefore kg/m^3. However, in chemistry, we generally use g/L or g/mL.

Density is a characteristic physical property of a substance (for a given temperature and pressure). Density does not depend on the amount of a substance present, so we call it an **intensive property**. Volume and mass do depend on the amount of a substance present, so we call them **extensive properties**. The density of water at ordinary pressure and room temperature is 1.00 g/mL. Substances that float in water will have densities less than 1.00 g/mL. Objects that sink in water (such as a nickel) will have densities greater than 1.00 g/mL. We say that muscle tissue is _more dense_ than fat. For a given volume, muscle tissue will weigh more than fat.

Density of a nickel:

A nickel weighs approximately 5 grams. Calculate the density of the nickel discussed above using the volume obtained by the water displacement method.

solution:

$$d = \frac{m}{V} = \frac{5\,g}{0.55\,cm^3} = 9\,g/cm^3$$

Let's compare this number to the density of elemental nickel. The Handbook of Chemistry and Physics lists the density as 8.90 g/mL. Our calculated number of 9 g/mL suggests that a nickel could be elemental nickel. A nickel is actually made from a mixture of 25% nickel and 75% copper. Copper has a density of 8.92 g/mL which is close enough to the density of nickel that our density determination is not accurate enough to distinguish pure nickel from a nickel/copper mixture.

Density of water:

Most cooks have used the relationship "a pint's a pound the world around" to convert between volume and mass. How close is 1 lb/pint to the density of water (1 g/mL)?

solution:

We will use the conversions:

$$1\,L = 1.06\,quart$$

$$1\,lb = 454\,gram$$

$$\frac{1\,lb}{1\,pint} \times \frac{454\,g}{1\,lb} \times \frac{2\,pint}{1\,quart} \times \frac{1.06\,quart}{1\,L} \times \frac{1\,L}{1000\,mL} = 0.96\,g/mL$$

The statement that "a pint's a pound" is a good approximation for the density of water (1.0 g/mL).

SECTION 1.4: Uncertainty in Measurement

Every measurement we make includes some uncertainty. When we weigh ourselves on a bathroom scale, we might tell someone our weight to within a pound, or perhaps, half a pound, but it would not be too believable to claim our weight was 145.259 pounds. An analytical balance in the lab is accurate to .0001 grams. It would be just as unbelievable to claim a reagent measured on this balance weighed 0.1357964 grams. We always estimate the rightmost digit when reading measuring devices, and we assume an uncertainty in this digit.

Significant digits are ones which we record. The exception to this is zeros that are used only to position the decimal point.

Significant digit examples:

How many significant digits are in each of the following?

1) 323 and 3.23

   Both numbers have 3 significant digits because <u>all non zero digits are significant</u>.

2) 3003 and 3.003

   Both numbers have 4 significant digits because <u>zeros between significant digits are significant</u>.

3) 0.00102

   This number has 3 significant digits because <u>zeros used to position the decimal point (those to the left of the first non zero digit) are not significant because they are not measured</u>.

4) 3.30 and 3.00

   Both numbers have 3 significant digits because <u>a zero that ends a number and is to the right of the decimal point is significant</u>.

5) 500

   This number may have 1, 2, or 3 significant digits. <u>A zero that ends a number and is to the left of the decimal point is significant only if it was actually measured</u>. This rule is the one which can cause confusion. We can clarify this problem by writing the number in scientific notation. 5×10^2 has 1 significant digit, 5.0×10^2 has 2 significant digits, and 5.00×10^2 has 3 significant digits. <u>For problems in this text, assume that any zero that ends a number is significant</u>. 500 will be assumed to have 3 significant digits.

Exact numbers are ones which have no uncertainty because they are not measured, but refer to a number of items, or are part of a definition: 4 oranges, 12 molecules, .9144 meters in a yard. <u>Exact numbers never limit the number of significant figures in a calculation because they can be assumed to have an infinite number of significant figures.</u>

Calculators are useful devices for plugging out answers, but too often, it is tempting to write down the whole string of numbers the calculator spews out without evaluating if all the digits are meaningful. Suppose you want to divide the weight of the gear for a camping trip equally among three people, and the total weight of the gear is 64 pounds according to your bathroom scale. You divide 64 by 3 on your calculator and announce that each person will carry 21.333333 pounds. Obviously this answer has exceeded a meaningful number of significant digits. The following rules tell how many significant figures to show in the final answer of a calculation.

Significant digits in calculations:

1) For multiplication and division: The answer contains the same number of significant figures as there are in the measurement with the fewest significant figures.

2) For addition and subtraction: The answer has the same number of decimal places as there are in the measurement with the fewest decimal places.

Rules for rounding off:

1) If the digit removed is more than 5, the preceding number increases by 1 (2.46 becomes 2.5).

2) If the digit removed is less than 5, the preceding number is unchanged (2.43 becomes 2.4).

3) If the digit removed is 5, the preceding number increases by 1 if it is odd and remains unchanged if it is even (2.45 becomes 2.4; 2.35 becomes 2.4)

4) Round off only the final answer in a calculation

Significant digits example problem:

Let's work through the significant digits in the calculation of the density of a nickel described above.

1) To find the volume of the nickel, you measure the volume of water it displaces in a graduated cylinder (initial volume = 15.50 mL, final volume = 16.05 mL).

$$\Delta V = \text{volume after} - \text{volume before}$$
$$= 16.05 \text{ mL} - 15.50 \text{ mL}$$
$$= 0.55 \text{ mL}$$

Both numbers in the calculation contain 2 places past the decimal, so your answer contains two decimal places past the zero.

2) You estimate the mass of a nickel to be 5 g. What is the density of the nickel?

$$d = \frac{m}{V} = \frac{5 \text{ g}}{0.55 \text{ mL}} = 9 \text{ g/mL}$$

Since you only know the mass of the nickel to one significant digit, this value limits the number of significant digits in your answer to <u>one</u> significant digit.

3) If you were to weigh the nickel on an analytical balance and obtain its mass to be 5.0012 g, your answer becomes:

$$d = \frac{m}{V} = \frac{5.0012 \text{ g}}{0.55 \text{ mL}} = 9.1 \text{ g/mL}$$

In this case, the fewest number of significant digits in the density calculation is 2 (the volume measurement), so your answer contains two significant digits.

It is important to be aware of the level of accuracy of measuring devices you use in the lab. What would be the problem with weighing an open beaker of water on an analytical balance? Analytical balances are accurate to 0.0001 grams, and are sensitive enough to measure weight loss due to water evaporation. They are also sensitive enough to measure weight gain due to fingerprints on glassware. Don't hope the weight gains and weight losses will cancel out! It is good practice to use covered flasks and a paper towel or tongs when handling glassware that will be weighed on an analytical balance.

Precision, Accuracy, and Instrument Calibration:

Precision and Accuracy are not synonyms. **Precision** refers to how close measurements in a series are to each other. **Accuracy** refers to how close a measurement is to the real value. You can have extremely precise measurements that are inaccurate. This situation usually occurs if there is a **systematic error** in the measurements that causes them to all read higher or lower than they should. Systematic error can be minimized through calibration of the measuring device.

Random error produces measurements that are both higher and lower than the average value. Random error always occurs, but its size depends on the instrument's precision and the skill of the measurer. Precise measurements have low random error. Accurate measurements generally have low random error as well. In some cases, even when there is high random error, the average value of the measurements may still be accurate.

Precision and Accuracy:

A nickel weighs 5.0012 g on a calibrated balance. Two students weigh the nickel five times and record these weights:

Student #1		Student #2	
#	weight	#	weight
1	4.8002	1	5.1020
2	4.8004	2	4.8882
3	4.8000	3	5.1095
4	4.8002	4	4.9055
5	4.8003	5	5.0012
Avg.	4.8002	Avg.	5.0013

Student #1 produced measurements that were precise, but not very accurate. A systematic error appears to have occurred, such as using a poorly calibrated balance. Student #2 did not have a good day. The measurements were not precise, and although reading #5 and the average were accurate, the size of the random error indicates a need for more work on weighing technique by the measurer.

Chapter 1 Exercises:

1.1) Nail polish remover, which is usually acetone, sometimes ethyl acetate (or a mixture of both), often with glycerol and various fragrances, has some characteristic properties. List some of the properties of nail polish remover and characterize them as chemical or physical properties. For the physical properties, state whether they are intensive or extensive.

1.2) Indicate whether the following processes involve a chemical change or a physical change.
      a. The melting of an ice cube on a hot griddle.
      b. The discharge of a dry cell battery as it runs down.
      c. The formation of carbon on adding sulfuric acid to table sugar.
      d. The cooling of skin as rubbing alcohol evaporates from it.
      e. The rising of bread dough.

1.3) a. Picture a pendulum swinging. At what point on the swing is the potential energy of the pendulum the greatest? At what point is the potential energy the lowest? At what point is the kinetic energy the greatest?

      b. When a glass vase is dropped to the floor, the glass breaks. Describe what happened to the potential energy of the vase.

1.4) The scientific method is an iterative approach to problem solving involving observation, hypothesis, prediction, experiment and observation. Explain how you could use the scientific method to determine why the electricity in your home has suddenly cut off.

1.5) a. If there are 5280 feet in one mile, 0.6214 mile in one kilometer, and 12 inches in one foot, how many inches are there in 10 km? How many strides, a stride being, e.g., 36"?

      b. Calculate the area in square meters of a swimming pool 150 feet long and 45.0 feet wide. (1 foot = 0.3048 meter exactly).

1.6) Calculate the volume in cubic centimeters of a piece of aluminum wire with a diameter of 3.00 mm and a length of 6.25 m.

1.7) How many milliliters are in a 12.0 oz can of soda?

1.8) The largest super tankers have the capacity to carry approximately 6.7×10^5 m^3 of oil. How many barrels of oil will the tanker carry? (Hint: 42 gallons = 1 barrel)

1.9) An empty bottle has a mass of 100 grams. When the bottle is filled with pure water (density 1.00 g/cm^3), the mass of the full bottle is 250 grams. When filled with an unknown liquid, the mass of the full bottle is 320 grams. What is the density of the unknown liquid?

1.10) a. The number presented on an electronic calculator as the answer to the problem 7.23 x 5.19 is 37.5237. How should this number be reported?

b. There are exactly 12 inches in one foot and exactly 2.54 cm in one inch. How many significant figures should be written in the answer to the following problem:

0.2252 ft x 12 inch/ft x 2.54 cm/inch

1.11)   The development of new materials exhibit superconductivity at temperatures that can be maintained with liquid nitrogen. The boiling point of liquid nitrogen is 77.0 K. What is the boiling point in Fahrenheit?

1.12)   Daniel Fahrenheit chose the lowest temperature he could achieve with a mixture of snow, salt, and water as the zero point of his scale. He chose body temperature as another point on his scale and gave it a value of 100 (Linus Pauling suggested, "perhaps he had a slight fever"). As a result, the freezing point of water on the Fahrenheit scale is 32°F, and the boiling point of water is 212°F. The difference between the freezing point and boiling points of water is 180 Fahrenheit degrees.

Suppose you developed a new temperature scale (°N), with a value of 0°N at the boiling point of liquid nitrogen (-196°C), and a value of 100 for the freezing point of water. What would normal body temperature (37°C) be on your new scale?

Chapter 1 Answers:

1.1)   A chemical property is a property that involves a chemical change.
An extensive property is a property that depends upon the amount of substance.
An intensive property is independent of the amount of substance.

Some of the properties that you might have thought of are:

Smell or odor	Physical	Intensive
Temperature	Physical	Intensive
Ability to dissolve nail polish	Physical	Intensive
Ability to dissolve varnishes and paints	Physical	Intensive
High volatility, evaporates easily	Physical	Intensive
Mass	Physical	Extensive
Volume	Physical	Extensive
Liquid state	Physical	Intensive
Translucent	Physical	Intensive
Density	Physical	Intensive
Ability to mix with water	Physical	Intensive
Flammable	Chemical	

1.2)   a. physical; b. chemical; c. chemical; d. physical; e. chemical

1.3)   a. The potential energy is the greatest at the top of the swing when the pendulum is momentarily stationary. The potential energy is the lowest at the bottom of the swing when the pendulum is moving fastest–this is where the kinetic energy is the greatest.

b. The potential energy is converted to kinetic energy as the vase falls. When the vase hits the ground, some of the kinetic energy is absorbed by the glass – so much so that some of the chemical bonds within the glass break and the vase breaks. Energy is also absorbed by the ground; the ground gets slightly warmer.

1.4)   The scientific method involves a cyclic reasoning process that might have proceeded as follows:

Observation:	lights go out; appliances don't work
Hypothesis:	electricity black-out in entire neighborhood
Prediction:	lights out in neighbors' houses
Experiment:	look at neighbors' houses
Observation:	lights are on
Hypothesis:	problem is in your own house
Prediction:	main fuse blown
Experiment:	replace main fuse
Observation:	lights still don't work; appliances don't work
Hypothesis:	electric line down outside house
Experiment:	...and so on

1.5)   a. 3.9×10^5 inches, 1.1×10^4 strides
b. 6.27×10^2 m^2

1.6)   177 cm^3

1.7)   354 mL

1.8)   4.2 x 10^6 barrels

1.9)   1.47 g/cm^3

1.10)   a. Electronic calculators often display as many digits in an answer to a problem as the display will accommodate. The calculator has no knowledge of the uncertainty in the data given to it. Each number in the product has 3 significant figures and the answer should be written with the same number of significant figures: 37.5.

       b. In this calculation the conversion factors are infinitely precise-there are *exactly* 12 inches in one foot, and *exactly* 2.54 cm in one inch. The answer should have the same number of significant figures as the least precise number involved in the calculation (0.2252)-i.e. four.

1.11)   -321°F

1.12)   Any two points, experimentally reproducible physical phenomena, can be used to define a temperature scale. The two points can be given any convenient numerical values. On your new scale (°N), the two points are: boiling point of liquid N$_2$ = 0°N (corresponds to -196°C), and freezing point of water = 100°N (corresponds to 0°C). The conversion factor for °C to °N is (100°N/196°C).

       An interval of 37°C gives 37°C x (100°N/196°C) = 19°N, so body temperature on the new scale is: 100°N + 19°N = 119°N

CHAPTER 2

The Components of Matter

TOPICS:

- Pure substances (elements and compounds)
- Impure substances (mixtures)
- Mass laws
- Model of an atom
- The Periodic Table
- Ionic and Covalent Bonding
- Naming compounds

SECTION 2.1: Elements, Compounds, and Mixtures

Pure substances:

Choose words from the list below to complete the following paragraph:

(a) atoms

(b) compounds

(c) copper (Cu)

(d) elements

(e) gold (Au)

(f) molecules

(g) same

(h) silver (Ag)

(i) water, (H_2O)

There are two kinds of pure substances. In one kind, the atoms are all the _____ (1). Such substances are called _____ (2). In the other kind of pure substance, all the _____ (3) are the same, but not all the _____ (4) in each molecule are the same. They are called _____ (5). Among the commonest examples of naturally occurring elements are the three "coinage metals": _____ (6) (the most common coinage metal), _____ (7) (the metal used in the manufacture of tableware, mirrors, and jewelry), and _____ (8) (one of the more precious metals, famous for not dissolving in nitric acid). The most common compound, _____ (9), is the major component of most living things.

answers: 1-g, 2-d, 3-f, 4-a, 5-b, 6-c, 7-h, 8-e, 9-i.

It is easy to confuse the terms molecule and compound. Molecule is the more general term that refers to the smallest particle of a substance that retains the properties of the substance, and is composed of one or more atoms. A compound is a special kind of molecule in which not all of its atoms are the same.

Molecules, Compounds, and Elements:

Classify the following as: a) molecules, b) compounds, or c) elements.

1) H_2O 4) S_8 7) N_2

2) O_2 5) Au 8) N_2O_4

3) CO_2 6) $C_{12}H_{22}O_{11}$

answers: 1-a,b; 2-a,c; 3-a,b; 4-a,c; 5-c; 6-a,b; 7-a,c; 8-a,b.

Impure substances:

Choose words from the list below to complete the following paragraph:

(a) aqueous   (e) mixture
(b) gaseous   (f) pure substances
(c) heterogeneous   (g) solution
(d) homogenous

A_____(1) is a combination of pure substances. There are two kinds of mixtures. A _____(2) mixture has visible boundaries; a _____(3) mixture has no visible boundaries. A homogeneous mixture is also called a_____(4). A substance dissolved in water is an_____(5) solution. Air is a _____(6) solution consisting of mainly oxygen and nitrogen molecules. We live in a world of predominately mixtures. Air, the oceans, soil, rocks, and living things all consist of _____(7) mixed together.

answers: 1-e, 2-c, 3-d, 4-g, 5-a, 6-b, 7-f.

Separation techniques:

Lab work often involves separating mixtures so pure substances can be identified. Mixtures can be separated by physical processes including filtration, crystallization, extraction, distillation, and chromatography. You are familiar with mixture separations even if you have never performed them in a chemistry lab.

Determine the type of separation technique used for each of the following processes:

A) Straining orange juice to remove the pulp
B) Pouring hot water over coffee grounds to make a cup of coffee
C) Pouring noodles and hot water in a colander
D) Producing moonshine
E) Making rock candy

answers: A) filtration; B) extraction; C) filtration; D) distillation; E) crystallization

Classification of matter:

Use the appropriate classifications of matter listed below (lettered a–g) to describe the substances (numbered 1–8) that follow:

a) element
b) molecule
c) compound
d) homogeneous mixture (solution)

e) heterogeneous mixture
f) pure substance
g) impure substance

1) Pure air (d,g)

Pure air is a <u>homogeneous mixture</u> or <u>solution</u> of primarily nitrogen and oxygen. (Air that is around us typically contains dust and other particles that cause air to be a <u>heterogeneous mixture</u>).

2) Stainless steel (d,g)

Stainless steel is a <u>solid solution</u> of iron with chromium and sometimes nickel or manganese.

3) Glass (d,g)

Glass is a <u>homogeneous mixture</u> (solution) of silica, or of oxides of boron or phosphorus and a stabilizer.

4) Wood (e,g)

Wood is a <u>heterogeneous mixture</u> of many substances; the proportions of these substances vary depending on the type of wood and where it grows.

5) Iron (a,f)

Iron is an <u>element</u>. It has the symbol Fe.

6) Water (b,c,f)

Pure water is a <u>compound</u> made up of hydrogen and oxygen in a fixed ratio (H_2O). (Drinking water is a <u>solution</u> containing dissolved minerals).

7) Milk (e,g)

Milk is a <u>heterogeneous mixture</u> or a colloidal suspension of many elements and compounds.

8) F_2, Cl_2, Br_2, I_2, H_2, N_2, O_2, P_4, S_8 (a,b,f)

All of the above are <u>elements</u> which exist in nature as <u>molecules</u>. The halogens (F_2, Cl_2, Br_2, I_2), hydrogen (H_2), nitrogen (N_2), and oxygen (O_2) exist as diatomic molecules. Some elements, such as phosphorus (P_4) and sulfur (S_8), exist as larger molecules.

SECTION 2.2: The Atomic View of Matter

Evidence for Atoms (Dalton's Induction):

1) Law of Mass Conservation:

Magnesium burns in oxygen to produce a white powder, magnesium oxide:

$$2Mg(s) + O_2(g) \rightarrow 2MgO(s)$$

The mass of the reactants (magnesium and oxygen) used up, equals the mass of the product (magnesium oxide) formed. Early scientists made many measurements on the masses of reactants and products of various reactions and found that <u>mass is always conserved in chemical transformations</u>. John Dalton (1766-1844) postulated that this would be the case if all matter were composed of indestructible particles he called "<u>atoms</u>" which in chemical reactions merely rearrange.

2) Law of Constant Composition:

Early scientists discovered that pure water is always 11.2 percent hydrogen and 88.8 percent oxygen by mass. Dalton postulated that this was further evidence for the existence of atoms, since the result could be explained <u>if all molecules of water have the same chemical formula and if each kind of atom has on the average the same mass</u>.

Law of Constant Composition:

1) What mass of hydrogen can be obtained from:

    a) 18.0 g of water; b) 1.00 g of water; c) 10.0 g water?

    <u>solutions</u>:

Water is always 11.2% hydrogen by mass so:

        a) mass of hydrogen in 18.0 g water = 0.112 x 18.0 g = 2.02 g

        b) mass of hydrogen in 1 g water = 0.112 x 1.00 g = 0.112 g

        c) mass of hydrogen in 10 g water = 0.112 x 10.0 g = 1.12 g

2) What mass of hydrogen can be obtained from:

    a) 18.0 lb of water; b) 1.00 lb of water; c) 10.0 lb of water?

    <u>solutions</u>:

        a) mass of hydrogen in 18.0 lb water = 0.112 x 18.0 lb = 2.02 lb

        b) mass of hydrogen in 1.0 lb water = 0.112 x 1.00 lb = 0.112 lb

        c) mass of hydrogen in 10 lb water = 0.112 x 10.0 lb = 1.12 lb

A test of the atomic model of matter (Dalton's deduction – the Law of Multiple Proportions):

Let: M_C = mass of 1 carbon atom

M_O = mass of 1 oxygen atom

Suppose carbon and oxygen form two compounds, as, in fact, they do. One compound might have the formula CO, e.g., and the other CO_2. If so, one would <u>predict</u> that:

$$\frac{\text{mass of oxygen in } CO_2 \,/\, \text{mass of carbon in } CO_2}{\text{mass of oxygen in CO} \,/\, \text{mass of carbon in CO}} = \frac{2M_O \,/\, M_C}{M_O \,/\, M_C} = \frac{\left[(16+16)+12\right]}{(16+12)} = \frac{2}{1}$$

which, in fact, is the case. This law is known as the **Law of Multiple Proportions**: <u>when two elements form a series of compounds, the ratios of the masses of the second element that combine with a fixed mass of the first element can always be reduced to small whole numbers.</u>

The Law of Multiple Proportions:

1) Hydrogen and oxygen form two compounds: water, H_2O, and hydrogen peroxide, H_2O_2. Evaluate the following ratio of ratios:

$$\frac{\text{mass of oxygen in } H_2O_2 \,/\, \text{mass of hydrogen in } H_2O_2}{\text{mass of oxygen in } H_2O \,/\, \text{mass of hydrogen in } H_2O}$$

<u>solution:</u>

$$\frac{2M_O \,/\, 2M_H}{M_O \,/\, 2M_H} = \frac{\left[(16+16)+(1+1)\right]}{\left[16+(1+1)\right]} = \frac{2}{1}$$

2) Given that:

$$\frac{\text{mass of H in hydrocarbon I} \,/\, \text{mass of C in hydrocarbon I}}{\text{mass of H in hydrocarbon II} \,/\, \text{mass of C in hydrocarbon II}} = \frac{2}{1}$$

Which of the following atomic models are consistent with the cited data:

	A	B	C	D	E
Hydrocarbon I	CH_2	CH_2	C_2H_4	CH_4	CH_4
Hydrocarbon II	CH	C_2H_2	C_2H_2	CH_2	C_2H_4

<u>solution:</u>

All 5 cases are consistent with the data. The case that Dalton discovered was E, CH_4 (methane, natural gas, or marsh gas) and C_2H_4 (ethylene).

Dalton's theory stated:

1. Each element is made up of tiny particles called atoms.

2. Atoms of one element cannot be converted into atoms of another element. Chemical reactions involve reorganization of the atoms, not changes in the atoms themselves (law of mass conservation).

3. Atoms of an element are identical, and are different from atoms of other elements.

4. Compounds form when atoms combine with each other. A given compound always has the same relative numbers and types of atoms (law of constant composition).

After Dalton published his theory, he prepared the first table of atomic masses. Because the masses of individual atoms are so small, he could only determine the relative masses of the elements. Dalton assigned an atomic mass of 1 to hydrogen, the lightest known substance.

Relative atomic masses:

It was known in Dalton's day that water was composed of hydrogen and oxygen and that every gram of hydrogen required 8 grams of oxygen to form water. Based on this knowledge, Dalton assigned the atomic mass of oxygen to be 8. What did Dalton assume was the molecular formula for water?

answer:

Dalton assumed that the molecular formula for water was HO. We know now that the molecular formula for water is H_2O, and the atomic mass of oxygen is 16. Dalton's billiard ball view of the atom was extremely useful for explaining many scientific observations, but it could not explain why elements combine in specific ratios of atoms to form compounds.

Avogadro's Hypothesis: Evidence for Molecules of Elements:

Experiments by Gay-Lussac on the volumes of gaseous reactants and products showed:

hydrogen(g) + chlorine(g) $\longrightarrow$ hydrogen chloride(g)

1 vol.   1 vol.   2 vol.

☐   +   ☐   =   ☐☐

One might have expected that the reaction of 1 volume of hydrogen gas with 1 volume of chlorine gas would yield 1 volume of hydrogen chloride gas:

hydrogen(g) + chlorine(g) $\longrightarrow$ hydrogen chloride(g)

1 vol.   1vol.   1 vol.

☐   +   ☐   =   ☐

To explain the confusing experimental results, Amadeo Avogadro made two proposals:

- equal volumes of gases contain equal number of gas particles. This equal volume – equal number hypothesis makes sense if the distances between gas particles is very great compared to the sizes of the particles so the <u>size</u> of the particles is insignificant, and the volume is determined by the <u>number</u> of molecules present.

- hydrogen and chlorine exist as **polyatomic molecules** which split apart and recombine to form gaseous hydrogen chloride molecules.

The evidence for diatomic hydrogen molecules:

Let's assume, as Avogadro did, that:

   A) equal number of gas particles take up equal volume

   B) atoms cannot be split

Which of the following possibilities for reaction of hydrogen with chlorine obey both of Avogadro's assumptions and yield the experimental result found above (1 vol. hydrogen + 1 vol. chlorine = 2 vol. hydrogen chloride)?

<u>Volumes of gases:</u>

Reaction	Volumes of gases:
$H + Cl \longrightarrow HCl$	1 vol. + 1 vol. = 1 vol.
$H_2 + Cl_2 \longrightarrow 2HCl$	**1 vol. + 1 vol. = 2 vol.**
$H_3 + Cl_3 \longrightarrow 3HCl$	1 vol. + 1 vol. = 3 vol.
$H_3 + Cl_3 \longrightarrow H_3Cl_3$	1 vol. + 1 vol. = 1 vol.
$H_4 + Cl_4 \longrightarrow 4HCl$	1 vol. + 1 vol. = 4 vol.
$H_4 + Cl_4 \longrightarrow 2H_2Cl_2$	**1 vol. + 1 vol. = 2 vol.**
$H_4 + Cl_4 \longrightarrow H_4Cl_4$	1 vol. + 1 vol. = 1 vol.
$H_6 + Cl_6 \longrightarrow 2H_3Cl_3$	**1 vol. + 1 vol. = 2 vol.**

<u>answer</u>:

The possibilities which give the experimentally found volumes of reactants and products are written in bold in the right column. Notice that the hydrogen and chlorine molecules which could produce 2 volumes of HCl all contain an **even number** of atoms.

One volume of hydrogen gas has never produced more than 2 volumes of gaseous product. Therefore, it is <u>not</u> necessary to assume more than 2 atoms per molecule of hydrogen. Chemists use the symbol H_2 for the hydrogen molecule.

Avogadro's concept of diatomic gaseous molecules preserved the idea that _____(1) are not created or destroyed in a chemical reactions. Instead, _____(2) are created and destroyed. (answers: (1) atoms, (2) molecules)

The Discovery of Electrons:

The first convincing evidence for subatomic particles came when experimenters observed rays of light when conducting electricity through gases at low pressures. J.J. Thomson (1856-1940) concluded that the rays appeared when negatively charged particles collided with the gas molecules in the tube. He named the particles **electrons**. Thomson determined the mass-to-charge ratio of the electron was the same regardless of what gas was in the tube, so he concluded that <u>the electron was a fundamental particle found in all atoms</u>. Since atoms were known to be electrically neutral, Thomson proposed that the negatively charged electrons were distributed randomly in a diffuse cloud of positive charge like raisins in a plum pudding.

Robert Millikan measured the charge of the electron by measuring the effect of an electrical field on the rate at which charged oil drops fell under the influence of gravity. He used this value and Thomson's mass/charge ratio to calculate an electron's mass to be 9.11×10^{-28} g. This mass is approximately 1/2000 the mass of hydrogen, the lightest element!

The evidence for a nucleus containing positively charged particles:

Ernest Rutherford established that an atom contains a tiny, positively charged center called an **atomic nucleus**. He bombarded a piece of thin gold foil with alpha particles (a helium nucleus, ^{4}He, which is emitted from certain radioactive materials). Since the mass of an alpha particle is 7300 times that of an electron, Rutherford expected all the alpha particles to smash right through the gold foil with, at most, very minor deflections in their paths. Unexpectedly, a tiny percentage of the alpha particles reflected back from the foil! The plum pudding model could not explain these results. Rutherford postulated that the alpha particles bounced off a tiny center of positive charge. He presented a new picture of an atom as mostly space occupied by electrons, but which contained a tiny region of concentrated positive charge which he called the **nucleus**. He postulated that the nucleus contained positively charged particles he called **protons** which accounted for most of the mass of the atom.

The Discovery of Neutrons:

Rutherford's work did not account for all the mass of an atom. Twenty years later, James Chadwick discovered the **neutron**. This uncharged particle is present in the nucleus with the protons and accounts for the remaining mass of the atom.

A Modern (ca. 1995) view of the atom:

Today we picture an atom as a positively charged nucleus surrounded by a negatively charged electron cloud. The electrons whiz around the nucleus, held there by the attraction of the nucleus. The nucleus is tiny and heavy - it is extremely dense! It contains protons (p^+) with a positive charge, and neutrons (n^0) with no charge. The protons and neutrons account for essentially all the mass of the atom. An atom is electrically neutral, so the number of protons and electrons in an atom are equal.

The Atomic Theory Today:

1. All matter is composed of atoms. Although atoms are composed of smaller particles (electrons, protons, and neutrons), <u>an atom is the smallest body that retains the unique identity of the element</u>.

2. Atoms of one element can be converted into atoms of another element <u>only</u> in nuclear processes. They are never transformed by a chemical reaction.

3. All atoms of an element have the same number of protons and electrons, which determines the chemical behavior of the element. Isotopes of an element differ in their number of neutrons and hence differ in their mass number. The mass of a sample of an element is the average of the masses of its naturally occurring isotopes.

4. Compounds are formed by the chemical combination of two or more elements in specific ratios as originally stated by Dalton.

Terminology for the Atom:

Atomic Number (Z): the number of protons in the nucleus of an atom.

$$Z = \#protons$$

Mass Number (A): The total number of protons and neutrons in an atom.

$$A = \#protons + \#neutrons$$

Each proton and neutron has one unit of mass. Notice that the mass number (A) minus the atomic number (Z) = the number of neutrons (N):

$$A - Z = (\#protons + \#neutrons) - \#protons = \#neutrons = N$$

Atomic Symbol: A symbol for each element based on its English, Latin, or Greek name. The convention is to write the atomic symbol with the atomic number as a left subscript, and the mass number as a left superscript. For atom X, we write:

$$_{Z}^{A}X$$

Isotopes: atoms with the same number of protons (the same element), but different numbers of neutrons and therefore different mass numbers.

Atomic mass unit (amu): 1/12 the mass of a carbon-12 atom. The mass of a carbon-12 atom is <u>defined</u> as exactly 12 atomic mass units.

Atomic mass: the average of the masses of an element's naturally occurring isotopes weighted according to their abundances.

Terminology for the Atom (exercise):

Choose words from the list below to complete the following paragraphs:

(a) atomic mass unit   (f) protons
(b) atomic number   (g) seven
(c) isotope(s)   (h) six
(d) mass number   (i) twelve
(e) neutrons

Elements are defined by the number of _____(1) in their nucleus, which is the element's_____(2). Every carbon atom contains _____(3) protons in its nucleus and therefore has an atomic number of _____(4). Essentially all the mass of an atom is contributed by its _____and _____(5 & 6), and the sum of these two gives the _____(7). The mass number varies depending on the number of _____(8) an atom contains in its nucleus (its _____(9)). A ^{13}C atom contains_____(10) protons and _____(11) neutrons in its nucleus. The chemical properties of an element are primarily determined by the number of electrons, so _____(12) have similar chemical behavior.

The mass unit which relates the relative masses of atoms is the _____(13). The mass of a ^{12}C atom is defined to be exactly _____(14) amu. The nucleus of a ^{12}C contains_____(15) protons and _____(16) neutrons. The atomic mass unit is based on this particular _____(17) of carbon, not on naturally occurring carbon which is a mixture of _____(18).

answers: 1-f, 2-b, 3-h, 4-h, 5-f, 6-e, 7-d, 8-e, 9-c, 10-h, 11-g, 12-c, 13-a, 14-i, 15-h, 16-h, 17-c, 18-c.

Atomic Mass Exercises:

1) Consider the lightest element, hydrogen. Hydrogen has an atomic mass of 1.008 amu. Why isn't it exactly 1.000?

solution:

Every hydrogen atom contains one proton in its nucleus (Z=1). Most naturally occurring hydrogen atoms (99.985%) contain no neutrons in their nucleus (A=1), and their atomic mass is 1 amu:

atomic mass of ^1H = 1.000 amu

A small number of naturally occurring hydrogen atoms contain 1 neutron in their nucleus (A=2), and their atomic mass is 2 amu:

$$\text{atomic mass of } ^2H = 2.000 \text{ amu}$$

It is this small number of hydrogen atoms containing a neutron that makes the average mass of naturally occurring hydrogen atoms greater than 1.000 amu.

2) When does atomic mass equal mass number?

solution:

Atomic mass equals the mass number when all atoms under consideration are the same isotope. In naturally occurring samples of an element where there is a mixture of isotopes, the atomic mass is the weighted average of the masses of all the isotopes.

Example: What is the atomic mass of carbon containing 98.9% ^{12}C and 1.1% ^{13}C?

$$\text{Atomic mass} = (.989 \times 12 \text{ amu}) + (.011 \times 13 \text{ amu}) = 12.011 \text{ amu}$$

SECTION 2.4: The Periodic Table

In 1871, Dmitri Mendeleev organized the elements by increasing atomic mass and arranged them so that elements with similar chemical properties would lie in the same column. The modern periodic table of the elements is based on Mendeleev's earlier version.

Periodic Table Exercise:

Choose words from the list below to complete the following paragraphs:

(a) atomic number
(b) group(s)
(c) inner transition
(d) letters

(e) main-group
(f) period(s)
(g) proton
(h) transition

The elements are arranged in order of increasing _____(1); each element takes the previous element and adds one _____(2) to its nucleus. The vertical columns are called_____(3) and are numbered 1 to 8 with either the letter A or B. The horizontal rows are called_____(4) and are numbered 1 to 7. A new system numbers the groups from 1 to 18 without_____(5). Elements in a _____(6) have similar chemical properties; elements in a_____(7) have different chemical properties.

The eight A-groups contain the _____(8) or representative elements. The 10 B-groups contain the _____(9) elements. The _____(10) elements are usually placed below the main body of the table, but actually fit between Ba and Hf (Lanthanides), and Ra and Rf (actinides).

answers: 1-a, 2-g, 3-b, 4-f, 5-d, 6-b, 7-f, 8-e, 9-h, 10-c.

Classification of the Elements:

Elements are classified as metals, non-metals, or metalloids depending on their physical and chemical properties. The group names are listed below in bold. Match the characteristics (a-n) to the classification (metal, non-metal, metalloid) that it best describes.

a) shiny solids at room temperature
b) group 8A elements, the relatively unreactive noble (or inert) gases
c) properties between those of metals and nonmetals
d) conduct heat and electricity poorly
e) group 7A elements, the highly reactive halogens
f) good conductors of heat and electricity
g) easily tooled into sheets and wires
h) group 1A elements, the highly reactive alkali metals
i) transition elements
j) dull, brittle solids or gases at room temperature
k) inner transition elements
l) elements in the upper right portion of the periodic table
m) elements that lie along the staircase line on the periodic table
n) group 2A elements, the reactive alkaline earth metals

answers:

Metals (a,f,g,h,i,k,n); Non-metals (b,d,e,j,l); Metalloids (c,m)

SECTION 2.5: Introduction to Bonding

Most elements don't occur naturally in elemental form, but in combination with other elements as compounds. Elements combine in two general ways, both of which involve the elements' electrons. In one case, electrons are transferred between atoms, in the other case, the electrons are shared between atoms.

Bonding Terminology:

Ionic compounds: form when an atom of one element transfers electrons to atoms of another element.

Covalent compounds: form when atoms share electrons.

Ion: a charged particle

Cation: a positively charged particle

Anion: a negatively charged particle

Monoatomic ion: an ion composed of a single atom

Polyatomic ions: two or more atoms bonded covalently and having a net positive or negative charge i.e.: NH_4^+, SO_4^{2-}

Ionic Bonding Exercise:

Choose words from the list below to complete the following paragraphs:

(a) anions	(f) gain	(j) neon
(b) argon	(g) ion	(k) Noble gases
(c) cations	(h) lose	(l) polyatomic
(d) Cl^-	(i) NH_4^+	(m) seventeen
(e) eleven		

When an atom, or small group of atoms, gains or loses electrons, it becomes a charged particle or an _____(1). Negatively charged particles called _____(2) attract positively charged particles called_____(3) to form ionic compounds. We can predict how many electrons an element is likely to gain or lose by examining its position in the periodic table. The elements in group 8A, the _____(4), are particularly stable elements. <u>Elements strive to obtain the same number of electrons as the nearest noble gas</u>. Elements on the left hand side of the periodic table, the metals, _____(5) electrons to become cations. Elements on the right hand side of the periodic table, the nonmetals, _____(6) electrons to become anions.

Table salt (sodium chloride) is a familiar ionic compound. We would predict that a sodium atom with_____(7) electrons would _____(8) an electron to obtain the same number of electrons as the noble gas _____(9), and that chlorine with _____(10) electrons would _____(11) an electron to obtain the same number of electrons as the noble gas _____(12).

Some ionic compounds are formed with_____(13) ions, ions consisting of two or more atoms covalently bonded which stay together as a unit during reactions. An example of an ionic compound formed with a polyatomic ion is ammonium chloride, NH_4Cl. _____(14) is the cation and _____(15) is the anion.

<u>answers</u>: 1-g, 2-a, 3-c, 4-k, 5-h, 6-f, 7-e, 8-h, 9-j, 10-m, 11-f, 12-b, 13-l, 14-i, 15-d.

Sodium chloride, written NaCl(s), is not a collection of pairs of sodium and chloride ions as the formula seems to imply, but is a three dimensional array of alternating Na^+ and Cl^- ions. The number of sodium ions must equal the number of chloride ions because ionic compounds have no charge. Although both sodium and chlorine are only one electron away from having the same number of electrons as a noble gas, some elements must lose or gain two or three electrons to obtain a noble gas electron configuration.

Covalent bonding exercise:

Choose words from the list below to complete the following paragraphs:

(a) anions   (e) O_2
(b) cations   (f) same
(c) H_2O (g) separate
(d) nonmetals   (h) share

Covalent compounds form when atoms _____ (1) electrons, which usually occurs between elements found in the upper right portion of the periodic table, the _____ (2). In the case of an oxygen molecule, two oxygen atoms share electrons to form the molecule _____ (3). In the case of water, two hydrogen atoms and an oxygen atom share electrons to form the molecule _____ (4).

Many molecules are not purely ionic, or purely covalent; their bonds possess some of each character. The purest type of covalent bond is one between two of the _____ (5) atoms since their nuclei attract the electrons to the same extent.

A covalent compound such as methanol, CH_3OH, consists of molecules which loosely associate with each other, but remain _____ (6). In contrast, an ionic compound such as magnesium chloride ($MgCl_2$), consists of a continuous array of ions which alternates positively charged _____ (7) and negatively charged _____ (8).

answers: 1-h, 2-d, 3-e, 4-c, 5-f, 6-g, 7-b, 8-a.

Types of formulas:

Chemical formulas use element symbols and numerical subscripts to show the type and number of each atom present in a compound. There are three types of chemical formulas:

1) **Empirical formula**: shows the relative minimum number of atoms of each element in a compound.

2) **Molecular formula** since it shows the _actual number_ of atoms of each element in the molecule.

3) **Structural formula**: shows the actual number of atoms of each element in a molecule and the bonds between them.

Ethylene has a molecular formula of C_2H_4. It's empirical formula is CH_2, and its structural formula shows the bonds between the atoms:

$$\begin{array}{ccc} H & & H \\ \diagdown & & \diagup \\ & C = C & \\ \diagup & & \diagdown \\ H & & H \end{array}$$

Methane has a molecular formula of CH_4. It's empirical formula is the same as its molecular formula, CH_4, and its structural formula is:

$$\begin{array}{c} H \\ | \\ H - C - H \\ | \\ H \end{array}$$

Ionic compounds such as sodium chloride generally use empirical formulas. NaCl is an empirical formula that tells us there is one sodium ion for every chloride ion.

SECTION 2.6: The Names of Simple Compounds

Ionic Compounds:

Let's consider the name for NaCl, <u>sodium chloride</u>. Notice that:

- The cation is named first followed by the anion.
- The name of the cation (sodium) is the same as the name of the neutral metal.
- The name of the anion (chloride) takes the <u>root</u> of the nonmetal name (chlorine) and adds the suffix "-ide".

Naming ionic compounds:

What are the names for the following ionic compounds:

       a) $MgCl_2$ c) CuCl

       b) CaO   d) $CuCl_2$

<u>answers:</u>

       a) magnesium chloride   c) copper(I) chloride

       b) calcium oxide   d) copper(II) chloride.

Notice that <u>two different compounds cannot have the same name</u>, so for metals (such as copper and iron) that can form cations with different charges, we indicate the metal's ionic charge by writing it in Roman numerals (in parentheses) after the metal ion's name. You will often see the common names for these compounds: the Latin root metal is followed by the suffix "-ous" for the ion with the lower charge and "-ic" for the ion with the higher charge.

CuCl = copper(I) chloride = cuprous chloride
CuCl$_2$ = copper(II) chloride = cupric chloride
FeCl$_2$ = iron(II) chloride = ferrous chloride
FeCl$_3$ = iron(III) chloride = ferric chloride

(Elements that form only one cation do not need to be identified by a roman numeral: group 1A elements form only 1+ ions, group 2A elements form only 2+ ions, and aluminum forms only Al^{3+}).

Ionic compounds containing polyatomic ions:

Polyatomic ions stay together as a charged unit. When two or more of the same polyatomic ion are present in the formula, the ion appears in parentheses with the subscript written outside.

Examples of ionic compounds containing polyatomic ions:

a) silver nitrate = $AgNO_3$

b) silver nitrite = $AgNO_2$

c) calcium nitrate = $Ca(NO_3)_2$

d) ammonium chloride = NH_4Cl

e) calcium phosphate = $Ca_3(PO_4)_2$

The naming convention for oxoanions:

When a polyatomic ion contains one or more oxygen atoms, we say it is an **oxoanion**. Often, there are families of oxoanions that differ only in the number of oxygen atoms.

A) With two oxoanions in the family, we use the suffixes

- "-ate" for the ion with more O atoms
- "-ite" for the ion with fewer O atoms.
  SO_4^{2-} is the sulfate
  SO_3^{2-} is sulfite

B) With four oxoanions in the family, in addition to the suffixes "-ate" and "-ite", we use the prefixes "per-" (more than) and "hypo-" (less than) to name the members of the series with the most and fewest oxygen atoms respectively.

ClO_4^- is perchlorate
ClO_3^- is chlorate
ClO_2^- is chlorite
ClO^- is hypochlorite

Naming acids:

An acid is a molecule with one or more H^+ ions associated to an anion. Acids dissolve in water to produce a solution containing free H^+ ions (protons). There are two classes of acids in terms of naming them:

A) If the anion does not contain oxygen, the acid is named with the prefix "hydro-" and the suffix "-ic". When gaseous HCl (Hydrogen chloride) dissolves in water, it forms **hydrochloric acid.**

B) If the anion contains oxygen, the names are similar to those of the oxoanions, except:

- the anion "-ate" suffix becomes an "-ic" suffix in the acid:

    H_2SC_4 = sulfuric acid

- the anion "-ite" suffix becomes an "-ous" suffix in the acid:

    H_2SC_3 = sulfurous acid

Binary Covalent Compounds:

Let's consider the name for CO_2, carbon dioxide. Notice that:

- The first element named (carbon) is the one in the lower group number in the periodic table. (An exception to this rule is when the compound contains oxygen and a halogen. In this case, halogen is named first). If both elements are in the same group, the one with the higher period number is named first.

- The second element (oxygen) is named with the suffix "-ide"

- Greek numerical prefixes indicate the number of atoms of each element in the compound. The first word has a prefix only when more than one atom of the element is present. The second word always has a numerical prefix.

Naming binary covalent compounds:

Give names for the following covalent compounds:

a) PCl_3 d) CO

b) NO_2 e) H_2O

c) Cl_2O_7 f) NH_3

answers:

a) phosphorus trichloride   d) carbon monoxide
b) nitrogen dioxide   e) dihydrogen monoxide
c) dichlorine heptaoxide   f) nitrogen trihydride.

These last two molecules are almost always referred to by their common names, water and ammonia.

SECTION 2.7: Molecular Masses from Chemical Formulas:

We can calculate the molecular mass of a compound by adding together its atomic masses.

Molecular mass = sum of atomic masses

Calculating molecular mass:

1) What is the molecular mass of an ammonia molecule?

solution:

• We calculate the molecular mass by adding together the atomic masses:

Molecular mass of NH_3 = (1 x atomic mass of N) + (3 x atomic mass of H)

= (14.01 amu) + (3 x 1.008amu) = 17.03 amu

2) What is the molecular mass of Magnesium nitrate?

solution:

• Again, we add together the atomic masses to calculate the molecular mass:

Molecular mass $Mg(NO_3)_2$

= (1 x atomic mass of Mg) + 2 x [(1 x atomic mass of N) + (3 x atomic mass of O)]

= 24.31 amu + 2 x [14.08 amu + (3 x 16.00 amu)]

= 148.47 amu

Notice that the atomic mass of Mg is used for the cation Mg^{2+} even though a magnesium ion is two electrons lighter than a magnesium atom The two electrons lost by Mg are gained by the anion, so electron masses are balanced in neutral compounds.

3) What is the molecular mass of sucrose, $C_{12}H_{22}O_{11}$?

solution:

Molecular mass $C_{12}H_{22}O_{11}$

= (12 x 12.01 amu) + (22 x 1.008 amu) + (11 x 16.00 amu)

= 342.3 amu

Chapter 2 Exercises:

2.1) Which of the following is a pure substance?

        seawater
        air
        methane gas
        orange juice
        homogenized milk
        a cup of coffee

2.2) An atom, ion, or molecule of a particular element X has the general symbol shown below. In this symbol various information is included at one or more of the four corners. What information is usually located at each corner?

$$_o^o X _o^o$$

2.3) a. The nucleus of a neutral atom contains 42 protons and 54 neutrons. What is the element?

b. A neutral atom contains 110 neutrons and 74 electrons. What is the element?

c. If a chlorine atom has a mass number of 37, how many neutrons are there in its nucleus?

d. One isotope of nitrogen has the symbol ^{14}N. How many protons, neutrons, and electrons are there in one ^{14}N atom?

2.4) Copper exists in nature as a mixture of two isotopes. One isotope contains 34 neutrons and has a relative abundance of 69.77%. The average atomic mass of copper is 63.55. How many neutrons must there be in the second isotope?

2.5) Alloys of zirconium, Zr, are used in the fabrication of nuclear reactors. The element has five naturally occurring isotopes:

        ^{90}Zr, 89.9043 amu (51.46%)
        ^{91}Zr, 90.9053 amu (11.23%)
        ^{92}Zr, 91.9046 amu (17.11%)
        ^{94}Zr, 93.9061 amu (17.40%)
        ^{96}Zr, 95.9082 amu (2.80%).

What is the average atomic mass of Zr?

2.6) Iridium has two naturally occurring isotopes. ^{191}Ir and ^{193}Ir have atomic masses of 190.9609 amu and 192.9633 amu respectively. The average atomic mass for iridium is 192.22 amu. What is the percent natural abundance for each isotope?

2.7) a. Which one of the following elements exists as a diatomic gas under normal conditions?

bromine	calcium	chlorine
sodium	argon	gallium

b. Which one of the following elements exists as a monatomic gas under normal conditions?

sulfur	krypton
potassium	bromine
oxygen	silicon

c. Atoms of which of the following elements form covalent bonds with themselves in molecules?

sodium	phosphorus
nitrogen	cadmium
neon	iodine

2.8)   What is the formula of the ionic compound...

   a. that contains only potassium and carbonate ions?
   b. that contains only calcium and carbonate ions?
   c. that contains only copper(II) and bisulfate ions?
   d. that contains only ammonium and phosphate ions?

2.9)   a. What is wrong with the following formulas?

$SrCl$ $Na_2(PO_4)_3$ $Ba(SO_4)$ $MgNO_3$ Li_2Br_2

   b. What is wrong with the following names?

$NaCl$	sodium monochloride
KNO_3	potassium nitrite
$CaBr_2$	dibromine calcium
$NaHCO_3$	sodium carbonate
$(NH_4)_2HPO_4$	diammonium hydrogen phosphate

2.10)   Name the following compounds.

a. NaI	e. Fe_2S_3
b. NH_4Cl	f. $PbCrO_4$
c. ZnO	g. $Mg(ClO_3)_2$
d. $Al(OH)_3$	h. H_3PO_3

2.11)   Give the formula for each of the following compounds.

a. nitrogen triiodide	e. sulfurous acid
b. copper(II) sulfate	f. sodium carbonate
c. diphosphorous pentoxide	g. hydrobromic acid
d. cobalt(II) nitrite	h. sodium hydride

2.12)   Which compounds contain only ionic bonds, only covalent bonds, and both covalent and ionic bonds?

$NaCl$ NH_4Br H_2O KNO_3 $(NH_4)_2HPO_4$ NH_3

2.13)   Calculate the molecular mass of:

nitric acid, HNO_3
carbon tetrachloride, CCl_4
acetic acid, CH_3CO_2H
toluene, $C_6H_5CH_3$

2.14)   Cinnabar is reddish ore from which mercury is derived. 8.62 g of Hg are in each 10.0 g of cinnabar. What is the mass fraction of mercury in cinnabar? What mass of cinnabar is necessary to yield 1.00 kg of Hg?

2.15)   a. What ending do the names of monatomic negative ions have?

b. What ending do the names of nonmetal positive ions such as H_3O^+ or NH_4^+ have?

c. If there are two possible polyatomic oxyanions of an element, what ending does the anion having the fewer oxygens have?

d. The maxiumum oxidation state of xenon is +8; other possible positive oxidation states are +2, +4 and +6. In accordance with the rules for assigning names to oxyanions, what would the name for the XeO_6^{4-} ion be?

Chapter 2 Answers:

2.1) methane gas (CH_4)

2.2) The atomic number, Z, is the number of protons in the nucleus. The mass number, A, is the protons and neutrons in the nucleus.

   upper left corner: the mass number of the atom; example ^4He
   lower left corner: the atomic number; example $_{27}Co$
   upper right corner: the charge, if any; example Zn^{2+}
   lower right corner: number of atoms in the molecule; example S_8

2.3) a. Mo; indicated by the number of protons, regardless of the number of neutrons
   b. W; the number of electrons equals the number of protons
   c. 20; the atomic number of chlorine is 17, therefore the number of neutrons is
      $37 - 17 = 20$.
   d. 7 protons, 7 neutrons, 7 electrons; the atomic number of nitrogen is 7, therefore
      there are 7 protons. There must then be 7 electrons. Since the mass number is
      14, there must also be 7 neutrons.

2.4) 36

2.5) 91.2 amu

2.6) 37.120% ^{191}Ir, and 62.880% ^{193}Ir

2.7) a. Chlorine. Argon is the only other gas in the list and is monatomic. Bromine is a
      liquid.

   b. Krypton. Oxygen, the other gas in the list, is diatomic O_2.

   c. Phosphorus forms P_4 molecules (and other allotropes)
      Iodine and nitrogen form diatomic molecules I_2 and N_2.
      Neon, the noble gas in the list, is monatomic. It does not form bonds.
      Sodium and cadmium are metals and form metallic solids.

2.8) a. K_2CC_3
   b. $CaCO_3$
   c. $Cu(HSO_4)_2$
   d. $(NH_4)_3PO_4$

2.9) a. Strontium is in Group 2 and invariably forms the ion Sr^{2+}; there should be two
      chloride ions to match the strontium ion: $SrCl_2$.

   The phosphate ion has a -3 charge, so the ratio of sodium ions to phosphate ions
   should be 3:1, not 2:3: Na_3PO_4.

   The ratio of atoms is correct in this formula but no parentheses are required around
   the sulfate ion since there's only one of them: $BaSO_4$.

Magnesium is in Group 2 and invariably forms the ion Mg^{2+}; there should be two nitrate ions for one magnesium ion: $Mg(NO_3)_2$.

The ratio of atoms is correct in this formula but the ratio should be expressed as simply as possible: LiBr.

b. sodium chloride; prefix mono– not required
potassium nitrate, not nitrite
calcium bromide; the metal is placed first in name; prefix di– not required
sodium bicarbonate; not carbonate
ammonium hydrogen phosphate; the number of hydrogen atoms still attached equals one and is specified; the number of ammonium ions follows automatically.

2.10) a. sodium iodide
b. ammonium chloride
c. zinc oxide
d. aluminum hydroxide
e. iron(III) sulfide
f. lead chromate
g. magnesium chlorate
h. phosphorous acid

2.11) a. NI_3
b. $CuSO_4$
c. P_2O_5
d. $Co(NO_2)_2$
e. H_2SC_3
f. Na_2CC_3
g. $HBr(aq)$
h. NaH

2.12) NaCl: ionic bond
NH_4Br: the ammonium ion contains covalent bonds; the bond between the NH_4^+ and the Br^- is ionic.
H_2O: covalent bonds
KNO_3: the anion NO_3^- contains covalent bonds; the bond between the potassium and the nitrate ion is ionic.
NH_4HSO_4: both the cation NH_4^+ and the anion HSO_4^- contain covalent bonds
NH_3: covalent bonds

2.13) Sum the atomic masses to obtain the molecular mass:

HNO_3 $1.008 + 14.01 + 3 \times 16.00 = 63.02$
CCl_4 $12.011 + 4 \times 35.45 = 153.8$
CH_3CC_2H $2 \times 12.011 + 4 \times 1.008 + 2 \times 16.00 = 60.05$
$C_6H_5CH_3$ $7 \times 12.011 + 8 \times 1.008 = 92.14$

The units of these molecular masses are amu (dalton) for the mass of an individual molecule, or grams for the mass of one mole of the molecules.

2.14) 0.862, 1.16 kg of cinnabar

2.15) a. Monatomic negative ions have the ending –ide.
b. Nonmetal positive ions have the ending –ium (hydronium, ammonium).
c. The ending of the anion having the fewer oxygens has the ending –ite.
d. The name for the XeO_6^{4-} ion is *perxenate*

CHAPTER 3

Stoichiometry

TOPICS:

- Units of Measure
- The Mole
- Molar Mass (m/n)
    - Empirical and Molecular Formulas
    - Chemical Composition from Combustion Analysis
- Balancing Chemical Equations
    - Limiting reactant, % Yield
- Concentration (n/V)

SECTION 3.1: Units of Measure

We saw in Chapter 1 that a physical quantity can be expressed by the product of a <u>pure number</u> and a <u>physical unit</u>:

$$\text{physical quantity} = \text{pure number} \times \text{physical unit}$$

Three physical quantities that we use to tell us how much of a substance there is are mass, volume, and amount:

<u>physical quantity</u>	<u>common physical unit</u>
mass (m)	gram
volume (V)	liter
amount (n)	mole

Using the term "amount" as a physical quantity can be confusing because we usually think of mass and volume as amounts too. It may be clearer to think of the physical quantity "amount" (n) as being a "population". It may also be confusing to think of the mole as a <u>unit</u> for amount because a mole stands for a number (6.02×10^{23}), and we don't normally think of a unit as being a pure number. When we use the mole as a unit, it stands for a population of "things" or "entities". In may be helpful to include "entity" as the unit. For example, if we have 12 eggs, we can say:

mass of eggs = 500 **grams**

amount, or population of eggs = 12 (entities), or 1 **dozen** (entities)

Notice that dozen is the <u>unit</u> for amount, and it is also a pure number (12). In this case, the entity is eggs. A mole is a chemist's dozen, and is used in the same way as a dozen. For example, if we have 1 mole of carbon atoms (6.02×10^{23} carbon atoms), we can say:

mass of atoms = 12.01 **grams**

amount, or number of atoms = 6.02×10^{23} (entities), or 1 **mole** (entities).

Problems are less likely to become confused if you include the entity that the mole is referring to: 1 mole H_2, 12.01 g C/ mole C, 5 mole electrons, etc.

Ratios of mass, volume, and amount:

Chemists often refer to ratios of these three units of measure. Match the ratio to the name given for it:

Names:	Ratios:
A) density (d)	1) m/V
B) molar mass (M)	2) V/n
C) concentration (c)	3) n/V
D) molar volume	4) m/n
E) specific volume	5) V/m
F) specific amount	6) n/m

<u>answers</u>:

A-1, B-4, C-3, D-2, E-5, F-6

Notice that IUPAC uses molar to mean "divide by n" and specific to mean "divide by m". Despite the fact that molar sounds like mole, <u>they have nothing to do with each other</u>, except that IUPAC uses a "mole" as a unit of n. (The IUPAC unit for mass is a kilogram, and we don't refer to specific volume as "kilogrammer volume").

We discussed density (mass/volume) in Chapter 1. In this chapter, we will first define a mole and then discuss **molar mass**, the mass per unit amount, and **concentration**, the amount per unit volume.

SECTION 3.2: The Mole

We learned in Chapter 2 that **atomic mass** is the average of the masses of an element's naturally occurring isotopes weighted according to their abundances. Atomic masses are given in **atomic mass units** (amu) defined as exactly 1/12 the weight of a ^{12}C atom.

mass of one ^{12}C atom = 12.000 amu.

Carbon found on the earth (natural carbon) does not have an atomic mass of 12.000 amu, but 12.011 amu because it is a mixture of the isotopes ^{12}C, ^{13}C, and ^{14}C. All three isotopes contain 6 protons, but they contain 6, 7, or 8 neutrons giving atomic masses of 12.000, 13.000, or 14.000 amu respectively. The average mass of these three isotopes (weighted for abundance) gives the atomic mass of naturally occurring carbon (12.011 amu). <u>No single carbon atom</u> has an atomic mass of 12.011, however for stoichiometric purposes, we consider carbon to be composed of only one type of atom with a mass of 12.011.

Atomic mass units allow us to know only the <u>relative</u> atomic masses of the elements. The periodic table lists these relative masses: a hydrogen atom (atomic weight = 1.008) is 1/12 as heavy as a carbon atom; a helium atom (atomic weight = 4.003) is 1/3 as heavy as a carbon atom (or 4 times as heavy as a hydrogen atom).

Relative atomic masses:

1) How many grams of iron would contain the <u>same number</u> of iron atoms as the number of carbon atoms contained in 1.000 g of carbon?

   <u>solution:</u>

   Iron has an atomic weight of 55.85 meaning that it is 55.85/12.01 times as heavy as carbon. If you have 1 gram of carbon, you would need to weigh out:

$$\frac{55.85}{12.01} \times 1.000 \text{ g} = 4.650 \text{ g of iron}$$

2) How many grams of iron would contain the <u>same number</u> of iron atoms as the number of carbon atoms contained in 12.01 g of carbon?

   <u>solution:</u>

$$\frac{55.85}{12.01} \times 12.01 \text{ g} = 55.85 \text{ g of iron}$$

The above example illustrates that <u>you can weigh out the atomic weight of any element (in grams) to give you the same number of atoms of that element</u>:

12.000 g carbon-12 = N carbon-12 atoms

1.008 g hydrogen = N hydrogen atoms

55.85 g iron = N iron atoms

107.868 g silver = N silver atoms

What is N? N has been found to be 6.022×10^{23}. This number (six hundred and two sextillion) is called a "**mole**". A mole is the number equal to the number of carbon atoms in exactly 12 grams of pure carbon-12.

```
1 mole =
        602,200,000,000,000,000,000,000
        602 sextillion
        6.02 x 10²³
        no. of atoms in 12 g of Carbon-12
        Avogadro's number
```

This huge number tells us <u>the number of atoms of an element it takes to weigh the element's atomic mass in grams</u>. It is also called Avogadro's number. It relates a <u>number</u> of atoms to a <u>mass</u>:

$$12.000 \text{ g } ^{12}C \approx 6.022 \times 10^{23} \text{ } ^{12}C \text{ atoms} = 1 \text{ mole } ^{12}C \text{ atoms}$$

$$1.008 \text{ g H} \approx 6.022 \times 10^{23} \text{ H atoms} = 1 \text{ mole H atoms}$$

$$55.85 \text{ g Fe} \approx 6.022 \times 10^{23} \text{ Fe atoms} = 1 \text{ mole Fe atoms}$$

$$107.868 \text{ g Ag} \approx 6.022 \times 10^{23} \text{ Ag atoms} = 1 \text{ mole Ag atoms}$$

Section 3.3: Molar Mass

The Atomic Point of View of Mass:

Chemists make atomic models of everything. We saw in Chapter 1 atomic models for solids, liquids, and gases, and we discussed in Chapter 2 models for molecules, elements, and compounds. Now we consider the atomic model chemists have of <u>mass</u>.

Chemists think of the total mass (m) being equal to the mass per unit population times the population:

Total mass = mass per unit population × population

Total mass = molar mass × amount

$$m = M \times n$$

Molar mass, M is therefore:

$$M = m/n$$

Common units for M:

$$\frac{amu}{atom} = \frac{g}{mole \text{ atoms}} \quad \text{(often written g/mol)}$$

Note: 1 amu × 1 mole = 1 g, therefore

$$1 \text{ amu} = \frac{1 \text{ g}}{1 \text{ mole}}$$

Atomic mass and molar mass:

Choose words from the list below to complete the following paragraph:

(a) amu/atom   (e) g/mole

(b) atomic mass   (f) molar mass

(c) atom   (g) mole

(d) average

12.000 grams of carbon-12 contains one _____(1) of carbon-12. Therefore, the _____(2) of carbon-12 equals 12 _____(3). 12.000 amu of carbon-12 contains one_____(4) of carbon-12. Therefore, the _____(5) of carbon-12 equals 12 _____(6). <u>The mass (in amu) of one atom of an element is numerically the same as the mass (in grams) of one mole of atoms of an element</u>. We work with grams and moles of atoms because we can measure these amounts in the laboratory.

Naturally occurring carbon is a mixture of predominately carbon-12 and carbon-13. The molar mass of naturally occurring carbon is 12.011 g/mole. No single carbon atom has an atomic mass of 12.011 amu/atom, but the _____(7) mass of the carbon atoms in a sample will equal 12.011 amu.

<u>answers</u>: 1-g, 2-f, 3-e, 4-c, 5-b, 6-a, 7-d.

Conversion between grams and amu's:

1) What is the mass of 6.02×10^{23} hydrogen atoms, H, in amu? In grams?

<u>solution</u>:

- The mass of 1 H atom = 1 amu, therefore:

    mass of 6.02×10^{23} H atoms = 6.02×10^{23} amu

- The mass of 6.02×10^{23} H atoms = 1 g, hydrogen's atomic mass in grams.

    The conversion between grams and atomic mass units is therefore:

$$6.02 \times 10^{23} \text{ amu} = 1 \text{ g}$$

We can use this equality to derive two useful conversion factors:

$$\frac{1 \text{ g}}{6.02 \times 10^{23} \text{ amu}} = 1 \quad \text{and} \quad \frac{6.02 \times 10^{23} \text{ amu}}{1 \text{ g}} = 1$$

2) What is the mass of a hydrogen atom in grams?

<u>solution</u>:

$$1\,H\,\text{atom} \times \frac{1.00\ gH}{6.02 \times 10^{23}\ \text{atoms}} = 1.66 \times 10^{-24}\ g$$

Since one hydrogen atom weighs one atomic mass unit, we can write:

$$1\ \textbf{amu} = \textbf{1.66} \times \textbf{10}^{-24}\ \textbf{g}$$

Additional exercises:

1) What mass of hydrogen can be obtained from:

    a) 18.0 g of water, b) 1.00 g of water, c) 10.0 g of water?

    solutions:

- The atomic mass of a H_2O molecule is:

$$2 \times \frac{1.008\ \text{amu}}{H\,\text{atom}} + \frac{16.00\ \text{amu}}{O\,\text{atom}} = \frac{18.02\ \text{amu}}{\text{water molecule}}$$

- The percent hydrogen by mass is:

$$\frac{2 \times 1.008\ \text{amu}/H\ \text{atom}}{18.02\ \text{amu}/\ \text{water molecule}} \times 100\% = 11.2$$

- Water is always 11.2% hydrogen by mass so:

    a) mass of hydrogen in 18 g water = 0.112 × 18.0 g = 2.02 g

    b) mass of hydrogen in 1.00 g water = 0.112 × 1.00 g = 0.112 g

    c) mass of hydrogen in 10.0 g water = 0.112 × 10.0 g = 1.12 g

2) What mass of hydrogen can be obtained from:

    a) 18.0 lbs of water, b) 1.00 lb of water, c) 10.0 lbs of water?

    answers:

    a) 2.02 lb; b) 0.112 lb; c) 1.12 lb

3) How many molecules of water are there in 18 amu of water? In 18 g of water? In 18 lb of water?

    solutions:

    a) The atomic mass of a H_2O molecule is 18.02 amu (from part 1).

      Therefore, 1 molecule of water weighs 18 amu

    b) The molar mass of a water molecule is:

$$2 \times \frac{1.008 \text{ g}}{\text{mole H atoms}} + \frac{16.00 \text{ g}}{\text{mole O atoms}} = \frac{18.02 \text{ g}}{\text{mole water molecules}}$$

Therefore 1 mole (6.02×10^{23} molecules) of water weighs 18 g

c) We know from part "b" that 6.02×10^{23} water molecules weigh 18 g. There are 454 g in 1 lb, therefore:

$$18 \text{ lb} \times \frac{6.02 \times 10^{23} \text{ H}_2\text{O molecules}}{18 \text{ g}} \times \frac{454 \text{ g}}{1 \text{ lb}} = 2.73 \times 10^{26} \text{ H}_2\text{O molecules}$$

4) "A pint of liquid water is a pound the world around". How many water molecules are there in a cup of water?

solution:

We know that:

1 pint water = 1 lb water

1/2 pint water = 1/2 lb water = 1 cup water (1 pint = 2 cups)

18 lb water = 2.73×10^{26} water molecules (from exercise 3c)

Therefore:

$$\tfrac{1}{2} \text{ lb water} = \frac{18 \text{ lb water}}{36} = \frac{2.73 \times 10^{26} \text{ H}_2\text{O molecules}}{36} = 7.58 \times 10^{24} \text{ H}_2\text{O molecules}$$

We could also work this problem by the factor label method:

$$n = 1 \text{ cup H}_2\text{O} \times \frac{1 \text{ pint}}{2 \text{ cups}} \times \frac{1 \text{ lb H}_2\text{O}}{1 \text{ pint H}_2\text{O}} \times \frac{454 \text{ g}}{1 \text{ lb}} \times \frac{6.02 \times 10^{23} \text{ amu}}{1 \text{ g}} \times \frac{1 \text{ H}_2\text{O molecule}}{18 \text{ amu}}$$

$$= 7.59 \times 10^{24} \text{ H}_2\text{O molecules}$$

Notice that if we substitute variables for the numbers in the above calculation, we obtain the equation:

$$n = \left(V_{H_2O} \right) \times 1 \times \left(d_{H_2O} \right) \times 1 \times 1 \times \left(\frac{1}{M_{H_2O}} \right)$$

$$n = \left(V_{H_2O} \right) \times \left(d_{H_2O} \right) \times \left(\frac{1}{M_{H_2O}} \right)$$

$$n = \frac{Vd}{M}$$

We can arrive at this same equation using the equations for molar mass and density:

$$M = m/n \rightarrow n = m/M$$
$$d = m/V \rightarrow m = Vd$$

Substituting m = Vd into the equation for n gives:

$$n = \frac{Vd}{M}$$

Converting from mass to atoms:

You weigh a nickel and find that it weighs 5.0012 grams. How many nickel atoms are in the nickel? A nickel is 25% elemental nickel and 75% copper.

<u>solution:</u>

• First calculate the mass of elemental nickel in a nickel:

$$0.25 \times 5.0012 \text{ g} = 1.2503 \text{ g}$$

• Next convert the mass of the nickel to an amount (moles of nickel):

$$\frac{1.2503 \text{ g}}{58.69 \text{ g/mole}} = 0.02130 \text{ moles}$$

• Finally, convert moles of nickel to atoms of nickel:

$$0.02130 \text{ moles Ni} \times 6.022 \times 10^{23} \text{ atoms/mole} = 1.283 \times 10^{22} \text{ atoms}$$

A nickel weighs less than 1/10th of a mole of nickel atoms (58.69 g), and our answer is a less than 1/10th of Avogadro's number, so it appears to be the right magnitude.

We could also work this problem by first converting grams to amu's, and then converting amu's to atoms:

$$1.2503 \text{ g} \times \frac{1 \text{ amu}}{1.661 \times 10^{-24} \text{ g}} = 7.527 \times 10^{23} \text{ amu}$$

$$7.527 \times 10^{23} \text{ amu} \times \frac{1 \text{ Ni atom}}{58.69 \text{ amu}} = 1.283 \times 10^{22} \text{ atoms}$$

This problem illustrates how molar mass (M, g/mole) relates a mass (in grams) to an amount (moles). Avogadro's number (6.022×10^{23} atoms/mol) relates an amount (moles) to a population of atoms.

Converting from mass to moles and molecules:

A typical baking powder contains 28% sodium bicarbonate, $NaHCO_3$. How many moles and molecules of sodium bicarbonate are in 1.00 gram of this baking powder?

solution:

- First we calculate the molar mass of $NaHCO_3$:

   M ($NaHCO_3$) = molar mass Na + molar mass H + molar mass C + (3 × molar mass O)

   = 23.00 g/mol + 1.01 g/mol + 12.01 g/mol + 3(16.00 g/mol)

   = 84.02 g/mole

- We use the molar mass to convert from grams of baking powder to moles.
  1.00 g of the baking powder contains: 0.28 × 1.00 g = 0.280 g $NaHCO_3$

$$\frac{0.280 \text{ g}}{84.02 \text{ g/mole}} = 3.33 \times 10^{-3} \text{ moles}$$

- Finally, we convert from moles to molecules:

$$3.33 \times 10^{-3} \text{ moles} \times \frac{6.022 \times 10^{23} \text{ molecules}}{\text{mole}} = 2.00 \times 10^{21} \text{ molecules}$$

Empirical and Molecular Formulas:

Empirical formulas show the relative number of atoms of each element in a compound using the smallest whole numbers possible. **Molecular formulas** show the actual number of atoms in a molecule. Water has a molecular formula of H_2O; its empirical formula is also H_2O. Hydrogen peroxide has a molecular formula H_2O_2. Its empirical formula is HO.

Determining Empirical and Molecular Formulas:

Ethanol contains the elements carbon, hydrogen, and oxygen. The weight percentages of each element was measured and found to be:

$$\%C = 52.1; \quad \%H = 13.1; \quad \%O = 34.7$$

Determine the empirical formula. The molar mass of ethanol is found to be about 46.1 g/mole. Determine its molecular formula.

solution:

- We need to determine the relative numbers of C, H, and O atoms. We'll assume, for convenience, we have a 100 gram sample of ethanol. In 100 grams of ethanol, there are 52.1 g C, 13.1 g H, and 34.7 g O. We use the molar masses to find the moles of each element:

$$\frac{52.1 \text{ g C}}{12.011 \text{ g/mole}} = 4.34 \text{ mole C}$$

$$\frac{13.1 \text{ g H}}{1.01 \text{ g/mole}} = 13.0 \text{ mole H}$$

$$\frac{34.7 \text{ g O}}{16.00 \text{ g/mole}} = 2.17 \text{ mole O}$$

• At this point in the problem we can write a preliminary equation:

$$C_{4.34}H_{13}O_{2.17}.$$

However, the empirical formula is written with the simplest <u>whole numbers</u>, so we need to divide each number of moles by the smallest number (2.17):

$$C: \frac{4.34}{2.17} = 2; \qquad H: \frac{13.0}{2.17} = 6; \qquad O: \frac{2.17}{2.17} = 1$$

• The empirical formula for ethanol is: C_2H_6O

• The molar mass of the empirical formula is:

$$M = (2 \text{ mol C} \times 12 \text{ g/mol}) + (6 \text{ mol H} \times 1 \text{ g/mol}) + (1 \text{ mol O} \times 16 \text{ g/mol})$$

$$= 46 \text{ g/mol}$$

Since the molar mass of ethanol equals the molar mass of the empirical formula, the molecular formula is the same as the empirical formula.

$$\text{molecular formula} = C_2H_6O$$

Combustion Analysis:

When a substance containing carbon and hydrogen is burned in pure oxygen, all the carbon in the combusted material converts to CO_2, and all the hydrogen converts to H_2O. If we collect the carbon dioxide and water vapor in separate containers, we can determine their masses, and calculate the masses of carbon and hydrogen in the combusted material. This technique is called combustion analysis.

Chemical composition from combustion analysis:

A 105.5 mg sample of a white substance, suspected of being cocaine ($C_{17}H_{21}NO_4$), forms 279.3 mg of CO_2 and 66.46 mg of H_2O on combustion. Chemical analysis shows that the compound contains 4.68% N by mass. Would you conclude that the white solid is cocaine?

<u>solution:</u>

- First calculate the mmol of CO_2 and H_2O formed:

$$\frac{279.3 \text{ mg } CO_2}{44.01 \text{ mg/mmole}} = 6.346 \text{ mmole } CO_2$$

$$\frac{66.46 \text{ mg } H_2O}{18.02 \text{ mg/mmole}} = 3.688 \text{ mmole } H_2O$$

- Since all the carbon in the unknown converts to CO_2, we know that
  mmol CO_2 = mmol C = 6.346 mmol C,

  and we can use the molar mass of carbon to calculate mg of carbon in the unknown:
  g C = 6.346 mmol × 12.01 mg/mmol = **76.22 mg C**

- Since all the hydrogen in the unknown converts to H_2O, we know that:
  mmol H_2O = 1/2 × mmol H in unknown = 7.376 mmol H,

  and we can use the molar mass of hydrogen to calculate mg of H in the unknown:
  g H = 7.376 mmol × 1.008 mg/mmol = **7.435 mg H**

- The compound contains 4.68% N by weight, so in 105.5 mg, there is:

  105.5 mg × 0.0468 = **4.937 mg N**

- The mass in the unknown sample from carbon, hydrogen, and nitrogen is:

  76.22 mg C + **7.435** mg H + **4.937** mg N = 88.592 mg

- The difference between the sum of the masses of C, H, and N and the total mass of the unknown must be due to the mass of oxygen: 105.5 mg – 88.592 = **16.9 mg O**

- Calculate the percent of each element by mass:

  C: $\dfrac{76.22 \text{ mg}}{105.5 \text{ mg}}$ × 100 = 72.25%

  H: $\dfrac{7.435 \text{ mg}}{105.5 \text{ mg}}$ × 100 = 7.047%

  N: $\dfrac{4.937 \text{ mg}}{105.5 \text{ mg}}$ × 100 = 4.680%

  O: $\dfrac{16.9 \text{ mg}}{105.5 \text{ mg}}$ × 100 = 16.0%

  Total = 100%

- Our last step is to compare this composition to that of cocaine ($C_{17}H_{21}NO_4$):

  M (cocaine) =

  (17 x 12.01 g/mol) + (21 x 1.008 g/mol) + 14.01 g/mol + (4 x 16.00 g/mol)

  = 303.4 g/mol

One mole of cocaine weighs 303.4 g and contains:

  (17 mol x 12.01 g/mol) = 204.2 g C

  (21 mol x 1.008 g/mol) = 21.17 g H

  (1 mol x 14.01 g/mol) = 14.01 g N

  (4 mol x 16.00 g/mol) = 64.00 g O

The percent of each element by mass is:

$$C: \frac{204.2 \text{ mg}}{303.4 \text{ mg}} \times 100 = 67.30\%$$

$$H: \frac{21.17 \text{ mg}}{303.4 \text{ mg}} \times 100 = 6.978\%$$

$$N: \frac{14.01 \text{ mg}}{303.4 \text{ mg}} \times 100 = 4.618\%$$

$$O: \frac{64.00 \text{ mg}}{303.4 \text{ mg}} \times 100 = 21.09\%$$

$$\text{Total} = 99.99\%$$

- The unknown is not cocaine because the composition by mass is not the same. (The unknown is $C_{18}H_{21}NO_3$, codeine).

Section 3.4: Balancing Chemical Equations

Chemical reactions occur when atoms recombine to form new molecules or compounds. Let's imagine oxygen and magnesium reacting to form magnesium oxide: diatomic O_2 molecules break apart, and the oxygen atoms bond to magnesium atoms to produce magnesium oxide molecules (MgO). The chemical equation is:

$$2Mg(s) + O_2(g) \rightarrow 2MgO(s)$$

A balanced equation has the same number of each type of atom on both sides of the equation. This equation tells us that 2 magnesium atoms react with 1 oxygen molecule to produce 2 magnesium oxide molecules, or 2 million Mg atoms react with 1 million O_2 molecules to produce 2 million MgO molecules, or 2 moles of Mg atoms react with 1 mole of O_2 molecules to produce 2 moles of MgO molecules.

Steps for balancing an equation:

1) Write chemical formulas for reactants to the left of an arrow, and the chemical formulas for the products to the right of the arrow.

2) Balance the number of each atom on the two sides of the equation by placing coefficients in front of chemical formulas; begin with the substance with the largest number of atoms.

3) Adjust the coefficients to the smallest whole number coefficients.

4) Specify the states of matter: solid (s), liquid (l), gaseous (g), aqueous solution (aq).

Balancing Combustion Reactions:

The combustion of a hydrocarbon (a compound containing carbon and hydrogen), or a carbohydrate (a compound containing carbon, hydrogen, and oxygen) in the presence of O_2, produces carbon dioxide and water. The easiest way to balance a combustion reaction is to balance the atoms in the order: carbon, hydrogen, and then oxygen. The oxygen is easily balanced at the end because the coefficients of the oxygen molecule can be changed without affecting the number of any other atoms.

Balancing a combustion reaction:

Balance the equation for the combustion of ethanol, C_2H_6O.

solution:

- First write the reactants and products:
$$C_2H_6O + O_2 \rightarrow CO_2 + H_2O$$

- Balance the carbon atoms by placing a 2 in front of CO_2:
$$C_2H_6O + O_2 \rightarrow 2CO_2 + H_2O$$

- Balance the hydrogen atoms by placing a 3 in front of H_2O:
$$C_2H_6O + O_2 \rightarrow 2CO_2 + 3H_2O$$

- Balance the oxygen atoms by placing a 3 in front of O_2:
$$C_2H_6O + 3O_2 \rightarrow 2CO_2 + 3H_2O$$

- Specify states of matter:
$$C_2H_6O(l) + 3O_2(g) \rightarrow 2CO_2(g) + 3H_2O(g)$$

- Check: 2 carbon atoms on both sides, 6 hydrogen atoms on both sides, and 7 oxygen atoms on each side.

> Remember that a coefficient operates on all the atoms in the formula.
>
> Add only __coefficients__ to balance equations, never add or change __subscripts__ (the chemical formulas themselves cannot be changed).

Limiting reactant:

In a chemical reaction, when one reactant is used up, the reaction stops. The **limiting reactant** is simply the reactant that runs out first. If the reactants react in a 1:1 mole ratio, the limiting reactant is the reactant with the fewest moles of material. We use the balanced equation to determine which reactant disappears first.

Limiting reactant:

Ammonia (17.03 g/mol) reacts with copper oxide (79.54 g/mol) to form nitrogen gas (28.02 g/mol), copper (63.55 g/mol), and water (18.02 g/mol). If we react 40.00 g of ammonia with 50.00 g of copper oxide, which is the limiting reactant? How much copper can be produced in this reaction?

__solution:__

• Write the balanced equation:

$$2NH_3(g) + 3CuO(s) \rightarrow N_2(g) + 3Cu(s) + 3H_2O(g)$$

• Calculate the number of moles of each reactant:

$$NH_3: \frac{40.0\ g}{17.03\ g/mole} = 2.349\ mole\ NH_3$$

$$CuO: \frac{50.00\ g}{79.54\ g/mole} = 0.6286\ mole\ CuO$$

• Using the coefficients from the balanced equation, determine how many moles of CuO would theoretically react with 1.468 moles of NH_3:

$$2.349\ mole\ NH_3 \times \frac{3\ mole\ CuO}{2\ mole\ NH_3} = 3.523\ mole\ CuO$$

There are only 0.6286 moles of CuO available, so it is the limiting reactant. We will base the amount of Cu formed on the limiting reactant, CuO:

$$g\ Cu\ produced = 0.6286\ mole\ CuO \times \frac{3\ mole\ Cu}{3\ mole\ CuO} \times 63.55\ g/mole = 39.95\ g\ Cu$$

The amount of product formed based on the limiting reactant is the **theoretical yield**, the yield that could be obtained if 100% of the limiting reactant becomes product.

Real life: Theoretical, actual, and percent yields:

The actual yield that you can obtain from a chemical reaction is never as high as the theoretical yield. Some loss of product always occurs either due to physical loss of material to laboratory equipment, or to competing side reactions. The amount of product that is actually obtained is the **actual yield**. The **percent yield** (% yield) is the ratio of actual to theoretical yield:

$$\% \text{ yield} = \frac{\text{actual yield}}{\text{theoretical yield}} \times 100$$

Percent yield:

If 5.00 g of each ammonia, oxygen, and methane react, how much hydrogen cyanide can be produced? If 2.10 g of HCN is produced, what is the percent yield?

<u>solution:</u>

• The balanced equation is:

$$2NH_3(g) + 3O_2(g) + 2CH_4(g) \rightarrow 2HCN(g) + 6H_2O$$

• Determine how much hydrogen cyanide could be produced if each reactant is completely used up:

$$NH_3 : \frac{5.00 \text{ g}}{17.03 \text{ g/mole}} \times \frac{2 \text{ mole HCN}}{2 \text{ mole NH}_3} = 0.294 \text{ mole HCN}$$

$$O_2 : \frac{5.00 \text{ g}}{32.00 \text{ g/mole}} \times \frac{2 \text{ mole HCN}}{3 \text{ mole O}_2} = 0.104 \text{ mole HCN}$$

$$CH_4 : \frac{5.00 \text{ g}}{16.04 \text{ g/mole}} \times \frac{2 \text{ mole HCN}}{2 \text{ mole CH}_4} = 0.312 \text{ mole HCN}$$

Oxygen produces the least amount of hydrogen cyanide, so it is the limiting reactant.

• The amount of HCN produced if all the oxygen reacts is (the theoretical yield):
0.104 mol HCN × 27.03 g/mol = 2.81 g HCN

• If 2.10 g of HCN is actually produced, the percent yield is:

$$\% \text{yield} = \frac{\text{actual}}{\text{theoretical}} \times 100 = \frac{2.10 \text{ g}}{2.81 \text{ g}} \times 100 = 74.7\%$$

When might an "actual" yield <u>appear</u> higher than the theoretical yield?

Section 3.5: Concentration

Concentration (c) is the amount per unit volume (n/V). For the <u>specificcase</u> when n = moles, and V = liters, concentration is given the special name molarity (M).

molarity = moles of solute/liters of solution

Solution stoichiometry: molarity

What is the molarity of a 750 mL solution containing 14.8 g of silver nitrate?

<u>solution</u>:

• The molar mass of silver nitrate, $AgNO_3$ = 169.9 g/mole, so moles of silver nitrate in 14.8 g is:

$$\frac{14.8 \text{ g}}{169.9 \text{ g/mole}} = 0.08711 \text{ mole}$$

• The molarity is:

$$\frac{0.08711 \text{ mole}}{0.750 \text{ liter}} = 0.116 \text{ M}$$

Remember that molarity refers to moles of a substance in a volume of <u>solution</u>, not a volume of <u>solvent</u>. If you were asked to make up a 0.116 M silver nitrate solution, you would add 14.8 g of silver nitrate to a volumetric flask and dilute to 750 mL with water. You would not add 750 mL of water because the total volume would be greater than 750 mL, and the solution would be too dilute.

Exercise:

How many grams of $ZnCl_2$ (100.83 g/mol) are present in 600 mL of a 0.500 M $ZnCl_2$ solution?

<u>solution</u>:

This problem is easily solved if you write down what is given, and watch units so they cancel to give your answer in grams of zinc chloride:

(0.500 mol/L)(0.600L)(100.83g/mol) = 30.2 g $ZnCl_2$

Limiting reactant for a reaction in solution:

The reaction of silver nitrate with potassium bromide is:

$$AgNO_3(aq) + KBr(aq) \rightarrow AgBr(s) + KNO_3(aq)$$

If you react 45.0 mL of a 1.25 M AgNO$_3$ solution with 75.0 mL of a 0.800 M KBr solution, what is the limiting reactant, and how much AgBr is produced?

<u>solution:</u>

• First calculate the moles of reactants:

AgNO$_3$: (0.0450L)(1.25 mol/L) = 0.0562 mol

KBr: (0.0750L)(0.800 mol/L) = 0.0600 mol

Since the reactant react in a 1:1 molar ratio, the reactant with the fewer number of moles (AgNO$_3$) is the limiting reactant.

• We use the amount of the limiting reactant, AgNO$_3$, to determine the theoretical yield of silver bromide:

$$0.0562 \text{ mole AgNO}_3 \times \frac{1 \text{ mole AgBr}}{1 \text{ mole AgNO}_3} \times 187.77 \text{ g/mole} = 10.6 \text{ g AgBr}$$

SUMMARY:

Units of measure:

mass (m), volume (V), number (n):

density (d) = mass/volume = m/V

concentration (c) = amount/volume = n/V
special case: molarity (M) = moles/L

molar mass (M) = mass/amount = m/n

SI unit for amount (n) = **mole**

1 **mole** = Avogadro's number

= 6.022 $\times$ 10^{23}

= the number of atoms in 12.000 g of carbon-12.

Balanced equations:

Equations are balanced when the same number of each atom type appears on both sides of the equation.

Combustion reactions of hydrocarbons or carbohydrates in the presence of oxygen always produce CO_2, and H_2O. The amounts of CO_2 and H_2O formed can be used to determine the composition of the combusted substance.

The atoms in combustion reactions should be balanced in the order: carbon, hydrogen, and then oxygen.

The **limiting reactant** is the one which is used up first in a reaction and causes the reaction to stop. The amount of the limiting reactant determines the **theoretical yield.**

$$\% \text{ yield} = \frac{\text{actual yield}}{\text{theoretical yield}} \times 100$$

Chapter 3 Exercises:

3.1)   a. How many moles of carbon atoms are there in 24 grams of carbon?
b. How many moles of oxygen atoms are there in 64 grams of oxygen?
c. How many moles of hydrogen molecules are there in 20 grams of hydrogen?

3.2)   a. What is the molar mass of methane CH_4 in g/mol?
b. What is the molar mass of potassium sulfide K_2S in g/mol?
c. What is the molar mass of neon gas in g/mol?

3.3)   a. How many grams of nitrogen contain the same number of moles of atoms as 24.00 grams of hydrogen?

b. Which sample contains the greatest number of atoms?
     One mole of hydrogen molecules H_2
     One mole of carbon atoms
     Two moles of helium atoms
     One mole of uranium atoms
     One-half mole of methane molecules CH_4

c. Which sample has the greatest mass?
     One mole of iron atoms
     One mole of sulfur atoms
     One mole of nickel atoms
     One mole of potassium atoms
     One mole of tin atoms

3.4)   a. How many moles of nitrogen are there in 12 moles of TNT (trinitrotoluene; $CH_3C_6H_2(NO_2)_3$) ?

b. How many moles of nitrogen are there in 3 moles of ammonium nitrate (NH_4NO_3) ?

3.5)   What is the percentage by mass of:
     a. sulfur in sulfur dioxide SO_2
     b. carbon in ethanol C_2H_5OH
     c. iron in limonite $2Fe_2O_3 \cdot 3H_2O$

3.6)   a. Malonic acid has the composition:
     Carbon 34.6%
     Hydrogen 3.9%
     Oxygen 61.5%
What is its empirical formula?

b. Exactly ten grams of an oxide of uranium is treated with fluorine to produce oxygen and 13.45 grams of uranium hexafluoride UF6. What is the empirical formula of the oxide?

3.7)   The elemental analysis of estrone, one of the female sex hormones, reported the compound to be 80.0% C, 8.15% H, and 11.85% O by mass. The mass spectrum shows the molar mass to be 270 g/mole. What are the empirical and molecular formulas for estrone?

3.8) Talc, the principle ingredient of baby powder, has the following elemental composition: 19.24% Mg, 0.53% H, 29.56% Si, and 50.67% O by mass. Further analysis shows that each formula contains two hydroxide groups. What is the empirical formula for talc?

3.9) Ibuprofen, an alternative analgesic to aspirin, contains only C, H, and O. It is sold under trade names such as Motrin and Advil. The combustion of 24.72 mg of Ibuprofen produced 68.64 mg of CO_2 and 12.96 mg of H_2O. What is the empirical formula for Ibuprofen?

3.10) a. What coefficient is necessary for nitric acid to balance the equation:

$$3Cu + HNO_3 \rightarrow 3Cu(NO_3)_2 + 2NO + 4H_2O$$

b. What coefficients are necessary to balance the equation:

$$Pb(NO_3)_2 \rightarrow PbO + NO_2 + O_2$$

3.11) a. When benzene C_6H_6 burns in air, carbon dioxide and water are produced. What is the ratio of the moles of water produced to the moles of carbon dioxide produced?

b. Aluminum carbide reacts with water as follows:

$$Al_4C_3 + 12H_2O \rightarrow 3CH_4 + 4Al(OH)_3$$

How many grams of methane will be produced if 12 g of aluminum carbide are used?

3.12) Sodium azide, NaN_3, is used to generate the gas which fills automobile air bags during a collision. What mass of N_2 gas is produced by the decomposition of 715 grams of NaN_3?

$$2NaN_3 \rightarrow 2Na + 3N_2$$

3.13) a. If 7.00 grams of ammonia and 14.0 grams of oxygen react to form nitric oxide and water as shown in the equation below, which one of the two reactants limits the amount of product formed?

$$4NH_3 + 5O_2 \rightarrow 4NO + 6H_2O$$

b. How much of the non-limiting reactant remains after all the limiting reactant has been used up?

c. How much nitric oxide is formed?

3.14) Phosphoric acid H_3PO_4 is manufactured by treating mineral phosphates with concentrated sulfuric acid H_2SO_4. A typical phosphate mineral, $Ca_5(PO_4)_3F$, reacts with sulfuric acid as shown by the equation:

$$Ca_5(PO_4)_3F + 5H_2SO_4 \rightarrow 3H_3PO_4 + 5CaSO_4 \cdot 2H_2O + HF$$

If 5.00 tons of this mineral is treated with 6.00 tons of sulfuric acid to produce 2.70 tons of phosphoric acid, what is the percent yield of phosphoric acid?

3.15) Liquid propane (LP gas) is the fuel used in outdoor gas grills and is commonly used for heating in rural areas. The use of liquid propane is a safe fuel as long as an abundant source of oxygen is available.

$$C_3H_8 + 5O_2 \rightarrow 3CO_2 + 4H_2O$$

If 55.0 g of propane burns in the presence of 192 g of oxygen, what is the limiting reactant? How much of the excess reactant remains?

3.16) a. When 8.0 grams of sodium hydroxide is dissolved in sufficient water to make 400 mL of solution, what is the molarity of the solution?

b. If this solution is then poured into a volumetric flask and made up to 1.0 L in volume, what would the molarity of the solution become?

3.17) a. How many moles of nitric acid are there in a 75 mL sample of a 0.60 M solution of nitric acid?

b. If this sample is made up to 2.0 L in volume, what would the molarity of the solution be?

3.18) A laboratory technician has been asked to prepare 750 mL of a 0.200 $M H_2SO_4(aq)$ solution. The only sulfuric acid solution the technician has available in the stockroom is 6.00 $M H_2SO_4(aq)$. Describe how to prepare the 0.200 $M H_2SO_4(aq)$ solution.

3.19) Mylanta, a commonly used over-the-counter antacid, is used to relieve heartburn. Each gelcap contains 311 mg of calcium carbonate and 232 mg of magnesium carbonate.

$$CaCO_3 + 2HCl(aq) \rightarrow CaCl_2(aq) + CO_2(g) + H_2O$$
$$MgCO_3 + 2HCl(aq) \rightarrow MgCl_2(aq) + CO_2(g) + H_2O$$

How many milliliters of stomach acid, 0.030 M HCl, will two Mylanta gelcaps neutralize?

3.20) 2.50 g of elemental phosphorous, P_4, was burned in an excess of oxygen to produce a compound with the elemental composition of 43.7% P and 56.3% O by mass. The mass spectrum of the product indicates it has a molar mass between 280 and 285 g/mol.
   a. What is the molecular formula of the product?
   b. Give the name of the product.
   c. Write a balanced chemical equation for the formation of the product.
   d. What mass of product was produced?

Chapter 3 Answers:

3.1)   a. 2.0 mol carbon atoms; b. 4.0 mol oxygen atoms; c. 10 mol hydrogen molecules

3.2)   a. 16.04 g/mol; b. 110.26 g/mol; c. 20.18 g/mol

3.3)   a. 333.5 g; b. One-half mole of methane (contains 2.5 mol atoms)
         c. Each sample contains one mole of atoms. The sample with the greatest mass is the
         substance with the greatest molar mass: one mole of tin atoms has a mass = 118.7 g.

3.4)   a. 36 mol; b. 6 mol

3.5)   a. 50.05%; b. 52.14%; c. 59.82%

3.6)   a. $C_3H_4O_4$; b. U_2O_3

3.7)   empirical formula: $C_9H_{11}O$; molecular formula: $C_{18}H_{22}O_2$

3.8) $Mg_3H_2Si_4O_{12}$

3.9) $C_{11}H_{10}O_2$

3.10)   a. 8; b. $2Pb(NO_3)_2 \rightarrow 2PbO + 4NO_2 + O_2$

3.11)   a. One mole of benzene will produce 6 moles of carbon dioxide on combustion, and 3
         moles of water. The ratio between the compounds, water and carbon dioxide, is:
         6 moles CO_2 : 3 moles H_2O, or 2:1.

         b. 4.0 grams CH_4.

3.12)   462 g N_2

3.13)   a. oxygen; b. 1.04 g NH_3 left over (0.061 mol NH_3); c. 10.5 g NO

3.14)   100% yield = 2.91 tons; Actual yield is 2.7 tons, or 92.8%

3.15)   Limiting reactant is O_2; 2.10 g C_3H_8 remains

3.16)   a. 0.50 M; b. 0.20 M

3.17)   a. 0.045 moles HNO_3
         b. The new solution contains 0.045 moles in 2.0 L. The molarity is
         0.045/2 = 0.023 M

3.18)   Add 25.0 mL of 6.00 M H_2SO_4 to 350 mL of distilled water, dilute to 750 mL, and mix.

3.19)   780 mL HCl

3.20)   a. P_4O_{10}
         b. tetraphosphorous decoxide
         c. $P_4 + 5O_2 \rightarrow P_4O_{10}$
         d. 5.73 g P_4O_{10}

The Major Classes of Chemical Reactions

TOPICS:

- Combination, Decomposition, Displacement Reactions
- Acid-Base Reactions
- Precipitation Reactions
- Oxidation Reduction (redox) Reactions
- Chemical Equilibrium

Chapter 4 presents an introduction to many concepts that will be explored in more depth in later chapters.

SECTION 4.1: Types of Chemical Reactions

Atoms and molecules are in constant motion. We can picture them as "sticky" electron clouds that bounce into each other and may "stick" together, or fall apart and stick to new atoms. These atomic dances are called chemical reactions. The historical classification of reactions as combination, decomposition, and displacement reactions does not require much knowledge of chemistry since chemists didn't know much chemistry at first. As scientists learned more about the chemistry involved in reactions, they were able to classify reactions according to how they behave. We will discuss acid-base, precipitation, and redox reactions in this chapter.

Historical classification of reactions:

1) Match the following reactions with their historical classification:

1) X + Y → Z	a) Single displacement
2) Z → X + Y	b) Combination
3) X + YZ → XZ + Y	c) Decomposition
4) WX + YZ → WZ + XY	d) Double displacement or Metathesis

answers: 1-b, 2-c, 3-a, 4-d

2) Choose words from the list below to complete the following paragraph:

(a) atoms	(d) decomposition	(g) elements
(b) combination	(e) displacement	(h) heat
(c) compound(s)	(f) electricity	(i) molecules

Just as the name implies, a _____(1) reaction is one in which two (or more) _____ or _____(2&3) stick together to form one substance. Combination

---

reactions can involve two elements, an element and a compound, or two compounds. When a reactant absorbs enough energy to break apart into two or more products, a _____(4) reaction occurs. Usually the starting material is a _____(5), and the products are smaller _____(6) and/or _____(7). Compounds are generally decomposed by _____(8) (thermal decomposition) or _____(9) (electrolytic decomposition). You can think of _____(10) reactions as molecules and atoms changing partners. They are often referred to as single-displacement and double-displacement (metathesis) reactions.

   answers: 1-b, 2-a, 3-i, 4-d, 5-c, 6-c, 7-g, 8-h, 9-f, 10-e.

Combination of two elements:

   As you look at a periodic table, it might appear that there are quite a few elements that could potentially combine together to form a compound. However, not all elements have the right arrangement of electrons to be reactive. Nonmetals, on the right hand side of the table (except for the noble gases in the last column), tend to be reactive.

Combination of Elements:

Reactions of the nonmetal, chlorine:

$$Cl_2(g) + 2Na(s) \rightarrow 2NaCl(s)$$
$$Cl_2(g) + Mg(s) \rightarrow MgCl_2(s)$$
$$3Cl_2(g) + 2Fe(s) \rightarrow 2FeCl_3(s)$$
$$Cl_2(g) + 3F_2(g) \rightarrow 2ClF_3(g)$$

Fill in the blanks based on the above reactions and the list of words below:

   (a) alkali metal   (d) reactive
   (b) alkaline earth metal   (e) transition metal
   (c) non-metal

   Chlorine, a nonmetal, reacts with sodium, an _____(1); chlorine reacts with magnesium, an _____(2); chlorine reacts with iron, a _____(3); and chlorine reacts with fluorine, another _____(4).

Conclusion: chlorine is _____(5).

   answers: 1-a, 2-b, 3-e, 4-c, 5-d

Combination of Elements:

Reactions of the nonmetal, oxygen:

Almost every element combines with oxygen to form an oxide.

A) Name the following oxides:

a) CO_2 d) CO

b) H_2O e) CaO

c) NO f) Fe_2O_3

answers:

a) carbon dioxide   d) carbon monoxide
b) water   e) calcium oxide
c) nitrogen monoxide   f) iron(III) oxide

B) Fill in the appropriate oxides from the list above:

Oxides surround us:

1) we exhale _____ and _____;

2) our cars exhale _____, _____, _____, and _____;

3) rust is _____;

4) lime is _____. 5).

Conclusion: oxygen is _____.

answers: 1-a,b; 2-a,b,c,d; 3-f; 4-e; 5-reactive.

Reactions of oxides:

Oxides may react with more oxygen (equation "a" below), with each other (equation b), or with water (equations c&d) to form larger compounds:

a) $2NO(g) + O_2(g) \rightarrow N_2O_4(g)$

b) $CaO(s) + SiO_2(s) \rightarrow CaSiO_3(s)$

c) $CaO(s) + H_2O(l) \rightarrow Ca(OH)_2(aq)$ (calcium hydroxide)

d) $SO_3(l) + H_2O(l) \rightarrow H_2SO_4(aq)$ (sulfuric acid)

Conclusion: oxides tend to be reactive.

Decomposition:

When compounds decompose thermally, they often release familiar gaseous products such as carbon dioxide (equation "a" below), oxygen (equation "b"), or water (equation "c").

a) $CaCO_3(s) \rightarrow$

b) $2HgO(s) \rightarrow$

c) $Mg(OH)_2(s) \rightarrow$

answers:

a) $CaO(s) + CO_2(g)$

b) $2Hg(l) + O_2(g)$

c) $MgO(s) + H_2O(g)$

Electrical energy can decompose compounds through the process of electrolysis. Water can be decomposed into the elements hydrogen and oxygen:

$$2H_2O(l) \xrightarrow{\text{electricity}} 2H_2(g) + O_2(g)$$

This reaction is useful for generating a potential fuel, H_2.

Displacement Reactions

1) Single Displacement Reactions

Metals may displace hydrogen gas from water or an acid. Metals may also displace other metal ions from solution. Halogens may displace another halide ion from solution.

Predict the products for the following reactions:

a) $Ba(s) + 2H_2O(l) \rightarrow$

b) $Cu(NO_3)_2(aq) + Zn(s) \rightarrow$

c) $2KBr(aq) + Cl_2(aq) \rightarrow$

answers:

a) $Ba(OH)_2(aq) + H_2(g)$

b) $Cu(s) + Zn(NO_3)_2(aq)$

c) $Br_2(l) + 2KCl(aq)$

2) Double Displacement Reactions

Three common double displacement reactions are:

a) precipitation of an insoluble ionic compound

b) neutralization of an acid and a base to form water

c) reaction of a carbonate, nitrite, or sulfite with an acid to form a gas.

Predict the products for the following reactions:

a) precipitation of the solid ionic compound, $AgBr(s)$:

$$AgNO_3(aq) + KBr(aq) \rightarrow$$

b) neutralization of an acid (HCl) and base (NaOH) to form water:

$$NaOH(s) + HCl(aq) \rightarrow$$

c) reaction of a carbonate with an acid to form a gas (CO_2):

$$MgCO_3(s) + 2HCl(aq) \rightarrow$$

answers:

a) $AgBr(s) + KNO_3(aq)$

b) $NaCl(aq) + H_2O(l)$

c) $MgCl_2(aq) + H_2O(l) + CO_2(g)$

We have been able to use the historical classification of reactions to predict products for reactions without understanding much about the chemistry. As you will see, as we learn more about the chemistry involved in reactions, we will be able to classify reactions according to how they behave.

SECTION 4.2: Water as a Solvent

Let's consider two types of solids both of which dissolve in water. An ionic solid breaks apart into charged ions as it dissolves. A covalent compound stays intact as uncharged molecules.

 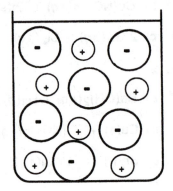

Which of the drawings represents a table salt (NaCl) solution and which represents a sugar solution? (Answer: the drawing on the right represents a sodium chloride solution with charged ions (Na^+ and Cl^-); the drawing on the left represents a sugar solution with intact $C_{12}H_{22}O_{11}$ molecules). We show that ions or molecules are in solution (or **solvated**) with the symbol "aq" in the chemical equation:

$$NaCl(s) \xrightarrow{H_2O} Na^+(aq) + Cl^-(aq)$$

$$C_{12}H_{22}O_{11}(s) \xrightarrow{H_2O} C_{12}H_{22}O_{11}(aq)$$

An ionic compound which is <u>insoluble</u> contains electrostatic attraction between its ions that are too strong for the water molecules to overcome. We saw above that silver bromide (AgBr) precipitates out in the double displacement reaction of silver nitrate ($AgNO_3$) with potassium bromide (KBr). Silver bromide is an insoluble compound often used in photographic films to record an image when exposed to light. If silver bromide were soluble, the silver bromide crystals, and hence the recorded image, would dissolve as soon as the film was immersed in aqueous developer solution. However, as we will see later, even compounds which are considered insoluble dissolve to a very small extent.

Ionic vs. molecular:

How could you tell a sugar and a salt solution apart? One way would be to taste them, but tasting is not recommended today for safety reasons. (Older chemists may attest that tasting used to be a common method for identification; we won't comment on the effects this method had on the chemists who used it). A safer method to tell a sugar and salt solution apart would be to immerse electrodes into the two solutions. The salt solution would conduct a current as its charged particles (ions) move towards oppositely charged electrodes. The sugar solution would not conduct a current because its molecules have no charge. A sodium chloride solution is called an **electrolyte** because it conducts a current, a sugar solution is called a **nonelectrolyte**.

Solubility of Ionic compounds in water

Why is water attracted to sodium and chloride ions? Water is a **polar** molecule; it has a partially negative pole which attracts cations, and a partially positive pole which attracts anions:

The oxygen atom attracts electrons more strongly than hydrogen, so each O-H bond in the water molecule is a **polar covalent bond**; the electrons spend more time near the oxygen atom than around the hydrogen atoms. The shape of the electron cloud around oxygen, pushes the hydrogen atoms away from a linear orientation to a bent configuration. It is this bent configuration which gives water its positive and negative poles.

Covalent compounds in water:

   The polar nature of a water molecule explains why some covalent compounds such as table sugar dissolve in water, and others such as octane, the major component in gasoline, do not dissolve in water. Sucrose contains polar covalent bonds which interact with those of water. Hydrocarbons such as octane do not contain polar bonds, and therefore do not dissolve in water. The general rule is that "like dissolves like".

Acids are a special case of covalent molecules which contain hydrogen and interact so strongly with water that they dissociate into ions. We show the dissociation of a fully dissociated "strong" acid such as nitric acid as:

$$HNO_3(aq) \xrightarrow{\ H_2O\ } H^+(aq) + NO_3^-(aq)$$

The H^+ ion is simply a <u>bare proton</u>. It is an extremely small volume of positive charge and associates strongly to the negative pole of water molecules. Often an aqueous H^+ ion is written as H_3O^+ to indicate this strong association:

$$HNO_3(aq) \xrightarrow{\ H_2O\ } H_3O^+(aq) + NO_3^-(aq)$$

SECTION 4.3:   Precipitation reactions

   When two aqueous ionic compounds are mixed, a precipitate may form. If we introduce silver and bromide ions into the same flask, a pale yellow solid (silver bromide) precipitates out of solution. Silver halide crystals used in photographic film are commonly produced by introducing silver nitrate and potassium bromide into an aqueous solution. Silver nitrate and potassium bromide are soluble in water which means they dissociate into solvated ions:

$$AgNO_3(s) \xrightarrow{\ H_2O\ } Ag^+(aq) + NO_3^-(aq)$$

$$KBr(s) \xrightarrow{\ H_2O\ } K^+(aq) + Br^-(aq)$$

The four ions are the potential reactants. The potential products are AgBr and KNO_3. It turns out that potassium nitrate (KNO_3) is soluble, and silver bromide (AgBr) is the insoluble precipitate which forms. There are three ways you may see this reaction written:

1) Total ionic equation:

$$Ag^+(aq) + NO_3^-(aq) + K^+(aq) + Br^-(aq) \rightarrow AgBr(s) + NO_3^-(aq) + K^+(aq)$$

Total ionic equations show <u>all the ions, even if they are not involved in a reaction</u>. In this equation, potassium and nitrate ions are not involved in the reaction (potassium nitrate is soluble), and they are referred to as **spectator ions**.

2) Net ionic equation:

$$Ag^+(aq) + Br^-(aq) \rightarrow AgBr(s)$$

Net ionic equations eliminate spectator ions and record only ions involved in precipitation.

3) molecular equation:

$$AgNO_3(aq) + KBr(aq) \rightarrow AgBr(s) + KNO_3(aq)$$

Molecular equations show all reactants and products as undissociated compounds. A molecular equation does not indicate which compounds are soluble and insoluble. Historically, this equation would have been called a double displacement reaction.

We can predict which combinations of ions will precipitate out from aqueous solutions by using solubility rules summarized below.

SOLUBILITY RULES FOR IONIC COMPOUNDS IN WATER

1. **Most nitrate (NO_3^-), sulfate (SO_4^{2-}) and perchlorate (ClO_4^-) salts are soluble**; exceptions are $CaSO_4$, $BaSO_4$, and $PbSO_4$ which are insoluble.

   Note: NO_3^-, SO_4^{2-}, and ClO_4^- are all anions of strong acids that lose H^+ readily to water (see the list of strong acids in the next section). These anions are able to spread their negative charge out over several oxygen atoms, and therefore have low charge densities which do not attract other cations strongly, so they tend to form soluble salts, especially with cations of low charge.

2. **Most sulfide (S^{2-}), carbonate (CO_3^{2-}), and phosphate (PO_4^{3-}) salts are only slightly soluble.**

   Note: S^{2-}, CO_3^{2-}, and PO_4^{3-} are anions of weak acids. These ions have higher charges spread over fewer oxygen atoms than the anions of strong acids. The higher charge densities attract cations, so they tend to form insoluble salts.

3. **Most hydroxide salts are only slightly soluble**; exceptions are NaOH, KOH, and $Ca(OH)_2$ which are soluble.

   Note: OH^- has only a -1 charge, but it is localized on one oxygen atom, and therefore attracts cations strongly, and tends to form insoluble salts.

4. **Most salts of Na^+, K^+, and NH_4^+ are soluble**

5. **Most chloride, bromide and iodide salts are soluble**; exceptions are AgCl, $PbCl_2$, and Hg_2Cl_2 which are insoluble.

SECTION 4.4: Neutralization reactions

For this chapter, we define an acid as a substance that produces protons (H^+ ions) when dissolved in water, and a base as a substance that produces hydroxide (OH^-) ions when dissolved in water.

Acids and bases:

Hydrochloric acid (a)and sodium hydroxide (b) are a common acid and base:

a) $HCl(aq) \rightarrow H^+(aq) + Cl^-(aq)$

b) $NaOH(s) \rightarrow Na^+(aq) + OH^-(aq)$

Choose words from the list below to complete the following paragraphs:

(a) dissociates   (e) weak
(b) low   (f) HF
(c) partly   (g) Mg^{2+}
(d) strong   (h) OH^-

Hydrochloric acid is a _____(1) acid because it _____(2) completely into ions. Likewise, sodium hydroxide is a _____(3) base because it _____(4) completely into ions. Most acids and bases are not strong, they dissociate only _____(5). Hydrogen fluoride (HF) is a weak acid; when it dissolves in water, it only partially ionizes; the concentration of H^+ and F^- ions is _____(6) compared to the concentration of undissociated _____(7) molecules.

Solubility is different than extent of dissociation. Magnesium hydroxide does not dissolve readily in water, but the molecules that do go into solution dissociate completely to _____(8) and _____(9) ions. In contrast, HF, a _____(10) acid, is completely soluble in water, but does not dissociate to a very great extent.

answers: 1-d, 2-a, 3-d, 4-a, 5-c, 6-b, 7-f, 8-g, 9-h, 10-e.

Not all bases contain OH^- as part of their structure. Bases may <u>remove</u> a proton from water producing hydroxide ion. Ammonia (NH_3) is a base because it obtains a proton from water leaving OH^- ions in solution:

$$NH_3(g) + H_2O(l) \rightarrow NH_4^+(aq) + OH^-(aq)$$

Soluble metal oxides act as strong bases in solution because the oxide ion reacts quickly with water to form hydroxide:

$$O^{2-}(aq) + H_2O(l) \rightarrow 2OH^-(aq)$$

THE STRONG ACIDS AND BASES

ACIDS   **BASES**

Hydrochloric acid   HCl   Hydroxides of the 1A and 2A metals:

Hydrobromic acid   HBr   LiOH, NaOH*, KOH*, RbOH, CsOH

Hydroiodic acid   HI $Mg(OH)_2$, $Ca(OH)_2$, $Sr(OH)_2$,

Nitric acid HNO_3 $Ba(OH)_2$

Perchloric acid $HClO_4$ Metal oxides

Sulfuric acid H_2SO_4 * most commonly used in the lab

Neutralization reaction:

Reaction of an acid and a base produces water and a salt (on evaporation) in a **neutralization** reaction. Write the 1) total ionic equation, 2) net ionic equation, and 3) molecular equation for the reaction of hydrochloric acid with sodium hydroxide.

<u>answers</u>:

1) total ionic equation:

$$H^+(aq) + Cl^-(aq) + Na^+(aq) + OH^-(aq) \rightarrow Na^+(aq) + Cl^-(aq) + H_2O(l)$$

2) net ionic equation: (Na^+ and Cl^- are the spectator ions and therefore not included)

$$H_3O^+(aq) + OH^-(aq) \rightarrow 2H_2O(l)$$

3) molecular equation:

$$NaOH(s) + HCl(aq) \rightarrow NaCl(aq) + H_2O(l)$$

Historically, this acid base reaction would have been written as the molecular equation above and would have been called a double displacement reaction.

Acid-Base titration

We can determine the concentration of an unknown acid (or base) by titration with a base (or acid) of known concentration. Titrations depend on **indicators** to indicate the point at which neutralization of the acid and base occurs. Indicators are themselves weak acids and are useful in that they are one color in their acidic form, and a different color in their basic form.

Let's suppose we want to know the concentration of acetic acid in vinegar. We would place a measured volume of the vinegar into a flask, and add a few drops of indicator solution that turns the solution pink (pink is the color of the acidic form of this indicator). We would then add a standardized basic solution dropwise into the acid, swirling the flask. The hydroxide ions from the base immediately react with protons from the acid to form water molecules. When all the moles of H^+ ion present in the vinegar react with an equivalent number of moles of OH^- ion added dropwise, we are at the **equivalence point**. A tiny amount of excess OH^- ions causes the indicator to become deprotonated to its basic form, and its color change indicates to us that we have reached the end point of the titration.

Acid base titration:

We find that 22.3 mL of 0.240 M NaOH is required to react with a 50.0 mL sample of vinegar. Calculate the concentration of acetic acid in the vinegar.

solution:

• First write down the equation for the reaction of OH^- with acetic acid:

$$HC_2H_3O\ (aq) + OH^-(aq) \rightarrow C_2H_3O^-(aq) + H_2O\ (l)$$

Notice that the reactants are consumed in a 1:1 mole ratio. Even though acetic acid contains more than one hydrogen atom, only one of the hydrogen atoms is "acidic". We write formulas for acids with the acidic hydrogens at the front.

• Now we will determine how many moles of hydroxide ions were required to reach the end point:

$$22.3 \text{ mL} \times \frac{1 \text{ L}}{1000 \text{ mL}} \times 0.240 \text{ mole/L} = 0.00535 \text{ mole}$$

• Since moles OH^- = moles HC_2H_3O, we can calculate the concentration of acetic acid:

$$\frac{0.00535 \text{ mole acetic acid}}{0.050 \text{ L}} = 0.107 \text{ M}$$

SECTION 4.5: Oxidation-Reduction Reactions

When two atoms stick together to form a chemical bond, one atom may attract the electrons more strongly than the other atom. The electrons spend more time around the atom with the stronger attraction, and if this change in the distribution of electrons occurs to a great enough extent, we can say electrons transfer from one reactant to another. These electron transfer reactions are called **oxidation-reduction** (or redox) reactions. Oxidation is the loss of electrons, reduction is the gain of electrons. "Oil rig" is a good mnemonic device for remembering this.

Oxidation and Reduction:

Choose words from the list below to complete the following paragraph:

(a) gains
(b) loses
(c) oxidized

(d) oxidizing agent
(e) reduced
(f) reducing agent

When sodium and fluorine react, sodium _____(1) an electron to become Na^+ and we say it's _____(2). Fluorine _____(3) an electron to become F^- and we say it's _____(4). Alternatively, we can say that sodium is the _____(5) because it causes fluorine to be reduced and fluorine is the _____(6).

<u>answers</u>: 1-b, 2-c, 3-a, 4-e, 5-f, 6-d.

Let's compare the reaction between hydrogen and fluorine and the reaction between sodium and fluorine. The reactions are:

$$1/2 H_2(g) + 1/2 F_2(g) \rightarrow HF(g)$$

$$Na(s) + 1/2 F_2(g) \rightarrow NaF(s)$$

In both reactions, there is an exchange of electrons between atoms, however there is a difference in the <u>extent</u> to which the electrons are transferred. In the case of sodium fluoride formation, we think of the electron as being completely transferred from sodium to fluorine to produce ions which are electrostatically attracted to each other to form an ionic solid. In the case of hydrogen fluoride formation, a hydrogen electron is displaced towards the fluorine atom but is shared with the hydrogen atom to form a covalent bond. Since the electrons in the H-F covalent bond are displaced strongly toward fluorine, for bookkeeping purposes, we think of fluorine as having gained an electron.

It is convenient to think of redox reactions as occurring in two steps, although it is important to remember that the oxidation and reduction occur simultaneously. In the above case with NaF formation, we can write:

Oxidation: $Na \rightarrow Na^+ + e^-$ (electron lost by sodium)

Reduction: $1/2 F_2 + e^- \rightarrow F^-$ (electron gained by F)

We can keep track of which atoms lose electrons and which gain them with **oxidation numbers**. <u>The oxidation number refers to the charge an atom would have if the electrons that bond two atoms together are assigned to the element which attracts the electrons more strongly</u>. For covalent compounds, the oxidation number is not an accurate description of the electron distribution because the electrons are only partially transferred from one atom to another and so do not carry a full positive or negative charge. However, for bookkeeping purposes, we assign oxidation numbers as whole numbers and keep in mind that it is only the atoms in ionic compounds which actually completely transfer electrons between atoms.

RULES FOR ASSIGNING OXIDATION NUMBERS

1) An atom in its elemental form has an oxidation number of 0.

   O.N. of N_2, H_2, and Na = 0.

2) A monatomic ion has an oxidation number equal to the ion charge.

   O.N. of Na^+ = +1; of Mg^{2+} = +2

3) The sum of oxidation numbers for the atoms in a compound equals 0.
   The sum of oxidation numbers for the atoms in a polyatomic ion equals the ion charge.

   For the compound NH_3 (ammonia), the oxidation numbers for nitrogen (-3) and 3 hydrogens (+1 each) add up to zero:

   $$(-3) + (3 \times 1) = 0.$$

   For the polyatomic ion, NH_4^+, the oxidation numbers for nitrogen (-3) and 4 hydrogens (+1 each) add up to +1, the charge on the ammonium ion:

   $$(-3) + (4 \times 1) = +1.$$

4) Some elements (listed below) have the same oxidation number in almost all their compounds:

   Group 1A elements = **+1**.

   Group 2A elements = **+2**.

   Fluorine = -1.

   Hydrogen usually has an oxidation number of **+1**; when it is in combination with metals, it has an oxidation number of -1.

   O.N. of hydrogen in:
   H_2O = +1
   NaH = -1.

   Oxygen usually has an oxidation number of **-2**; it has an oxidation number of -1 in peroxides, and -1/2 in superoxides.

   O.N. of oxygen in:
   H_2O = -2
   H_2O_2 (hydrogen peroxide) = -1
   KO_2 (a superoxide) = -1/2

   Group 7A elements usually have an oxidation number of **-1**; notable exceptions are when these elements are in combination with oxygen.

   O.N. of Cl in:
   HCl = -1
   $HAuCl_4$ = -1
   $HClO_4$ = +7

Most elements are more versatile than the ones listed above. Transition metals may have as many as five possible oxidation numbers. Metals tend to lose electrons, so their oxidation numbers are positive. With a few exceptions, an element's group number in the Periodic Table is its highest possible oxidation number. Once we understand oxidation numbers, the words oxidation and reduction make more sense: <u>an atom which is oxidized undergoes an increase in its oxidation number</u>; <u>an atom which is reduced undergoes a decrease (or reduction) in its oxidation number</u>.

Determining oxidation numbers:

Determine the oxidation number of each element in the following compounds:

      a) phosphorus trichloride

      b) sulfuric acid

      c) perchloric acid.

<u>answers</u>:

a) PCl_3: Halides usually have an oxidation number of -1, so we assign a -1 oxidation number to chlorine. Since there are three chloride ions, phosphorus must have a +3 oxidation number to make the compound neutral.

b) H_2SO_4: Hydrogen in combination with nonmetals has an oxidation number of +1. Oxygen has an oxidation number of -2 in all compounds except peroxides, superoxides and with fluorine. The four oxygen atoms give a -8 charge, and the two hydrogen atoms give a +2 charge, so to make the compound neutral, we assign an oxidation number of +6 to sulfur.

c) $HClO_4$: First we'll assign hydrogen a +1 oxidation number, and oxygen a -2 oxidation number. Cl must have a +7 oxidation number to make the compound neutral.

In redox equations, all electrons must be accounted for. Redox reactions can be balanced using the method summarized below.

Rules for Balancing Redox equations:

1) Assign oxidation numbers to all atoms.

2) Identify the oxidized and reduced atoms (atoms whose oxidation numbers change).

3) Make the number of electrons lost equal the number of electrons gained by multiplying by appropriate factors; use the factors as balancing coefficients.

4) Complete the balancing by inspection.

Balancing a redox equation:

Balance the equation for the reaction which occurs between chloride and permanganate ions in acidic solution. The equation is:

$$MnO_4^-(aq) + H^+(aq) + Cl^-(aq) \rightarrow Mn^{2+}(aq) + Cl_2(g) + H_2O(l)$$

- First we assign oxidation numbers to all atoms using the rules listed above. In the permanganate ion, MnO_4^-, oxygen is assigned an oxidation number of -2 giving a negative charge of 4 x -2= -8. The charge on the polyanion is -1, so we know that manganese must be contributing a +7 charge.

Reactants		Products	
Atom	Oxidation number	Atom	Oxidation number
Mn	+7	Mn	+2 (reduced)
O	-2	O	-2
H	+1	H	+1
Cl	-1	Cl	0 (oxidized)

- Manganese is reduced since its oxidation number decreases from +7 to +2; chlorine is oxidized since its oxidation number increases from -1 to 0.

- The number of electrons lost by chlorine = 1; the number of electrons gained by manganese = 5. To make electrons lost = electrons gained, we multiply electrons lost by chlorine by 5 and use this factor as a balancing coefficient:

$$MnO_4^-(aq) + H^+(aq) + 5Cl^-(aq) \rightarrow Mn^{2+}(aq) + 5/2Cl_2(g) + H_2O(l)$$

- Multiply through by two to eliminate fractional coefficients:

$$2MnO_4^-(aq) + 2H^+(aq) + 10Cl^-(aq) \rightarrow 2Mn^{2+}(aq) + 5Cl_2(g) + 2H_2O(l)$$

- Now we balance the remaining elements. We balance oxygen atoms with water molecules, and hydrogen atoms with H^+ ions. The oxygen is balanced first. Since $2MnO_4^-$ contains 8 oxygen atoms, there must be eight water molecules each containing one oxygen atom. Then to balance the hydrogen, 16 H^+ must be present on the left:

$$2MnO_4^-(aq) + 16H^+(aq) + 10Cl^-(aq) \rightarrow 2Mn^{2+}(aq) + 5Cl_2(g) + 8H_2O(l)$$

Redox Titrations:

Analogous to acid base titrations described above, we can determine an unknown concentration of a reducing agent by titrating with a known concentration of an oxidizing agent (or vice versa).

SECTION 4.6:   Reversible Reactions

If X reacts with Y to produce Z, we can write this as:

$$X + Y \rightarrow Z$$

The above reaction suggests that the reaction continues until either X or Y is consumed. Many reactions stop before either X or Y is consumed. The reason for this is that at the same time X and Y are combining to form Z, a certain percent of Z is decomposing to reform reactants X and Y. When the forward reaction occurs at the same rate as the reverse reaction, there is no further change in the amounts of the reactants or products and we say that the system is in **equilibrium**. We use double arrows to indicate a reaction at equilibrium:

$$X + Y \rightleftharpoons Z$$

If the product, Z, is removed from the system, because, for example, it is a gas which escapes, or it is a precipitate, the reaction will continue to completion.

Weak acids and bases dissociate to only a small extent because dissociation quickly becomes balanced by reassociation. HF molecules dissociate in water to H^+ and F^- ions, but as these ions form, many of them react together to reform HF molecules. At any given time, only a small percentage of the HF molecules are dissociated, and so we say that the equilibrium lies to the left:

$$HF(aq) \rightleftharpoons H^+(aq) + F^-(aq)$$

Chapter 4 Exercises:

4.1) Which solution has the greatest number of ions or molecules?

    a. 400 mL of 1.20 M $Ni(NO_3)_2$

    b. 700 mL of 2.00 M $C_{12}H_{22}O_{11}$

    c. 200 mL of 1.50 M $Fe_2(SO_4)_3$

4.2) Classify the following reactions as either combination, decomposition, or displacement:

    a. $2K + Cl_2 \rightarrow 2KCl$

    b. $3Mg + N_2 \rightarrow Mg_3N_2$

    c. $Zn + 2HCl \rightarrow H_2 + ZnCl_2$

    d. $SO_2 + H_2O \rightarrow H_2SO_3$

    e. $2H_2O \rightarrow 2H_2 + O_2$

    f. $2KClO_3 \rightarrow 2KCl + 3O_2$

    g. $SiO_2 + 2C \rightarrow Si + 2CO$

4.3) Balance the following reactions and classify each by its historical classification.

    a. $Al_2O_3 + H_2O \rightarrow Al(OH)_3$

    b. $NH_4NO_3 \rightarrow N_2 + O_2 + H_2O$

    c. $MgSO_4 \cdot 7H_2O \rightarrow MgSO_4 + H_2O$

    d. $SCl_2 + Cl_2 \rightarrow SCl_6$

    e. $V + HCl \rightarrow VCl_3 + H_2$

    f. $Zn(NO_3)_2 + (NH_4)_2S \rightarrow ZnS + NH_4NO_3$

    g. $Ca_3(PO_4)_2 + H_2SO_4 \rightarrow Ca(H_2PO_4)_2 + CaSO_4$

4.4) Which of the following substances would you expect to be insoluble in water?

barium hydroxide	lead iodide
hydrochloric acid	ammonium acetate
lithium sulfate	potassium cyanide
ammonium nitrate	lead acetate
silver chloride	strontium hydroxide
lithium carbonate	ammonium nitrate
calcium carbonate	silver nitrate
barium sulfate	cadmium acetate

4.5) Write net ionic equations for:

a. the precipitation reaction $Pb(NO_3)_2 + 2KI \rightarrow PbI_2(s) + 2KNO_3$

b. the neutralization of a strong acid by a strong base.

c. the evolution of sulfur dioxide when sodium sulfite is treated with a dilute solution of hydrochloric acid.

4.6) a. If the concentration of sodium ions in an aqueous solution of sodium sulfate is 2.4×10^{-2} mol/liter, how many moles of sodium sulfate must have been dissolved per liter to make the solution?

b. How many moles of ammonium ions are there in a 200 mL sample of 0.10 M ammonium phosphate?

4.7) a. If it requires 50 mL of 0.030 M HCl to titrate a 100 mL sample of potassium hydroxide to the equivalence point, how many moles of potassium hydroxide are in the sample?

b. What was the original concentration of the potassium hydroxide?

c. What is the concentration of the potassium chloride salt at the equivalence point?

4.8) 37.12 mL of 0.1151 M NaOH is required to neutralize a 20.00 mL sample of citric acid $(H_3C_6H_5O_7)$.

$$H_3C_6H_5O_7 + 3NaOH \rightarrow 3H_2O + Na_3C_6H_5O_7$$

What is the molarity of the citric acid solution?

4.9) Lactic acid, $HC_3H_5O_3$, is responsible for the sour taste in out-of-date milk. 57.04 mL of 0.09827 M NaOH is required to neutralize a 50.00 mL lactic acid solution.

$$HC_3H_5O_3 + NaOH \rightarrow H_2O + NaC_3H_5O_3$$

What is the molarity of the lactic acid solution?

4.10) In the following reactions, identify the element oxidized, the element reduced, the oxidizing agent, and the reducing agent.

a. $Sn + 2NaOH \rightarrow Na_2SnO_2 + H_2$

b. $2KCl + MnO_2 + 2H_2SO_4 \rightarrow K_2SO_4 + MnSO_4 + Cl_2 + 2H_2O$

4.11) When heated in the presence of base, an aqueous solution of chlorine disproportionates to chloride ion and chlorate ion. What ratio of chlorate ion to chloride ion is produced in this reaction? The unbalanced redox reaction is:

$$Cl_2 \rightarrow Cl^- + ClO_3^-$$

4.12) Copper metal reacts with concentrated nitric acid to produce nitric oxide and copper nitrate. Calculate the mass of nitric oxide produced when 5.00 g of copper is consumed. The unbalanced reaction is:

$$Cu(s) + HNO_3(aq) \rightarrow Cu^{2+}(aq) + NO_3^-(aq) + NO(g) + H_2O$$

4.13) Solutions of hydrogen peroxide are sold over-the-counter as topical antiseptics. 30.0 mL of H_2O_2 is first acidified and then titrated to a pale pink with 26.5 mL of 0.200 M $KMnO_4$ solution.

$$H_2O_2 + MnO_4^- + H^+ \rightarrow O_2 + Mn^{2+} + H_2O$$

a. Balance the net oxidation-reduction reaction above.

b. What is the molarity of the H_2O_2 solution?

c. If the density of the solution is 1.00 g/mL, what is the percent H_2O_2 by mass of the hydrogen peroxide solution?

4.14) Liquid bleach is normally sold as 5% solutions of sodium hypochlorite, NaOCl. Quality control analysis of the NaOCl content of bleach solutions can be carried out using an iodometric titration. The hypochlorite solution is treated with an excess of iodide solution:

$$OCl^- + I^- + H^+ \rightarrow Cl^- + I_2 + H_2O$$

The resulting iodine is titrated with a standard sodium thiosulfite, $Na_2S_2O_3$ solution:

$$S_2O_3^{2-} + I_2 + H_2O \rightarrow I^- + SO_4^{2-}$$

a. Balance each of the oxidation-reduction reactions above.

b. A 100 mL bleach sample is treated with an iodide solution and then titrated with 47.24 mL of a 0.347 M $Na_2S_2O_3$ solution. What is the molarity of the bleach solution?

c. Assume the density of the solution is 1.00 g/mL, what is the percent NaOCl by mass in the bleach solution?

4.15) A metallurgist determines the amount of manganese in stainless steel by using the method of Volhard oxidation-reduction titration. A 0.5229 g sample of stainless steel is dissolved in nitric acid and titrated to a pale pink endpoint with 33.81 mL of 1.296×10^{-3} M $KMnO_4$ solution.

$$Mn^{2+} + MnO_4^- + H_2O \rightarrow MnO_2 + H^+$$

a. Balance the net oxidation-reduction reaction above.

b. What is the mass of manganese in the stainless steel sample?

c. What is the percent Mn by weight in the steel sample?

4.16) How many milliliters of a 0.10 M potassium dichromate solution are required to titrate 50 mL of an acidic 0.10 M iron(II) chloride solution? The unbalanced chemical reaction is:

$$Fe^{2+} + H^+ + Cr_2O_7^{2-} \rightarrow Cr^{3+} + Fe^{3+} + H_2O$$

Chapter 4 Answers:

4.1)   c

4.2)   a, b, and c are combination reactions; d and e are decomposition reactions; f and g are displacement reactions

4.3)   a. $Al_2O_3 + 3H_2O \rightarrow 2Al(OH)_3$ (combination)

b. $2NH_4NO_3 \rightarrow 2N_2 + O_2 + 4H_2O$ (decomposition)

c. $MgSO_4 \cdot 7H_2O \rightarrow MgSO_4 + 7H_2O$ (decomposition)

d. $SCl_2 + 2Cl_2 \rightarrow SCl_6$ (combination)

e. $2V + 6HCl \rightarrow 2VCl_3 + 3H_2$ (single displacement)

f. $Zn(NO_3)_2 + (NH_4)_2S \rightarrow ZnS + 2NH_4NO_3$ (double displacement)

g. $Ca_3(PO_4)_2 + 2H_2SO_4 \rightarrow Ca(H_2PO_4)_2 + 2CaSO_4$ (double displacement)

4.4)   All the substances are soluble <u>except</u> lead iodide, silver chloride, calcium carbonate, and barium sulfate.

4.5)   a. First, write all components in the form in which they predominantly exist under the conditions of the reaction. Strong electrolytes in solution are in ionic form, weak electrolytes in solution are written in molecular form, evolved gases and precipitates are written in molecular (or formula unit form):

$$Pb^{2+}(aq) + 2NO_3^-(aq) + 2K^+(aq) + 2I^-(aq) \rightarrow PbI_2(s) + 2K^+(aq) + 2NO_3^-(aq)$$

Eliminate spectator ions to obtain the net ionic equation:
$$Pb^{2+}(aq) + 2I^-(aq) \rightarrow PbI_2(s)$$

b. The net ionic equation for the neutralization of any strong acid by any strong base is:
$$H_3O^+(aq) + OH^-(aq) \rightarrow 2H_2O$$

The cation of the strong base, and the anion of the strong acid are both spectator ions of the neutralization reaction and do not appear in the net ionic equation.

c. $SO_3^{2-}(aq) + 2H^+(aq) \rightarrow H_2O + SO_2(g)$

4.6)   a. 1.2×10^{-2} mol sodium sulfate; b. 0.060 mol

4.7)   a. 0.0015 mol; b. 0.015 M ; c. 0.010 M

4.8)   0.07120 M

4.9)   0.1121 M

4.10) a. Tin is oxidized (from 0 to +2); Hydrogen is reduced (from +1 to 0)
Hydrogen of the hydroxide is the oxidizing agent, it oxidizes the tin
Tin is the reducing agent, it reduces the hydrogen

b. Chlorine is oxidized (from -1 to 0); Manganese is reduced (from +4 to +2)
Manganese of the manganese dioxide is the oxidizing agent, it oxidizes the chlorine
Chlorine (as chloride) is the reducing agent, it reduces the manganese

4.11) A disproportionation reaction is a redox reaction in which the same species is both oxidized and reduced. In this reaction, chlorine is oxidized and reduced. In any redox reaction, the oxidation and reduction must be balanced. Compare the decrease in oxidation number of the chlorine that is reduced to the increase in the oxidation number of the chlorine that is oxidized:

Reduction: $Cl_2 \rightarrow Cl^-$ (change of -1)
Oxidation: $Cl_2 \rightarrow ClO_3^-$ (change of +5)

In order to balance the two processes, five times as much chlorate ion as chloride ion must be produced.

4.12) The balanced redox reaction is:

$$3Cu(s) + 8HNO_3(aq) \rightarrow 3Cu^{2+}(aq) + 6NO_3^-(aq) + 2NO(g) + 4H_2O$$

The stoichiometry indicates that 2 moles of NO is produced for every 3 moles of copper consumed. This same information could have been derived from the fact that the copper metal is oxidized to Cu^{2+} (a change of +2), and the nitrate ion is reduced to nitric oxide (a change from +5 to +2, or -3). Mass of NO produced = 1.57 grams

4.13) a. $5H_2O_2 + 2MnO_4^- + 6H^+ \rightarrow 5O_2 + 2Mn^{2+} + 8H_2O$; b. 0.442 M; c. 1.50%

4.14) a. $OCl^- + 2I^- + 2H^+ \rightarrow Cl^- + I_2 + H_2O$

b. $S_2O_3^{2-} + 4I_2 + 5H_2O \rightarrow 8I^- + 2SO_4^{2-} + 10H^+$

c. 0.656 M

d. 4.88%

4.15) a. $3Mn^{2+} + 2MnO_4^- + 2H_2O \rightarrow 5MnO_2 + 4H^+$; b. 0.003609 g; c. 0.6902%

4.16) Iron(II), Fe^{2+}, is oxidized to iron(III), Fe^{3+}: a change of +1
The dichromate ion, $Cr_2O_7^{2-}$, is reduced to chromium(III), Cr^{3+}: a total change of +6 (+3 for each Cr in the dichromate ion).

One mole of dichromate can oxidize six moles of iron(II). Alternatively, the equation can be balanced and the mole ratio derived directly from the stoichiometry:

$$6Fe^{2+} + 14H^+ + Cr_2O_7^{2-} \rightarrow 2Cr^{3+} + 6Fe^{3+} + 7H_2O$$

Since the concentrations of the two solutions are equal (both 0.10 M), 50 mL of dichromate solution requires only 50 x 1/6 mL of iron(II) solution (or 8.3 mL).

CHAPTER 5

Gases and the Kinetic-Molecular Theory
"Molecules in Motion"

TOPICS:

- Gas properties

- Ideal gases:
  - Gas laws
  - Kinetic molecular theory
  - Effusion and diffusion

- Real gases

INTRODUCTION:

Hydrogen and helium are both gaseous elements that are less dense than air, but hydrogen balloons aren't found at typical birthday parties. Hydrogen is extremely flammable when activated, while helium is inert at all temperatures and, in fact, is the most inert substance in the universe. In spite of their chemical differences, the _physical_ behavior of hydrogen and helium is similar. For example, balloons filled to the same volume with hydrogen and helium at the same temperature and pressure, contain the same _number_ of hydrogen molecules as helium atoms, even though a helium atom is about twice the mass of a hydrogen molecule. Chapter five explores the similar **physical** behavior of gases:

SECTION 5.1: The Physical States of Matter

In Chapter 1 we saw that, generally, there are three ways molecules may be arranged:

Touching & ordered
(solid)

Touching & disordered
(liquid)

Not touching
(gas)

Elemental solids, liquids, and gases:

Choose words from the list below to complete the following paragraph:

(a) bromine
(b) chlorine
(c) helium
(d) hydrogen
(e) mercury

(f) neon
(g) nitrogen
(h) oxygen
(i) solids

Most elements are _____(1) at ambient temperature and pressure. The only two elements that are liquids at ambient conditions are, _____(2), a metal, and _____(3), a nonmetal. Some familiar elemental gases are: _____(4), the lightest; _____(5), an inert, nonflammable gas used for balloons and airships; _____(6), the major component of air; _____(7), the most abundant element on earth which supports combustion; _____(8), a greenish-yellow poisonous gas; and _____(9), used in tubes to produce colored light.

answers: 1-i, 2-e, 3-a, 4-d, 5-c, 6-g, 7-h, 8-b, 9-f.

The drawing on the previous page representing gas molecules indicates that gas molecules are not touching.

- at normal conditions, gases are chiefly empty space
- mechanically and thermally, all gases behave the same way: they exhibit the same compressibility, and the same thermal expansion.

The same volume of allgases at the same temperature and pressure contain the same number of gas molecules. Experiments have shown that one mole of molecules of any gas at 273 K and 1 atmosphere occupies 22.4 L (the standard molar volume, V_m).

$$V_m(1 \text{ atm, } 273K) = 22.4 \text{ L/mol molecules} \quad \text{(all gases)}$$

We define the conditions, t = 0°C (273.15 K) and P = 760 torr as **standard temperature and pressure (STP)**. Notice that STP involve the usual pressure, but not the usual temperature in laboratories.

By contrast, the volume of solids and liquids, where molecules are touching, depend on the size of the molecules, their interactions, and, in the case of solids, the type of crystal lattice they form.

Comparison of liquid and gaseous volumes:

1) How much volume does a mole of water molecules in the gaseous state at 273 K and 1 atm occupy?

2) How much volume does this same amount of liquid water occupy (where the molecules are touching one another)?

answers:

1) One mole of any gas (including water) at 273 K and 1 atm. takes up **22.4 L.**

2) 6.02×10^{23} water molecules (1 mole) weigh 18 grams; the density of water is 1 g/mL. A mole of water molecules in the liquid state occupies:

$$6.02 \times 10^{23} \text{ H}_2\text{O molecules} \times \frac{18 \text{ amu}}{\text{H}_2\text{O molecule}} \times \frac{1 \text{ g}}{6.02 \times 10^{23} \text{ amu}} \times \frac{1 \text{ mL}}{1 \text{ g}} = 18 \text{ mL}$$

A drawing illustrates the difference in volumes of a mole of gaseous water and a mole of liquid water:

```
┌─────────────────┐
│  V = 22.4 L     │
│  (273 K, 1 atm) │
│                 │
│                 │                    ▯  V = 18 mL
│                 │
│                 │
└─────────────────┘
```

　　　1 mole gaseous water molecules　　　　　　　　1 mole liquid water

We can conclude that gases are mostly empty space.

Since gases are mostly space, they are easily compressed. They also have a low viscosity (they flow freely), and they typically have low densities. Densities of gases are often given in grams per liter instead of grams per milliliter because gases contain so much empty space and occupy such a large volume compared to solids and liquids.

SECTION 5.2: Pressure

Gases expand to fill a container. As gas particles collide with the walls of a container, they exert some force on the walls. The force per unit area that a gas exerts on its container is pressure (P):

$$P = \frac{f}{a}$$

You are probably already aware of the definition of pressure. Think, for example, why you might choose to lie down on thin ice instead of standing on it. The gravitational force exerted by your body to the ice is the same whether you are standing or lying, but by increasing the area of your body in contact with the ice when you lie down, you decrease the pressure at any point. The same concept applies to wearing snow shoes, or being able to lie (somewhat comfortably) on a bed of nails.

Atmospheric Pressure:

Air is a gaseous mixture containing mainly nitrogen and oxygen, but also a small amount of other gases. Gravity holds these gases around the earth in a layer about 50 miles thick. Our atmosphere exerts a pressure of about 15 pounds per square inch at the earth's surface. The atmospheric pressure varies depending on the altitude and weather conditions. Sometimes it is useful to picture the atmosphere as a weight exerting force on a solution surface or on a container.

$P \approx 15$ lb/sq. in.

atm

It is easy to forget about atmospheric pressure since its source, air, is invisible, but it is critical to remember its presence when working problems that involve gases. Expanding gases do work as they push back the weight of the atmosphere. In order to exhale, you must overcome atmospheric pressure.

What would happen if you filled a tube (closed at one end) with mercury and inverted it into a dish of mercury?

atmospheric pressure

760 mm

atmospheric pressure

Start   End

As the diagram shows, mercury flows out of the tube until the pressure of the column of mercury on the surface of the mercury in the dish is equal to the pressure the <u>atmosphere</u> exerts on the dish of mercury. Typically, the atmosphere exerts enough pressure to hold up a column of mercury about 760 mm high.

A barometer is a device, such as the one pictured above, used to measure atmospheric pressure. If a student in Denver (altitude 1524 m or 5000 ft.) made a barometer, the mercury in the column would be only about 700 mm high. Denver is almost a mile above sea level and therefore the blanket of gases forming the atmosphere is thinner and exerts less pressure than is typical at sea level. The atmospheric pressure might become that low at sea level during a hurricane.

A manometer measures the pressure exerted by a gas in a closed container. Manometers work by comparing the pressure of a gas in a flask to a known pressure, either the atmosphere (barometric pressure) or a vacuum (0 pressure).

Choice of liquid for a barometer:

Why do barometers use the poisonous liquid mercury? Let's suppose we wanted to make a barometer containing water. How high would the column need to be to hold a column of water supported by atmospheric pressure?

Hint: use the pressure of the atmosphere as 14.7 lbs/in^2.

<u>solution:</u>

- What do we know about water? It's <u>density</u> is 1.0 g/mL; does this information, along with the information given about atmospheric pressure, suggest a way to obtain a unit of length for the column height?

- Let's rewrite the density of water as 1.0 g/cm^3, and convert units of pressure to similar units (g/cm^2) using the equalities 1 lb = 454 g and 1 in = 2.54 cm:

$$P_{atm} = 14.70 \frac{lb}{in^2} \times 454 \frac{g}{lb} \times \left(\frac{1 \, in}{2.54 \, cm}\right)^2 = 1034 \frac{g}{cm^2}$$

- Let's consider a column of water which is a cm^2 in cross section. How high a column (h), of water in this column weighs 1034 g?

The volume of the above column is: $V = 1 \, cm^2 \times h$

We know that volume x density = mass, so

$$(1 \, cm^2 \times h) \times 1 \, g/cm^3 = 1034 \, g$$

$$h = 1034 \, cm \text{ or } \textbf{33.9 feet!}$$

- How high a column of mercury in the above column weighs 1034 g?

The density of mercury is 13.59 g/cm^3, so:

$$(1 \, cm^2 \times h) \times 13.59 \, g/cm^3 = 1034 \, g$$

$$h = 76.1 \, cm \text{ or } \textbf{761 mm (2.5 feet)}$$

Two and a half feet is a more convenient height for a barometer column than 34 feet! Mercury's high density makes it a useful liquid for barometers.

The ratio of <u>heights</u> of the liquid columns in a barometer is inversely related to the <u>densities</u> of the liquids for a given pressure. For the above example we could write:

$$\frac{h(H_2O)}{h(Hg)} = \frac{d(Hg)}{d(H_2O)}$$

$$\frac{1034 \text{ cm}}{76.1 \text{ cm}} = 13.6 = \frac{13.59 \text{ g/cm}^3}{1.0 \text{ g/cm}^3}$$

People in the 1600's knew that they couldn't lift water more than about 33 feet with a suction pump, but they didn't know why at first. An Italian scientist, Torricelli, recognized that the maximum height of the water column was a measure of the atmospheric pressure, and that once a pump lowered the gas pressure at the top of the pipe to zero, the water in the pipe would stop rising, no matter how fast the pump worked.

Units of Pressure:

Most chemists express pressure in **mm Hg** or **atmospheres (atm)**. A mmHg is sometimes called a **torr** in honor of the scientist Torricelli who invented the barometer.

1 atmosphere = 760 mm Hg = 760 torr (Hg at 0°C).

The SI unit for pressure uses the SI unit for force (Newton, N) and area (m^2) and is called a **pascal** (pa). A pascal is small compared to an atmosphere or a torr and is not used often in the United States.

$$1 \text{ Pa} = 1 \text{ N/m}^2$$

$$1 \text{ atm} = 101{,}325 \text{ Pa} = 760 \text{ mm Hg}$$

Pressure of a Gas in a Container:

The pressure of a gas inside a container (which is not pressurized or evacuated) equals the pressure of the atmosphere pressing against the container; if this were not the case, the container would collapse or explode.

SECTION 5.3: The Gas Laws

Consider a cylinder of gas equipped with a piston initially under atmospheric pressure, P_{atm}.

What could we do to change the volume of the gas? If we increase the <u>pressure</u> on the cylinder, we would compress the gas. We could also raise the <u>temperature</u> which increases the molecular motion of the gas molecules, and therefore increases the volume that the gas molecules occupy. Likewise, the volume increases if we increase the <u>number</u> of molecules. These relationships are summarized in the **gas laws**.

The Gas Laws:

1) **Boyle's Law:** at constant temperature, the volume of a fixed amount of gas is inversely proportional to the applied pressure. As the piston goes down, pressure increases as volume decreases:

$$V = \frac{k}{P} \quad (\text{T and n fixed})$$

where V = volume, P = pressure, T = temperature, n = moles, and k = a constant. We can use Boyle's law to solve problems that compare two sets of volume and pressure values:

$$V_1 = \frac{k}{P_1} \quad \text{and} \quad V_2 = \frac{k}{P_2}$$

$$k = P_1 V_1 = P_2 V_2$$

$$\text{or} \quad V_2 = V_1 \times \frac{P_1}{P_2}$$

We can see from this last equation that when $P_2 > P_1$, $V_2 < V_1$, or, as pressure increases (P_2 increases), volume decreases (V_2 decreases).

2) **Charles's Law:** at constant pressure, the volume of a fixed amount of gas is directly proportional to its absolute temperature. As a gas is heated and molecular motion increases, the piston goes up as the volume increases.

$$V = kT \quad (\text{P and n fixed})$$

To compare two sets of volume and temperature values, we can use the relationship:

$$\frac{V_1}{T_1} = \frac{V_2}{T_2} \quad \text{or,} \quad V_2 = V_1 \times \frac{T_2}{T_1}$$

3) Avogadro's Law: at fixed temperature and pressure, equal volumes of any ideal gas contain equal numbers of particles.

$$V \propto n \quad \left(T \text{ and } P \text{ fixed}\right)$$

To compare two sets of volume and number of moles, we can use the relationship:

$$\frac{V_1}{n_1} = \frac{V_2}{n_2}$$

$$\text{or } V_2 = V_1 \times \frac{n_2}{n_1}$$

Combining the Gas Laws:

1) What is the volume (V) of 1 mole of a gas at 1 atm and 273 K?

   answer:

   n = 1 mole   T = 273 K   P = 1 atm

   V = 22.4 L (from Section 1)

2) What is the volume of 2 moles of a gas at 1 atm and 273 K?

   answer:

   n = 2 moles   T = 273 K   P = 1 atm

   $$V = 22.4L \times \frac{2 \text{ mol}}{1 \text{ mol}} = 44.8L$$

   (common sense: twice the number of molecules takes up twice the volume)

3) What is the volume of n moles of a gas at 1 atm and 273 K?

   answer:

   n = n   T = 273 K   P = 1 atm

   $$V = 22.4L \times \frac{n}{1 \text{ mol}}$$

   (general case of exercise #2)

4) What is the volume of 1 mole of a gas at 2 atm and 273 K?

   answer:

   n = 1 mole   T = 273 K   P = 2 atm

   $$V = 22.4L \times \frac{n}{1 \text{ mol}} \times \frac{1 \text{ atm}}{2 \text{ atm}} = 11.2L$$

   (Boyle's Law: twice the pressure decreases volume to 1/2)

5) What is the volume of n moles of a gas at pressure P and 273 K?

answer:

$n = n$ $\qquad T = 273\ K$ $\qquad P = P$

$$V - 22.4L \times \frac{n}{1\ mol} \times \frac{1\ atm}{P}$$

(general case of exercise #4)

6) What is the volume of 1 mole of a gas at 1 atm and 546 K?

answer:

$n = 1\ mole$ $\qquad T = 546\ K$ $\qquad P = 1\ atm$

$$V - 22.4L \times \frac{n}{1\ mol} \times \frac{1\ atm}{P} \times \frac{546\ K}{273\ K} - 448L$$

(Charles' Law; twice the temperature doubles the volume)

7) What is the volume of n moles of a gas at pressure P and temperature T?

answer:

$n = n$ $\qquad T = T$ $\qquad P = P$

$$V - 22.4L \times \frac{n}{1\ mol} \times \frac{1\ atm}{P} \times \frac{T}{273\ K}$$

(general case of exercise #6)

8) Rewrite the expression for volume from exercise 7 with numbers (constants) and variables gathered together.

answer:

$$V - \underbrace{\frac{22.4\ L \times 1\ atm}{1\ mol \times 273\ K}}_{R} \times \frac{nT}{P}$$

The constants in this equation are called "R", the gas law constant. What is the value for the constant, R, in this equation?

$$R - \frac{PV}{nT} - 22.4\ L \times \frac{1\ atm}{1\ mol} \times 273\ K = 0.0821\ \frac{L \cdot atm}{mol \cdot K}$$

We can rewrite the equation for R above as:

$$PV = nRT$$

This equation, which contains the gas laws of Avogadro, Boyle, and Charles, is called the Ideal Gas Law.

Remember that a law <u>generalizes</u> behavior observed in many experiments, and may not apply to every situation. In fact, real gases <u>always</u> break the gas laws! The gas laws apply strictly only to <u>ideal</u> gases. We assume ideal gas particles have no volume and no interactions. Real gases behave much like ideal gases under certain conditions (low pressure, high temperature), and in these cases, the gas laws provide a close approximation for the behavior of real gases.

Gas law problem:

Suppose we fill an uninflated balloon with 15.0 g of dry ice (solid CO_2). The balloon's capacity is 7.5 L. Is the balloon likely to pop as the dry ice vaporizes? Assume the temperature is 298 K.

<u>solution:</u>

To work this problem, we need to first calculate the volume of 15.0 g of gaseous CO_2 (M = 44 g/mole) and compare the value to 7.5 L (the balloon capacity).

• What do we know?

$T = 298K$ $P = 1.0$ atm

$n = \dfrac{15.0 \text{ g}}{44.0 \text{ g/mol}} = 0.341$ mol $R = 0.0821$ L·atm/mole·K

• Solve for volume, V:

$$V = \frac{nRT}{P} = \frac{0.341 \text{ mol} \times 0.0821 \frac{\text{L·atm}}{\text{mol·K}} \times 298 \text{ K}}{1.0 \text{ atm}} = 8.3 \text{ L}$$

8.3 L is larger than the balloon's capacity, so BLAM!...the balloon will pop.

Density of an ideal gas:

Use the molar mass of helium and its standard molar volume to determine the density of helium at STP. Derive an equation for the density of a gas under any conditions.

<u>solution:</u>

The molar volume of any gas at STP, $V_m = 22.4$ L/mole

The molar mass of helium is M = 4.003 g/mole

Therefore,

$$d_{STP} = \frac{m}{V_m} = \frac{m/n}{V_m/n} = \frac{4.003 \text{ g/mol}}{22.4 \text{ L/mol}} = 0.179 \text{ g/L}$$

We know from the ideal gas law that $V = nRT/P$. Substituting this expression into the equation for density above gives:

$$d_{STP} = \frac{m}{V_m} = \frac{m}{nRT/P} = \frac{MP}{RT}$$

where M = molar mass, m/n.

At constant T and P, the density of a gas is directly proportional to its molar mass. The heavier the atom or molecule, the more dense it will be. Xenon has a greater mass than argon, so xenon is more dense than argon. This relationship does not hold with solids and liquids. Compare carbon (M = 12.0 g/mole) in its diamond form to sodium (M = 23.0 g/mole). A sodium atom is heavier than a carbon atom, but it is less dense. Diamonds sink in water (density = 3.51 g/mL) while sodium floats on water (as it reacts vigorously, density = 0.97 g/mL). The densities of solids and liquids depend both on their molar mass and on their volume which in turn depends on interactions between molecules.

Density of gases:

We know from experience that helium balloons rise in air. Will a nitrogen (N_2) filled balloon float or sink? What about balloons filled with oxygen (O_2), carbon dioxide (CO_2), methane (CH_4) and propane (C_3H_8)? (Air is approximately 78% nitrogen and 21% oxygen).

solution:

Since the density of a gas is directly proportional to molar mass, we will compare molar masses of the gases in the balloons to that of air to determine if the balloon will sink or float.

• The molar mass of air (78% nitrogen and 21% oxygen) is approximately:

$$M = (.78 \times 28) + (.21 \times 32) = 28.6 \text{ g/mole}$$

• The molar mass of nitrogen is 28 g/mole which is less than the molar mass of air, so it would seem the balloon should rise slightly in air. However, the weight of the balloon may prevent the balloon from rising.

• The molar mass of oxygen (32 g/mole), carbon dioxide (44 g/mole) and propane (44 g/mole) are all larger than the molar mass of air, so a balloon filled with those gases would sink.

• The molar mass of methane is 16 g/mole. A large balloon filled with methane will rise. A small one might not rise since the weight of the balloon might prevent it from doing so.

Density and the ideal gas law:

What is the density in g/L of radon at 25°C and 1 atm?

answer:

We can substitute m/M for n in the ideal gas law, and then we solve for m/V (density):

$$PV = \left(\frac{m}{M}\right)RT$$

$$d = \frac{m}{V} = \frac{PM}{RT} = \frac{1\,atm \times 222.02\ g/mol}{0.0821\frac{L \cdot atm}{mol \cdot K} \times 298K} = 9.07\ g/L$$

Radon is a very dense gas. Radon detectors should be placed in the basement of a house to determine if house levels are high. The equation used to work this problem indicates that the density of a gas is inversely proportional to the temperature. As temperature increases, the space between molecules increases, volume increases, and density decreases.

The ideal gas law and reaction stoichiometry:

We can use the ideal gas law to determine the volumes of gases involved in chemical reactions under given conditions. For example, how many liters of pure oxygen, measured at 740 mm Hg and 24°C, would be required to burn 1.00 g of octane, $C_8H_{18}(l)$ to carbon dioxide and water?

solution:

- First, we write the balanced equation for the reaction:
$$C_8H_{18}(l) + \frac{25}{2}O_2(g) \rightarrow 8CO_2(g) + 9H_2O(l)$$

- Using the reaction stoichiometry, determine how many moles of O_2 are required to burn 1.00 g of octane:

$$\frac{1\,g\,C_8H_{18}}{114\,g/mol} \times \frac{\frac{25}{2}\ mol\,O_2}{1\,mol\,C_8H_{18}} = 0.110\ mol\,O_2$$

- Now use the gas law to calculate the volume of O_2: (first convert pressure to atm, and temperature to Kelvin since R is in units of L atm/mole K)

$$P = 740\ mm\,Hg \times \frac{1\,atm}{760\,mm\,Hg} = 0.974\,atm; \quad T = 24 + 273 = 297K$$

$$V = \frac{nRT}{P} = \frac{0.110 \text{ mol} \times 0.0821\frac{L \cdot atm}{mol \cdot K} \times 297 \text{ K}}{0.974 \text{ atm}} = 2.75 \text{ L}$$

This reaction is analogous to the one which occurs in an automobile engine. A vehicle that gets 25 miles per gallon at 60 miles per hour would use 1 gram of gasoline almost every half second. Air is only 21% oxygen by volume, so the volume of air that your car engine would need to completely combust the octane would be approximately 13 L every half second! Hence the need for air filters.

Mixtures of Gases: Dalton's Law of Partial Pressures:

In a mixture of unreacting gases, the total pressure is the sum of the partial pressures of the individual gases:

$$P_{total} = P_1 + P_2 + P_3...$$

Dalton's law is useful when a gas is collected by displacement of water because the collected gas is a mixture of the gas produced by a chemical reaction, and water vapor. Water molecules escape from the surface of the water and collect in the space above the liquid. The pressure exerted by these water molecules on the container is called the **vapor pressure**. When the rate of escape of the water molecules from the water surface equals the rate of return to the water, the vapor pressure remains constant.

Partial pressures:

When aluminum reacts with concentrated hydrochloric acid, hydrogen gas evolves. The hydrogen is collected by displacement of water at 22°C at a total pressure of 750 torr. If the volume of the collected gas is 365 mL, how many moles of hydrogen were collected? The vapor pressure of water at 22°C is 21 torr.

<u>solution:</u>

• The collected gas is a mixture of hydrogen and water vapor. By Dalton's Law:
$P_{tot} = P_{H_2} + P_{H_2O}$. The partial pressure of water in the sample is its vapor pressure.

The partial pressure of hydrogen gas is: $P_{H_2} = P_{tot} - P_{H_2O}$

$$= 750 \text{ torr} - 21 \text{ torr} = 729 \text{ torr}.$$

$$= 729 \text{ torr} \times \left(\frac{1 \text{ atm}}{760 \text{ mm Hg}}\right) = 0.959 \text{ atm}$$

• Use the ideal gas law to calculate moles of hydrogen:

$$n_{H_2} = \frac{P_{H_2}V}{RT} = \frac{0.959 \text{ atm} \times 0.365 \text{L}}{0.0821\frac{L \cdot atm}{mol \cdot K} \times 295 \text{K}} = 0.0144 \text{ mol } H_2 \text{ gas}$$

SECTION 5.4: The Kinetic-Molecular Theory

Chemists theorize about what happens at the molecular level to explain observed macroscopic behavior. The kinetic-molecular theory proposes a model for molecular behavior that explains the observedmacroscopic behavior of ideal gases. This theory is widely accepted because it can be used with the principles of physics to derive an expression for pressure that agrees with the ideal gas law. In addition, it correctly predicts the relative effusion rates of gases summarized by Graham's law. This theory can be stated in terms of four postulates:

Postulates of the Kinetic-Molecular Theory:

1) **Particle size:** Gas molecules are so small compared to the space between molecules that their volume is assumed to be zero.

2) **Particle interaction:** Gas molecules exert no forces on each other; they are assumed neither to attract nor to repel each other.

3) **Particle motion:** Gas molecules are in constant random motion. It is the collisions of the particles with the walls of the container that are the cause of the pressure exerted by the gas. The collisions are elastic, which means that no energy is lost to frictional forces (their total kinetic energy remains constant).

The kinetic molecular theory's explanation for ideal gas behavior:

The kinetic molecular theory states that collisions of gas particles with the walls of its container cause gas pressure. The kinetic molecular theory can be combined with the laws of mechanics to predict that the pressure of a gas in a container will depend on the speed and mass of the particles.

Pressure:

$$P = \frac{\frac{1}{3} mu^2}{V} = \frac{nRT}{V}$$

kinetic molecular theory   experimental results
molecular behavior   macroscopic behavior

where u is the is the average molecular speed of the gas particles.

We will use the above equations to obtain equations for a) the average kinetic energy of gas particles and, b) the average velocity of the gas particles.

Kinetic energy:

Use the equations above, and the equation for kinetic energy (K.E. = $1/2mu^2$), to obtain an equation for the average kinetic energy of gas molecules moving through space.

<u>answer:</u>

- We know $P = \dfrac{\frac{1}{3} mu^2}{V}$ (kinetic molecular theory) and, K.E. = K.E. = $\frac{1}{2} mu^2$

- We can multiply each side of the K.E. equation by 2/3 to obtain:

$$\frac{2}{3} K.E. = \frac{1}{3} mu^2$$

and substitute this into the equation for pressure to give:

$$P = \dfrac{\frac{2}{3} K.E._{avg}}{V} \quad \text{or} \quad K.E._{avg} = \frac{3}{2} PV$$

- Using the ideal gas law:

$$K.E._{avg} = \frac{3}{2} nRT$$

The kinetic energy of a mole of gas is directly proportional to temperature and <u>has the same value for all gases</u>. Heavier molecules do not bombard container walls with more energy than lighter gas molecules because the kinetic energy of an object is related to its mass <u>and</u> its speed. For two gases at the same temperature, the gas molecule with the higher mass will move more slowly than the lighter gas molecule.

Average velocity of gas molecules:

Rearrange the equations for pressure of gas particles above to calculate the average speed of gas molecules.

<u>answer:</u>

$$P = \dfrac{\frac{1}{3} mu^2}{V} = \dfrac{nRT}{V} \quad \text{(from the kinetic molecular theory and the ideal gas law)}$$

therefore, $\frac{1}{3} mu^2 = nRT$

Solving for u gives:

$$u = \left(\dfrac{3nRT}{m} \right)^{\frac{1}{2}}$$

$$u = \left(\frac{3RT}{M}\right)^{\frac{1}{2}}$$

The gas constant R is not useful with the units liter·atm/mole·K for problems such as these. R can be expressed as 8.31×10^7 ergs/mole·K (1 erg = 1 g·cm^2/s^2) which will provide speed in cm/s.

Average velocity of nitrogen:

Find the average speed of a nitrogen molecule in air at room temperature (25°C).

<u>answer:</u>

$$u = \left(\frac{3RT}{M}\right)^{\frac{1}{2}}$$

$$u = \left(\frac{3 \times (8.31 \times 10^7 \text{ erg/mol·K}) \times (25 + 273) \text{ K}}{28.0 \text{ g/mol}}\right)^{\frac{1}{2}}$$

$$= 5.15 \times 10^4 \text{ cm/sec} = 515 \text{ meter/sec}$$

If we convert our answer to miles per hour (1 m/sec = 2.24 miles/hr, a slow walk), we can see that the average speed of a nitrogen molecule at room temperature is about 1150 miles per hour!

In the above problem, we solved for u, the <u>average</u> speed of a nitrogen molecule. In fact, the many collisions among gas particles produces a wide <u>distribution</u> of speeds, and most nitrogen molecules will not be moving at the average speed, but will be moving either faster or slower than the average speed. As temperature increases, average velocity increases and the range of velocities becomes much larger.

We can compare this very high average speed of molecules to the speed of sound in air. Molecular motion propagates sound waves, so the speed of sound and the speed of the gas molecules through which it passes should be approximately equal. The speed of sound in air is about 800 miles an hour, which is close to the average speed of an oxygen or nitrogen molecule at room temperature.

Molar energy of gases:

What is the energy of one mole of a gas at room temperature? Use R with units of energy, 1.987 cal/mole K.

solution:

$$K.E._{avg} = \frac{3}{2} nRT$$

$$K.E._{avg} = \frac{3}{2} \times 1 \text{ mol} \times 1.987 \frac{cal}{mol \cdot K} \times 298K = 888 \text{ cal/mol}$$

The kinetic energy of a mole of molecules of <u>any gas</u> at room temperature is 888 cal.

Values for R:

0.0821 L atm/mole K	(useful in gas law problems)
8.31 x 10[7] ergs/mole K	(useful for determining speed)
1.987 cal/mole K	(useful for determining energy in calories)

SECTION 5.5: Effusion and Diffusion

We saw in Section 4 that the kinetic molecular theory could be used to derive an expression for pressure that agrees with the ideal gas law. The kinetic molecular theory also satisfactorily explains the observed behavior of diffusion and effusion. You are probably familiar with the word diffusion, but may not be as familiar with the word effusion – we hope not to cause confusion!

Diffusion is the movement of one gas through another; **effusion** is the movement of a gas into a vacuum. Effusion is a fast process, diffusion happens more slowly. When we open a perfume bottle, it takes some time for every one in the room to smell it. Why is this the case if, as we have seen, gas particles at room temperature move at very high speeds (nitrogen has an average velocity of 1150 miles per hour or 515 m/s)? The answer is that the path of a gas particle is extremely convoluted because of collisions with nitrogen and oxygen in the air so most likely, a gas molecule travels a long distance before it reaches the opposite side of the room. Effusion happens quickly because there are no molecules in a vacuum for the gas molecules to collide with.

The average distance a molecule travels between collisions is called its **mean free path.** For a nitrogen molecule in air at room temperature, it travels, on average, 6.6×10^{-8} m before colliding with another molecule or atom. This distance is equal to about 180 molecular diameters between collisions.

Collision frequency:

Use the average speed (515 meter/second) of a nitrogen molecule in air at room temperature and its mean free path (6.6×10^{-8} m) to calculate its collision frequency (number of collisions per second).

answer:

Collision frequency can be calculated by taking the ratio of the average speed (distance per second) and the mean free path (distance per collision).

$$\text{collision frequency} = \frac{\text{distance/second}}{\text{distance/collision}} = \frac{\text{collisions}}{\text{second}}$$

$$= \frac{515 \text{ m/s}}{6.6 \times 10^{-8} \text{ m/collision}} = 7.7 \times 10^{9} \text{ collisions/s}$$

Graham's Law of Effusion: the rate of effusion of a gas is inversely proportional to the square root of its molecular weight:

$$\text{rate} = \frac{c}{\sqrt{M}}$$

The ratio of the effusion rates of two gas molecules such as helium and argon is:

$$\frac{\text{rate}_{Ar}}{\text{rate}_{He}} = \frac{\sqrt{M_{He}}}{\sqrt{M_{Ar}}}$$

The lighter gas (He) effuses faster than the heavier gas (Ar).

Graham's Law also applies to diffusion. The ratio of the diffusion of two gases moving through another gas or a mixture of gases such as air is:

$$\frac{\text{time}_{A}}{\text{time}_{B}} = \frac{\text{rate}_{B}}{\text{rate}_{A}}$$

Graham's Law and Molar Mass:

Graham's Law is generally used to determine the molar mass of an unknown gas. In an effusion experiment, it requires 68 seconds for a sample of N_2 to effuse down a capillary, and 85 seconds for a second gas to effuse under the same conditions of temperature and pressure. Was the gas ethane (C_2H_6) or propane (C_3H_8)?

solution:

• The measured times will vary inversely with the rates of effusion (a high rate of effusion = short time):

$$\frac{time_A}{time_B} = \frac{rate_B}{rate_A}$$

• Substituting into Graham's law gives:

$$\frac{time_{unk}}{time_{N_2}} = \frac{\sqrt{M_{unk}}}{\sqrt{M_{N_2}}}$$

$$M_{unk} = M_{N_2} \times \left(\frac{time_{N_2}}{time_{unk}}\right)^2$$

$$M_{unk} = 28 \text{ g/mol} \times \left(\frac{85 \text{ sec}}{68 \text{ sec}}\right)^2 = 44 \text{ g/mol}$$

The unknown gas is propane which has a molecular weight of 44 g/mole.

SECTION 5.6: Real Gases: Deviations from Ideal Behavior

Real Gases:

Choose words from the list below to complete the following paragraphs describing the behavior of real gases.

(a) container	(f) interact	(k) speed
(b) distance	(g) low	(l) temperature
(c) free	(h) molecules	(m) volume
(d) high	(i) **pressures**	(n) **volumes**
(e) ideal	(j) Real	

No gas <u>exactly</u> follows the _____(1) gas law. Ideal gas behavior is approached by real gases under certain conditions. The kinetic molecular theory successfully explains ideal behavior by making the assumptions that gas particles do not _____(2) and have no _____(3). _____(4) gases have a finite volume and exhibit interparticle attractive forces. This last section discusses how the ideal gas law can be changed to more closely reflect the behavior of real gases.

At_____(5) temperatures (close to a gas's condensation point), and _____(6) pressures (above 10 atm), gas particles begin to _____(7) appreciably. As applied pressure increases, the average _____(8) between molecules becomes smaller, and attraction to nearby molecules decreases the _____(9) of a gas particle, and therefore the force with which it hits the container walls. Lowering the

_____(10) has the same effect because it slows the molecules and attractive forces become more significant. **Real gases exhibit lower** _____**(11) than the ideal gas law would predict.**

At low pressures, we assume that the _____(12) volume inside the container equals the _____(13) volume (we assume that the gas particles themselves do not contribute significantly to the volume). At _____(14) pressures, this assumption no longer holds. As the pressure increases and the free volume decreases, eventually, the volume of the _____(15) themselves takes up a significant proportion of the container volume and the container volume is significantly larger than the free volume. **At very high pressures, real gases exhibit larger** _____**(16) than the ideal gas law would predict based on free volume inside the container.**

answers: 1-e, 2-f, 3-m, 4-j, 5-g, 6-d, 7-f, 8-b, 9-k, 10-l, 11-i,
12-c, 13-a, 14-d, 15-h, 16-n.

Initially, as pressure increases, interactive forces become important and PV/RT decreases below the ideal value. As the pressure increases further, however, the volume of the molecules becomes significant, and outweighs the importance of the intermolecular attractions so PV/RT increases above the ideal value.

An equation for real gases was developed in 1873 by Johannes van der Waals Equation which takes into account molecular interactions and molecule volume:

$$\left(P + n^2 \frac{a}{V^2}\right)(V - n b) = n R T$$

$$\begin{array}{cc} \text{adjusts} & \text{adjusts} \\ \text{P up} & \text{V down} \end{array}$$

where a and b are van der Waals constants, experimentally determined positive numbers that depend on the particular gas.

SUMMARY:

1) The physical attributes of gases:

- Gases are mostly empty space at normal conditions

    - easily compressed
    - low viscosities
    - low densities

- Mechanically, all gases behave the same way

    1 mole of any gas at STP = 22.4 L (standard molar volume)

- Standard temperature and pressure (STP) conditions: T = 0°C (273 K), P = 1 atm.

- The density of a gas is directly proportional to its molar mass.

2) Pressure:

- Definition: $P = \dfrac{f}{a}$
- A barometer measures atmospheric pressure (typically 760 mm Hg)
- Units of pressure: 1 atm = 760 mm Hg = 760 torr

3) The Gas Laws:

- Boyle's Law: $V = k/P$ (T and n fixed) $P_1V_1 = P_2V_2$
- Charles's Law: $V = bT$ (P and n fixed) $\dfrac{V_1}{T_1} = \dfrac{V_2}{T_2}$
- Avogadro's Law: $V \propto n$ (T and P fixed) $\dfrac{V_1}{n_1} = \dfrac{V_2}{n_2}$
- Ideal Gas Law: $PV = nRT$ (R = 0.0821 L·atm/mole K)

  The ideal gas law may also be written in the form: $m/V = d = PM/RT$

4) Dalton's Law of Partial Pressures:

For a mixture of gases, $P_{total} = P_1 + P_2 + P_3 ...$

For a gas, X, collected over water, $P_X = P_{tot} - P_{H2O}$

5) The Kinetic-Molecular Theory:

- Assumptions:
  - gas particles have no volume
  - gas particles do not interact
  - gas particles are in constant motion
  - The average kinetic energy of a gas molecule is directly proportional to the absolute temperature of the gas.

- Average speed of gas molecules: $u = \left(\dfrac{3RT}{M}\right)^{\frac{1}{2}}$

  u = 1150 mi/hr for a nitrogen molecule at room temperature

- Molecular energy:

  $KE_{(avg)} = 3/2\ RT$

  $KE_{(avg)} = 888$ cal/mole for any gas at room temperature

6) Graham's Law of Effusion and Diffusion: $\dfrac{rate_A}{rate_B} = \dfrac{\sqrt{M_B}}{\sqrt{M_A}}$

7) Real gases: van der Waals Equation: $\left(P + n^2\dfrac{a}{V^2}\right)(V - nb) = nRT$

Chapter 5 Exercises:

5.1) A U-tube manometer is open to the atmosphere on the right and connected to a sealed glass vessel on the left. If the atmospheric pressure is 760 torr, and the difference in the levels of mercury in the two sides is 15 cm, so that the level on the side open to the atmosphere is the higher, calculate the pressure of the gas in the sealed vessel on the left.

5.2) a. If a mixture of 4.0 mol of nitrogen and 1.0 mol of oxygen is placed in a 5.0 liter container at 27°C, what pressure is exerted by the gas mixture?

b. At a constant temperature, how many liters will a gas occupy at a pressure of 3.00 atm if it occupies 8.0 liters at a pressure of 4.5 atm?

5.3) Scuba divers use compressed air tanks for air sources while under water. A scuba tank contains 12.00 L of compressed air at 200 atm. What is the volume of air within the tank at 1.25 atm?

5.4) A bicycle pump contains a column of air 40.0 cm long with a 5.00 cm diameter at 1.00 atm pressure. How far must the handle be compressed to create a pressure of 3.00 atm within the bicycle pump?

5.5) Expandable balloons are used in weather and atmospheric research. Such an instrument balloon, at takeoff, has a volume of 14.0 cubic meters at 748 mm Hg and 18°C. As the balloon rises through the atmosphere, it expands to 900 cubic meters when the pressure drops to 8.72 mm Hg. What is the temperature at this point?

5.6) Methane gas, CH_4, contributes to the process of global warming. An average dairy cow produces 24.0 pounds of methane gas each year. What volume of CH_4 is produced at 25°C and 750 mm Hg?

5.7) A 10 liter vessel of nitrogen gas at 1 atm pressure is heated from a temperature of 200K to a temperature of 300K. The volume of the container does not change. How much of the nitrogen gas would have to be released from the container in order to reduce the pressure back to 1 atm while maintaining the temperature at 300K?

5.8) a. A 7.20 liter flask contains 20 grams of an unknown gas at a pressure of 1.02 atm. and a temperature of 87°C. What is the molar mass of the gas?

b. What is the density of ammonia gas at 2.00 atm pressure and 25.0°C?

5.9) The molar volume of any gas at standard temperature and pressure is 22.4 L/mol.

a. What is the density of PF_3 at STP?
b. If the pressure is held constant at 1.00 atm, what is the density of PF_3 at 25°C? at 40°C?
c. How does temperature affect the density of a gas?
d. If the temperature is kept constant at 273 K, what is the density of PF_3 at 2.00 atm? at 5 atm?
e. How does pressure affect the density of a gas?

5.10)   a. What is the partial pressure of helium gas in a mixture of 24 grams of oxygen and 24 grams of helium held in a vessel at a pressure of 1350 torr?

b. A 2.0 liter vessel of hydrogen gas at 1.0 atm and a 3.0 liter vessel of nitrogen gas at 1.5 atm are connected by a valve. If the valve between the flasks is opened and the gases are allowed to mix at constant temperature, what is the final pressure in the apparatus?

5.11)   Halothane, C_2HF_3BrCl, is a commonly used gaseous anathesia. One suggested dosage is 3.0% halothane by volume. What is the partial pressure of halothane if air administered to the patient is at a pressure of 775 mm Hg?

5.12)   834 mL of O_2 was collected over water at 24°C and 726 mm Hg. What is the mass of the dry oxygen gas collected?

5.13)   If an unknown gas, which creates copious white fumes in the presence of hydrochloric acid, diffuses 3 times faster than carbon tetrachloride CCl_4, what might the gas be?

5.14)   The early enrichment of uranium-235 involved the slightly different rates of diffusion between UF_6 prepared with uranium-238 and UF_6 prepared with uranium-235. What is the relative rate of diffusion for $^{235}UF_6$ to $^{238}UF_6$?

5.15)   2.50 L of CO_2 and 4.50 L of NH_3 are injected into a reaction chamber at 11 atm pressure and heated to 200°C to produce urea.

$$2NH_3 + CO_2 \rightarrow CO(NH_2)_2 + H_2O$$

What mass of urea can be produced from this reaction?

5.16)   A hot air balloon is filled with 1500 L air at 51°C and 785 mm Hg. What is the density of the air inside the balloon? Assume the air is 80% N_2 and 20% O_2 by mass.

5.17)   21.0 L of N_2 gas and 50.0 L of H_2 gas is mixed and pumped at 5.00 atm into a reaction column containing a catalyst which is heated to 500°C. Ammonia was produced with a 87.0% yield. The unreacted N_2 and H_2 are recycled back to the reaction column.

$$N_2 + 3H_2 \rightarrow 2NH_3$$

What is the volume of NH_3 produced during the first pass through the reaction column?

5.18)   Which of the following gases would you expect to be most ideal in its behavior? Which one would you expect to deviate the most from ideal behavior?

nitrogen	hydrogen chloride
hydrogen sulfide	ammonia
carbon dioxide	methane
nitric oxide	hydrogen
water	argon

5.19)   If the temperature of an ideal gas is increased while keeping the volume constant, the mean free path of the gas molecules remains unchanged. Explain, on a qualitative bases, how this can be.

Chapter 5 Answers:

5.1) If the level of mercury on the atmosphere side is higher, then the gas inside the glass vessel is pushing down with greater pressure than the atmosphere; the pressure in the glass vessel must be higher than atmospheric pressure.

Pressure inside vessel = atmospheric pressure + difference in the two levels in mm
$$= 760 \text{ torr} + 150 \text{ mm Hg}$$
$$= 910 \text{ torr}$$

5.2) a. You assume the gases behave ideally and use the ideal gas law, $PV = nRT$
$$P = 5.0 \text{ mol} \times 0.08206 \text{ L atm K}^{-1}\text{mol}^{-1} \times 300.15 \text{ K}/5.09 \text{ L} = 25 \text{ atm.}$$

b. Use Boyle's Law: $P_1V_1 = P_2V_2$
$$V_1 = 4.5 \text{ atm} \times 8.0 \text{ L}/3.0 \text{ atm} = 12 \text{ L}$$

5.3) 1920 L air

5.4) 13.3 cm

5.5) 219 K

5.6) 1.68×10^4 L of methane

5.7) You can use the relationship $n_1 T_1 = n_2 T_2$, so that
$$n_2 = n_1 \times T_1/T_2 = n_1 \times 200/300$$
One-third of the nitrogen gas must be removed to leave 2/3 behind.

5.8) a. density = PM/RT; M, molar mass = 80 g/mol
b. density = 1.39 g/L

5.9) a. 3.92 g/L at STP
b. 3.59 g/L at 25°C, and 3.42 g/L at 40°C
c. An increase in temperature decreases the density
d. 7.84 g/L at 2.0 atm, and 19.6 g/L at 5 atm
e. An increase in pressure will increase the density

5.10) Both problems involve Dalton's Law of Partial Pressure. The partial pressure of a gas in a mixture of gases is the contribution that gas makes to the total pressure exerted by the mixture. The partial pressure is a molar quantity; it depends upon the mole fraction of the gas in the mixture.

a. mole fraction of helium = 6/6.75 = 0.89
   partial pressure of helium = 0.89 × 1350 torr = 1200 torr

b. A useful way to express Dalton's law is to say that the partial pressure of a gas is the pressure that the gas would exert if it alone occupied the vessel.
   partial pressure of hydrogen gas = 0.40 atm
   partial pressure of nitrogen gas = 0.90 atm
   total pressure is the sum of the partial pressures = 0.40 + 0.90 = 1.30 atm.

5.11) 23 mm Hg halothane

5.12) 1.02 g O_2

5.13) The molar mass of carbon tetrachloride is 153.8 g/mol.
Graham's Law of Diffusion states that the relative rates of diffusion of two gases equals the square root of the inverse ratio of their molar masses. If one gas diffuses three times faster than another, then its molar mass must be nine times (3^2) less than the molar mass of the other gas. The molar mass of the unknown gas must be about 153.8 g/mol / 9 = 17.09 g/mol. This value suggests the gas is ammonia NH_3.

5.14) 1.0043

5.15) 38.1 g urea

5.16) 1.12 g/L

5.17) 29.0 L NH_3

5.18) The predominant reason for deviation from ideal behavior is intermolecular attraction. The greater the attraction between molecules, the greater the deviation. The secondary reason for deviation is the size of the molecules. First look at intermolecular forces: hydrogen bonding (the strongest, e.g. water, ammonia), dipole-dipole attraction (molecules with permanent dipole moments, e.g. hydrogen chloride, hydrogen sulfide, nitric oxide), London forces (the weakest, carbon dioxide, methane, hydrogen, argon, nitrogen).

Most ideal gas: hydrogen (very weak intermolecular attraction and low molar mass).
Least ideal gas: water (strong hydrogen bonding; in fact it's a liquid at room temperature and 1 atm pressure).

5.19) If the temperature of an ideal gas is increased, then the kinetic energy of the gas molecules increases, i.e. the molecules move faster. This suggests that the distance traveled between collisions should increase. However, increasing the temperature at constant volume increases the pressure of the gas and the collision frequency increases.

The velocity, or distance/time, increases
The collision frequency, or collisions/time increases.

Dividing one by the other, distance/collision, or mean free path, remains constant.

CHAPTER 6
Thermochemistry

TOPICS:
- Energy
- Heat and Work
- Enthalpy
- Calorimetry
    - Heat capacity
    - Coffee cup calorimeter
    - Bomb calorimeter
- Heat of Reaction
    - Heat of formation
    - Hess's Law of Heat Summation

SECTION 6.1: Energy

All mechanical energy can be classified as kinetic energy (energy of motion), or potential energy (energy of position). When we hold a ball above the ground, it contains potential energy which is converted into kinetic energy as the ball falls to the ground. At any given time, the sum of the potential and kinetic energies are a constant:

$$E_{kinetic} + E_{potential} = constant \quad \text{(purely mechanical system)}$$

where $E_{kinetic} = 1/2mv^2$ and $E_{potential} = mgh$.

Units of energy:

The SI unit for energy is the joule (J), $1 \text{ J} = 1 \dfrac{kg \, m^2}{s^2}$

Units for a joule:

Use (a) the potential energy equation, and (b) the kinetic energy equation to obtain the units for a Joule ($kg \, m^2/s^2$).

solution:

If we substitute kg for mass, m/s^2 for g, m for height, and m/s for velocity we obtain:

$$(a) \quad E_{pot} = mgh = kg \times \frac{m}{s^2} \times m = \frac{kg \, m^2}{s^2}$$

$$(b) \quad E_{kin} = \frac{1}{2}mv^2 = \frac{1}{2} \times kg \times m \times \frac{m}{s^2} = \frac{1}{2} \frac{kg \, m^2}{s^2}$$

You may be more familiar with the term calorie which is about 4 times larger than a joule:

$$1 \text{ cal} = 4.184 \text{ J}$$

A calorie is the amount of energy needed to raise 1 gram of water 1°C. A nutritional Calorie (note the capital C) is equivalent to a 1000 calories, or 1 kilocalorie.

$$1 \text{ Cal} = 1 \text{ kcal} = 1000 \text{ cal}$$

Demonstration that $E_{kin} + E_{pot}$ = constant:

1) If a ball (mass 20 g) is dropped and falls for 1.5 seconds before it hits the ground, what is its potential energy before it falls, and what is its kinetic energy when it reaches the ground? The distance the ball drops from its initial height (h_i) to its final height (h_f) is $(h_i - h_f) = \Delta h = 1/2gt^2$. The value for g is 9.8 m/s^2.

solution:

- In order to solve for potential energy, (E_{pot} = mgh), we need to know the ball's initial height:

$$h = h_i$$
$$\Delta h = h_i - h_f$$
$$h = h_f = 0$$

- We know from physics that the distance the ball will fall in 1.5 seconds is:

$$\Delta h = h_i - h_f = \frac{1}{2} \times 9.8\frac{m}{s^2} \times \left(1.5 \text{ s}\right)^2 = 11.0 \text{ m}$$

$$E_{pot} = mgh = 0.020 \text{ kg} \times 9.8\frac{m}{s^2} \times 11.0 \text{ m} = 2.2\frac{kg\,m^2}{s^2} = 2.2 \text{ J}$$

- In order to solve for kinetic energy (E_{kin} = 1/2mv^2), we need to know the ball's velocity as it reaches the ground:

$$\text{velocity, } v_f = gt = 9.8 \text{ m/s}^2 \times 1.5 \text{ s} = 14.7 \text{ m/s}$$

$$E_{kin} = \frac{1}{2}mv^2 = \frac{1}{2}\left[0.020 \text{ kg} \times \left(14.7 \text{ m/s}\right)^2\right] = 2.2\frac{kg\,m^2}{s^2} = 2.2 \text{ J}$$

Notice that $E_{kin} = E_{pot} = 2.2$ J

2) Use algebra to show: $E_{kin} + E_{pot} = $ constant

   solution:

   - We start with the equation for the distance, Δh, an object falls:

   $$\Delta h = h_i - h_f = \frac{1}{2} g t^2,$$

   and substitute for t, $t = \frac{v_f}{g}$ ($v_i = 0$):

   $$\Delta h = h_i - h_f = \frac{1}{2} g \left(\frac{v_f}{g} \right)^2 = \frac{1}{2} \frac{v_f^2}{g}$$

   - Multiplying through by g gives:

   $$g h_i - g h_f = \frac{1}{2} v_f^2$$

   - Multiplying both sides of the equation by m gives:

   $$m g h_i - m g h_f = \frac{1}{2} m v_f^2$$

   $$-\Delta \left(E_{pot} \right) = \Delta \left(E_{kin} \right)$$

   $$\Delta \left(E_{pot} \right) + \Delta \left(E_{kin} \right) = 0$$

   $$\Delta \left(E_{pot} + E_{kin} \right) = 0$$

   $$E_{pot} + E_{kin} = \text{constant}$$

Conservation of Energy

Let's go back to the ball in free-fall. We know from experience that eventually the ball will come to rest. Does this mean that the kinetic and potential energy are lost? No. Experience teaches us that energy is not lost, it is converted to other forms. As the ball bounces on the ground, frictional forces cause the mechanical energy to be converted to thermal energy. When one form of energy disappears, equivalent amounts of other types of energy always appear. Experience tells us:

ENERGY CANNOT BE CREATED OR DESTROYED,

ONLY TRANSFORMED

This observation is the Law of Conservation of Energy or the 1st law of thermodynamics.

We cannot in practice measure absolute energies, but we can measure energy changes. To measure a change in energy (ΔE), we measure the difference between a system's internal energy before some change, and its energy after the change:

$$\Delta E = E_{final} - E_{initial}$$

A <u>system</u> consists of the components whose change we wish to measure. How do we measure the energy change of a system? By looking at the <u>surroundings</u>. We can measure the energy change of the surroundings to infer the energy change of the system. We divide the surroundings into two types of energy changes: mechanical and thermal. We measure ΔE of the <u>mechanical surroundings</u> with meter sticks and stopwatches, and we use a thermometer to determine ΔE of the <u>thermal surroundings</u>. We obtain ΔE of the system from ΔE of the thermal plus mechanical surroundings.

In the case of the ball in free fall, the ball is the system. In a chemical reaction, the system is often the reactants. The <u>surroundings</u> are everything else that is affected by the change. The system can gain energy from the surroundings or lose energy to the surroundings. Since we know the total amount of energy cannot change, we can write:

$$\Delta E_{total} = \Delta E_{sys} + \Delta E_{surr} = 0$$

$$E_{total} = E_{system} + E_{surr} = constant$$

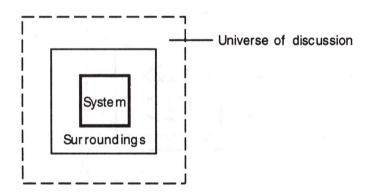

The figure shows the universe of discussion enclosed with a dotted line. The central box represents the <u>system</u> which contains the components we wish to study. The outer box represents the <u>surroundings</u> which contains everything else that is affected by the change. In a chemical reaction, if a gas is produced, it does work by pushing back the atmosphere (mechanical change of the surroundings.). If it absorbs, or gives off heat, we can measure the thermal changes of the surroundings to determine the energy changes of the system.

Calorimeters are useful devices because they enclose the system in an insulated container. We use a thermometer to measure the change in thermal energy of the surroundings (often water) to determine the energy changes from a chemical reaction (the system).

SECTION 6.2: Heat and Work

Heat and work defined:

Choose words from the list below to complete the following paragraph:

    (a) energy   (c) temperature
    (b) heat   (d) work

Energy can be transferred from one system to another by either heat flow or by work. Heat is energy transferred due to differences in _____(1) between the system and the surroundings. Energy transferred when an object is moved by a force is _____(2). We saw above that a bouncing ball causes its surroundings to _____(3) up. It could also have done _____(4) by causing some of the particles on the ground to move. When we eat food, we input calories (_____(5)) into our bodies which is converted to _____(6) (to keep our body temperature at approx. 98.6°F), and into _____(7) (to propel our bodies around if we choose to do so).

<u>answers</u>: 1-c, 2-d, 3-b, 4-d, 5-a, 6-b, 7-d.

Let's consider a reaction run in a flask exposed to the constant pressure of the atmosphere. We can picture the system and surroundings as follows:

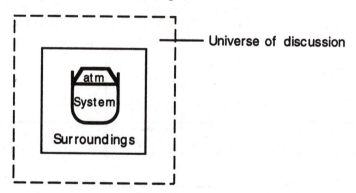

If a reaction produces a gas, it does <u>work</u> by pushing back the atmosphere. If the system is at a different temperature from the surroundings, energy is transferred as <u>heat</u>. We must consider both the energy change of the <u>mechanical surroundings</u> (the atmosphere pushing down on the reaction), and the <u>thermal surroundings</u>. We can write this as:

$$\Delta E_{sys} + \Delta E_{surr} = 0$$

$$\Delta E_{sys} = -\Delta E_{surr}$$

$$\Delta E_{sys} = -(\Delta E_{thermal.\ surr} + \Delta E_{mech\ surr})$$

$$\Delta E_{sys} = q + w$$

where:

- $q = -\Delta E_{thermal\ surr}$, and is defined as energy transferred between the system and surroundings due to a temperature change

- $w = -\Delta E_{mech.\ surr}$, and is defined as energy transferred by other mechanisms. Many different kinds of work are possible, but we will be concerned only with mechanical work which is due to volume changes against an external pressure ($P\Delta V$ work).

- The equation is written from the system's perspective so q is positive when the system <u>gains</u> heat from the surroundings, and w is positive when work is done <u>on</u> the system by the surroundings.

Notice that the equations in bold above are simply different ways to express that the total amount of energy in the universe is constant:

$$\Delta E_{sys} = q + w$$

$$\Delta E_{sys} - q - w = 0 = \Delta E_{total}$$

$$\Delta E_{sys} + \Delta E_{therm.\ surr} + \Delta E_{mech.\ surr} = 0 = \Delta E_{total}$$

$$E_{sys} + E_{thermal\ surr} + E_{mech.\ surr} = constant = E_{total}$$

Conversion of energy into heat:

If you weigh 65 kg (143 lb.), and eat a 400 Calorie breakfast, what would be your body temperature if the energy were all converted to heat? (Assume a starting body temp. of 98.6°F or 37°C, and that, for simplicity's sake, you are 100% water).

<u>solution:</u>

- 400 nutritional Calories = 400 kcal or 400,000 cal.

- 1 gram of water requires 1 calorie to raise its temperature 1°C. It would therefore take 65,000 calories to raise 65,000 g (65 kg) of water 1°C or 65,000 cal/°C.

- Our breakfast supplied us with 400,000 calories of energy, so 400,000 calories would raise 65,000 g by:

$$\frac{400,000\ cal}{65,000\ cal/°C} = 6.15°C$$

- Our body temperature would be:

$$37°C + 6.15°C = 43.1°C \text{ or } 109.6°F \quad \text{(probably fatal)}.$$

We metabolize our food so that not all the heat is released at one time.

Conversion of energy into work:

Assuming you convert all the chemical potential energy from a 400 Calorie breakfast into work with 100% efficiency, how high would you have to climb to work off this energy?

solution:

- $E_{potential} = mgh$ (m = 65 kg, g = 9.8 m/s^2)

    Solving for h gives:

    $h = E_{pot}/mg$

- $E_{potential}$ from breakfast = 400 Cal = 400,000 cal

    400,000 cal x 4.184 J/cal = 1,673,600 J

    (Remember: $J = \dfrac{kg\,m^2}{s^2}$)

- Substituting into the equation for h gives:

    $$h = \frac{1,673,600 \; kg\,m^2/s^2}{65\,kg \times 9.8\,m/s^2} = 2600\ m \ \ (about\ 1.6\ miles)$$

Note: we cannot convert much of our food calories into work because most of the energy is used to keep our body temperatures around 98°F.

A note about q:

Heat and temperature are not synonyms. Heat transfer is measured by a change in temperature. Unlike temperature, heat depends upon the amount of material involved. A five gallon container of gasoline sitting in a garage has the same temperature as a cupful of gasoline from the container, but the "heat content" of the five gallon container is much larger than the cupful, as one could easily demonstrate by lighting the two with a match...

A note about w:

In this chapter we are concerned with work done by expanding gases. The useful equation for pressure-volume work which can be derived from the definition of work is:

$$w = -P\Delta V$$

- the system loses energy when it does work ($\Delta V > 0$), (indicated by the negative sign).

- this equation assumes constant pressure conditions, in this case the constant pressure of the atmosphere.

Using $w = -P\Delta V$, we can write:

$$\Delta E_{sys} = q + w$$

$$\Delta E_{sys} = q_p - P\Delta V \quad \text{(constant p, only } P\Delta V \text{ work done)}$$

The subscript "p" of q_p reminds us that this equation is written for constant pressure conditions.

Note: when the volume is constant, $\Delta V = 0$, so $P\Delta V = 0$, and

$$\Delta E = q_v \quad \text{(constant volume)}$$

State Functions

A property that is independent of the pathway is a <u>state function</u>. E is a state function because the difference in E_{final} and $E_{initial}$ is independent of the path taken between them. Another way to think about this is that the same ΔE can occur through any combinations of q and w. After we eat a 400 Calorie breakfast, we can either use the energy by taking a nap, in which case much of the energy is slowly(!) converted to heat, or we can climb 8000 ft. to the top of a mountain, in which case some of the energy ($\approx$ 25%) is converted into our ability to do work.

Another example of a state function is the altitude gain on a mountain climbing expedition. If you and a friend climb Long's Peak in Colorado, and one decides to scale the vertical east face, while the other takes the more circuitous, but decidedly safer "keyhole" route, the change in altitude from the trail head to the peak is a state function: it is the same regardless of the route taken.

SECTION 6.3:   Enthalpy

For a universe containing a system with its mechanical and thermal surroundings we have seen that:

$$\Delta E = q + w$$

When the system is at constant pressure and only $P\Delta V$ work is possible, we can write:

$$\Delta E_{sys} = q_p - P\Delta V_{sys}$$

We can solve this equation for q_p to give:

$$q_p = \Delta E + P\Delta V$$

We can substitute $\Delta E = E_{final} - E_{initial}$ and $\Delta V = V_{final} - V_{initial}$:

$$q_p = (E_{final} - E_{inital}) + P(V_{final} - V_{initial})$$

and rearrange the equation to give:

$$q_p = (E_{final} + PV_{final}) - (E_{initial} + PV_{initial})$$

We can simplify the above equation by <u>defining</u>:

$$H = E + PV$$

We call H **enthalpy**. Enthalpy simplifies the writing of equations when working under constant pressure conditions, the conditions you normally encounter in the laboratory. Enthalpy, like energy, is a state function and so a change in enthalpy, ΔH, is dependent only on the initial and final states of the system. Substituting H into the equation for q_p above gives:

$$q_p = H_{final} - H_{initial} = \Delta H$$

We can substitute ΔH into our equation for ΔE ($\Delta E = q_p - P\Delta V$) to give:

$$\Delta E = \Delta H - P\Delta V \textbf{ (constant p)}$$

ΔH and ΔE differ only by the $P\Delta V$ term, which for most chemical reactions is insignificant, so **usually, $\Delta H \approx \Delta E$**. $P\Delta V$ may become significant when:

1) There is a change in the number of moles of gases between reactants and products (the change in volume term, ΔV, may become significant). The volume change when dealing with liquids and solids is too small to cause the ΔV term to be significant.

2) Pressures are approximately those reached in the interior of the earth (the pressure term, P, becomes significant) - not likely a concern with reactions carried out in a laboratory!

Determining when $P\Delta V$ is significant:

1) When 1 mole of water is vaporized at 1 atm pressure at 100°C, **$\Delta H = 40.6$ kJ**. Would you expect ΔE to be larger, smaller, or approximately the same as ΔH?

solution:

Because there is a large increase in volume between a mole of water and a mole of steam, the work term ($P\Delta V$) is significant (3.1 kJ). Since the system does work against the atmosphere as water is vaporized, it loses energy and ΔE is smaller than ΔH:

$$\Delta E = 40.6 \text{ kJ} - 3.1 \text{ kJ} = \textbf{37.5 kJ}.$$

2) When 1 mole of water freezes, would you expect ΔE to be larger, smaller, or approximately equal to ΔH?

solution:

The volume change between water and ice is small, so the work term ($P\Delta V$) is insignificant. In fact, when a mole of water freezes, ΔE and ΔH differ by only 0.00017 kJ and:

$$\Delta H \approx \Delta E.$$

Note on enthalpy: when a system is at constant pressure, and only $P\Delta V$ work is possible, enthalpy represents the heat liberated. It is merely a useful way to simplify the expression for q_p.

H, like E, is a <u>state function</u>; it is independent of the route used to reach the products.

$$\Delta H = H_{products} - H_{reactants}$$

- If $H_{products} > H_{reactants}$, $\Delta H > 0$ and the reaction is endothermic (absorbs heat from the surroundings.)

- If $H_{reactants} > H_{products}$, $\Delta H < 0$, and the reaction is exothermic (gives off heat to the surroundings).

SECTION 6.4:   Calorimetry

How do we measure ΔH or ΔE in the laboratory? We measure a change in enthalpy or energy by measuring heat released or absorbed by the system from the thermal surroundings. We can't <u>directly</u> measure heat flow, but we can easily measure temperature changes which are proportional to the amount of heat transferred. We use a calorimeter to measure these temperature changes. The amount the temperature changes will depend on the material's heat capacity.

Heat Capacity:

Heat capacity (C) is the amount of heat required to raise the temperature of a substance 1 K.

Two specific types of heat capacity are:

Specific heat capacity (c): the amount of heat required to raise the temperature of **1 gram** of a substance by 1 K.

Molar heat capacity (C_m): the amount of heat required to raise the temperature of **1 mole** of a substance by 1 K.

Notice that specific heat capacity and molar heat capacity specify an amount of material; heat capacity is a term which does <u>not</u> specify an amount of material (its units are J/T). You cannot look up heat capacity in a table since a number would be meaningful only if it were defined for a specific amount of a system.

Water has a specific heat capacity of 4.184 J/g K or 1 calorie/g K (remember a calorie is the amount of heat required to raise 1 gram of water 1 K). Water has a high heat capacity! For comparison, metals typically have specific heat capacities of less than 1 J/g K.

Specific heat capacity:

What is the amount of energy (in joules) required to bring a cup of water to boiling (100°C) from room temperature (25.0°C)?

<u>solution:</u>

• Specific heat capacity of water = 4.184 J/g°C. The amount of water is:

$$1 \text{ cup} \times \frac{1 \text{ pint}}{2 \text{ cups}} \times \frac{1 \text{ lb}}{1 \text{ pint}} \times \frac{454 \text{ g}}{1 \text{ lb}} = 227 \text{g}$$

• The temperature change is:

$$\Delta T = 100°C - 25.0°C = 75.0°C$$

• Therefore, the amount of energy required is:

$$4.184 \text{ J/g°C} \times 227 \text{g} \times 75.0°C = 71,200 \text{ J} \quad \text{or} \quad 71.2 \text{ kJ}$$

Mathematically we write:

$$q = c \times m \times \Delta T$$

As long as we remember the definition for specific heat capacity (c), we can remember this equation by using unit analysis. The specific heat capacity must be multiplied by terms which provide q with energy units (typically joules):

$$J = \frac{J}{g \, °C} \times g \times °C$$

It can be confusing to keep heat capacity, specific heat capacity and molar heat capacity straight unless you always include units in your equations. It is also very helpful to notate in the algebraic equations what type of heat capacity you are using. In this study guide, we use a capital C to notate heat capacity, a small c for specific heat capacity and C_m for molar heat capacity.

Note also that J/g K = J/g°C since the magnitude of a degree Celsius and a degree Kelvin are equivalent.

The amount of heat evolved or absorbed in a process can be measured using a calorimeter. A coffee cup calorimeter is a simple apparatus that constructs surroundings so that water captures and retains the heat released or absorbed by the system. The water in a Styrofoam cup is defined to be the thermal surroundings. A thermometer, lowered into the water through the cover of the calorimeter, monitors the temperature change of the surroundings (ΔT). We can weigh the amount of water in the coffee cup, and we know that the specific heat capacity of water is 4.184 J/g K. We can then calculate q_{surr}:

$$q_{surr} = c \times m \times \Delta T$$

Since heat gained by the surroundings = heat lost by the system (or vice versa), we can calculate q_{sys} by:

$$q_{surr} = -q_{sys}$$

Coffee cup calorimeter (Determination of ΔH):

A 45.0 g spoon is heated to 100°C in boiling water and is dropped into a coffee cup calorimeter containing 50.0 g of water at 25°C. If the final water temperature is, 28.6°C, calculate the specific heat capacity of the spoon. Is the spoon stainless steel or silver?

solution:

First we calculate the heat absorbed by the water. Next we set up an expression for the heat released by the spoon leaving the specific heat capacity of the spoon (c_{spoon}) as the variable we want to solve for. Finally, we use the relationship:

$$q(spoon) = -q(water)$$

to solve for c_{spoon} and compare the value with the known specific heat capacities of steel and silver.

• Heat absorbed by the water:

$$q(water) = c \times m \times (T_{final} - T_{initial})$$

$$= 4.184 \ J/g°C \times 50.0 \ g \times \left(28.6°C - 25°C\right)$$

$$= 753.1 \ J$$

• Heat released by the spoon:

$$q(spoon) = c_{spoon} \times m \times (T_{final} - T_{initial})$$

$$= c_{spoon} \times 45.0g \times (28.6°C - 100°C)$$

$$= -3210 \ g°C \times c_{spoon}$$

• To solve for c_{spoon} we use the relationship:

$$\Delta E_{sys} = -\Delta E_{surr} \text{ or, } q(spoon) = -q(water)$$

• Substituting into this equation we obtain:

$$-3210 \text{ g°C} \times c_{spoon} = -753.1 \text{ J}$$

$$c_{spoon} = 0.234 \text{ J/g°C}$$

The specific heat capacity for steel is 0.45 J/g°C and for silver is 0.237 J/g°C; therefore, the spoon is most likely silver.

A note about heat capacity calculations: the problems are not difficult to set up algebraically, but they must be solved carefully so that minus signs are not misplaced. Remember that ΔT is always defined as $T_{final} - T_{initial}$.

Comments about the coffee cup calorimeter:

• We measure ΔT of the surroundings to calculate q_{surr}. Then we can solve for q_{sys} since we know:

$$q_{sys} = -q_{surr}$$

• We assume that the water in the coffee cup is the only element of the surroundings that changes temperature.

• It is a constant <u>pressure</u> system since the contents of the coffee cup are exposed to constant atmospheric pressure, so $q_p = \Delta H$ (by definition).

A coffee cup calorimeter is a simple calorimeter that you can easily make, but it isn't effective for reactions that involve gases, or ones that produce high temperatures. A **bomb calorimeter** can obtain precise heat flow measurements for these reactions.

In a bomb calorimeter, weighed reactants are sealed in a rigid steel container called the "bomb" (and they wonder why students are sometimes nervous in chemistry labs!). The bomb is immersed in a water filled container equipped with a stirrer and thermometer. Electrical ignition starts the reaction in the bomb, and if the reaction is exothermic, it gives off heat. The volume inside the bomb does not change ($\Delta V = 0$). The surroundings which absorb the heat are not only the water, but the water, the bomb, and the other parts of the calorimeter. Before a bomb calorimeter is useful for calculating energy changes, therefore, we must know the heat capacity of the <u>entire calorimeter</u>. We measure the heat capacity of the calorimeter by running a reaction in which a known amount of heat will be released.

Calibration of a bomb calorimeter:

It is known that 1 gram of benzoic acid evolves 26.42 kJ of heat when combusted. If one gram of benzoic acid is burned in a bomb calorimeter containing 2.95 kg of water, and the temperature rises from 24.33°C to 26.67°C, what is the heat capacity of the bomb (in kJ/°C)?

solution:

- We know q and ΔT:

  $q = 26.42$ kJ/g x 1 g = 26.42 kJ

  $\Delta T = T_{final} - T_{initial} = 26.67°C - 24.33°C = 2.34°C$

- We can solve for heat capacity using the relationship:

  $q =$ heat capacity x ΔT

  heat capacity $= \dfrac{q}{\Delta T} = \dfrac{26.42 \text{ kJ}}{2.34 °C} = 11.3$ kJ/°C

Note that the heat capacity value for the bomb calorimeter doesn't specify an amount of material, so it is only useful if the calorimeter is set up with the same amount of water each time.

Now that we know the heat capacity of the bomb calorimeter, we can run a combustion reaction and determine the heat evolved.

Bomb calorimeter experiment:

When hydrogen (H_2) burns in oxygen, it produces only water as a product. Hydrocarbons which contain carbon and hydrogen produce water and carbon dioxide. CO_2 is an undesirable product since it is one of the "Greenhouse" gases that contributes to global warming. We can compare the heats of combustion of H_2 and methane (CH_4) using a bomb calorimeter.

When a 1.5 g sample of methane gas is burned with excess oxygen in the calorimeter, the temperature increases by 7.3°C. When a 1.15 g sample of hydrogen gas is burned, the temperature increase is 14.3°C. Calculate the heats evolved when 1 gram of hydrogen and when 1 gram of methane are burned.

solution:

- We know from our experiment with benzoic acid (above) that the heat capacity of our bomb calorimeter is 11.3 kJ/°C

For methane, we calculate q:

$q = C_{tot} \times \Delta T = 11.3$ kJ/°C x 7.3°C = 82 kJ for 1.5 g CH_4

For 1 gram of CH_4, $q = \dfrac{82 \text{ kJ}}{1.5 \text{ g}} = 55$ kJ / g

For hydrogen we calculate q:

$$q = 11.3 \text{ kJ/°C} \times 14.3°C = 162 \text{ kJ for } 1.15 \text{ g } H_2$$

$$\text{For 1 gram of } H_2, \quad q = \frac{162 \text{ kJ}}{1.15 \text{ g}} = 141 \text{ kJ/g}$$

More energy is released when 1 gram of hydrogen is combusted than when 1 gram of methane is combusted. Hydrogen is a potentially useful and clean fuel. It is not a practical fuel yet because it is difficult to store.

Comments about the bomb calorimeter:

- Like the coffee cup calorimeter, we measure the temperature change of the surroundings to calculate q_{surr} and q_{sys}:

$$q_{surr} = C \times m \times \Delta T$$

$$q_{sys} = -q_{surr}$$

- Unlike the coffee cup calorimeter, we do NOT assume that water is the only element of the surroundings that changes temperature. We take into account the temperature change of the entire calorimeter. The heat capacity of the calorimeter must be calculated in a separate experiment. In the coffee cup calorimeter, the temperature change of the Styrofoam cup is assumed to be negligible.

- A bomb calorimeter is a constant <u>volume</u> system: $\Delta V = 0$ and $P\Delta V = 0$. Since there is no work term, $q_v = \Delta E$

In summary:

Energy:

- Heat flow at constant volume is an <u>energy</u> change: $\Delta E = q_v$

  A bomb calorimeter measures ΔE.

Enthalpy:

- Heat flow at constant pressure is an <u>enthalpy</u> change: $\Delta H = q_p$

  A coffee cup calorimeter measures ΔH.

SECTION 6.5: Heat of Reaction, ΔH_{rxn}.

The heat of reaction, ΔH_{rxn}, is dependent on the amount of substances reacting. ΔH_{rxn} is equal in magnitude but opposite in sign for ΔH of the reverse reaction. Processes which are frequently studied are given special names.

Frequently studied ΔH reactions:

- ΔH_{comb} = Heat of Combustion
  1 mole of a substance combines with oxygen in a combustion reaction.

$$CH_4(g) + 2O_2(g) \rightarrow CO_2(g) + 2H_2O(g) + Heat \qquad \Delta H = -802 \text{ kJ}$$

- ΔH_f = Heat of formation
  1 mole of a compound is formed from its elements

$$Mg(s) + \tfrac{1}{2}O_2(g) \rightarrow MgO(s) + Heat \qquad \Delta H = -602 \text{ kJ}$$

- ΔH_{fus} = Heat of fusion
  1 mole of a substance melts

$$H_2O(s) + Heat \rightarrow H_2O(l) \qquad \Delta H = 6 \text{ kJ}$$

- ΔH_{vap} = Heat of vaporization
  1 mole of a substance vaporizes

$$H_2O(l) + Heat \rightarrow H_2O(g) \qquad \Delta H = 41 \text{ kJ}$$

- ΔH^o_{rxn} = Heat of reaction measured with all substances in their standard states. The standard states are defined to be:

  -for a gas: 1 atm.
  -for an aqueous solution: 1 M concentration
  -for an element or compound: the most stable form at 1 atm. pressure
   and the temperature of interest, usually 25°C (298 K).

Heat of Formation (ΔH_f)

Consider the equation for magnesium burning in oxygen:

$$2 \, Mg(s) + O_2(g) \rightarrow 2 \, MgO(s); \quad \Delta H^o_{rxn} = -1203 \text{ kJ}$$

The negative ΔH tells us it's an exothermic reaction, that heat is given off. The superscript zero on the ΔH tells us the reactants and products are in their standard states. The amount of heat given off is related to the quantity of material. If we want to find the heat of <u>formation</u> of MgO from the elements, we would need to write the equation as:

$$Mg(s) + \frac{1}{2}O_2(g) \rightarrow MgO(s); \quad \Delta H_f^o = \frac{-1203 \text{ kJ}}{2} = -601.6 \text{ kJ}$$

We must divide the value for ΔH_{rxn}^o by two because ΔH_f^o is defined as formation of <u>1 mole</u> of the compound.

Two Comments about Heats of Formation:

- An element in its standard state is assigned a $\Delta H_f^o = 0$
- Most compounds have a negative ΔH_f^o.

Hess's Law of Heat Summation:

Sometimes it may be difficult to measure ΔH for a chemical reaction directly because, for example, the change is very slow. Since H is a state function (it does not depend on the path), we can add together ΔH's for individual steps to obtain the ΔH in question:

We use Hess's law of heat summation to write: $\Delta H_1 = \Delta H_2 + \Delta H_3 + \Delta H_4$

Hess's Law:

The standard enthalpy of combustion of elemental carbon to form carbon dioxide is -393.7 kJ/mol carbon. The standard enthalpy of combustion of carbon monoxide to form carbon dioxide is -283.3 kJ/mol CO. Use these data to calculate ΔH^o for the reaction:

$$2C(s) + O_2(g) \rightarrow 2CO(g)$$

<u>solution:</u>

- The balanced equation for the combustion of carbon is:

$$C(s) + O_2(g) \rightarrow CO_2(g)$$

$$\Delta H^o = -393.7 \text{ kJ/mol carbon}$$

- The balanced equation for the combustion of carbon monoxide is:

$$CO(g) + \tfrac{1}{2}O_2(g) \rightarrow CO_2(g)$$

$$\Delta H° = -283.3 \text{ kJ/mol CO}$$

- Manipulation of equations to allow us to obtain the equation in question:
We want 2 moles of carbon to appear on the left hand side of the final equation, so we multiply the equation for the combustion of C by 2:

$$2C(s) + 2O_2(g) \rightarrow 2CO_2(g) \qquad \Delta H° = -787.4 \text{ kJ}$$

- We need 2 moles of CO to be a product, so we write the reverse of the equation for the combustion reaction for CO and multiply it by 2. We inverse the sign of $\Delta H°$, and multiply it by 2:

$$2CO_2(g) \rightarrow 2CO(g) + O_2(g) \qquad \Delta H° = 566.6 \text{ kJ}$$

- Adding the two equations and corresponding $\Delta H°$ values gives:

$$2C(s) + 2O_2(g) \rightarrow 2CO_2(g) \qquad \Delta H° = -787.4 \text{ kJ}$$
$$2CO_2(g) \rightarrow 2CO(g) + O_2(g) \qquad \Delta H° = 566.6 \text{ kJ}$$

$$2C(s) + O_2(g) \rightarrow 2CO(g) \qquad \Delta H° = -220.8 \text{ kJ}$$

Another example where Hess's law is frequently used is in using bond energy data to calculate heats of reaction when ΔH_f^o data is unavailable. We can think of a reaction as a process in which reactant bonds are broken and product bonds are formed. Breaking bonds requires energy, and forming bonds releases energy. We can calculate ΔH_{rxn} by determining the energy required to break bonds in the reactants and the energy released as bonds are formed in the products.

For the following reaction, we will use bond energy data to obtain ΔH_{rxn}.

$$\underset{H}{\overset{H}{H-\underset{|}{\overset{|}{C}}-H}} + Cl-Cl \longrightarrow \underset{H}{\overset{H}{Cl-\underset{|}{\overset{|}{C}}-H}} + H-Cl$$

	ΔH_{sys} (kJ/mol bonds)	
1. One H-C bond is broken	414.2	ΔH_1
2. One Cl-Cl bond is broken	242.7	ΔH_2
3. One C-Cl bond is formed	-330.5	ΔH_3
4. One H-Cl bond is formed	-431.0	ΔH_4

Using Hess's law we write:

$$\Delta H_{rxn} = \Delta H_1 + \Delta H_2 + \Delta H_3 + \Delta H_4$$

$$\Delta H_{rxn} = 414.2 + 242.7 - 330.5 - 431.0 = -104.6 \text{ kJ.}$$

The reaction is exothermic because more energy is released forming the C-Cl and H-Cl bonds than is needed to break the H-C and Cl-Cl bonds.

Determining the Standard Heat of Reaction from the Standard Heats of Formation:

We can view any reaction as a two step process in which the first step is breaking the reactants down into their elements ($-\Delta H_f^o$), and the second step is combining the elements to form the products (ΔH_f^o). Using Hess's Law, we find that the standard heat of reaction for any reaction is the sum of the standard heats of formation of the products minus the sum of the standard heats of formation of the reactants. This is written mathematically as:

$$\Delta H_{rxn}^o = \sum m\Delta H_{f(products)}^o - \sum n\Delta H_{f(reactants)}^o$$

where $\sum$ means "the sum of", and m and n are the coefficients of the reactants and products from the balanced equation.

Standard Enthalpy of Combustion:

Ethanol (C_2H_5OH) is sometimes used in gasoline to decrease the amount of undesirable combustion products. The standard heat of formation of ethanol is -279 kJ/mol. Compare the standard enthalpy of combustion per gram of ethanol to that per gram of octane (C_8H_{18}), a major component of gasoline. ΔH_f^o for octane is -269 kJ/mol, ΔH_f^o for $CO_2(g)$ is -394 kJ/mol, and ΔH_f^o for $H_2O(l)$ is -286 kJ/mol.

solution:

• First we write balanced equations for the combustion of octane and ethanol:

$$C_8H_{18}(l) + \frac{25}{2}O_2(g) \rightarrow 8CO_2(g) + 9H_2O(l)$$

$$C_2H_5OH(l) + 3O_2(g) \rightarrow 2CO_2(g) + 3H_2O(l)$$

• Calculate ΔH_{comb}^o (per mole and per gram of octane and ethanol):

$$\Delta H_{comb}^o = \sum m\Delta H_{f(products)}^o - \sum n\Delta H_{f(reactants)}^o$$

$$\Delta H^o_{comb}(octane) = [8 \times \Delta H^o_f\, CO_2(g) + 9 \times \Delta H^o_f\, H_2O(l)] - [\Delta H^o_f\, C_8H_{18}(l)]$$

$$= [(8 \times -394\ kJ/mol) + (9 \times -286\ kJ/mol)] - [1 \times (-269 kJ/mol)]$$

$$= -5457\ kJ/mole\ of\ octane$$

- Molar mass of octane = 114 g/mol so:

$$\frac{-5457\ kJ/mol}{114\ g/mol} = -47.9\ kJ\ per\ gram\ of\ octane$$

- $\Delta H^o_{comb}(ethanol) = [2 \times \Delta H^o_f\, CO_2(g) + 3 \times \Delta H^o_f\, H_2O(l)] - [\Delta H^o_f\, C_2H_5OH(l)]$

$$= [(2 \times -394\ kJ/mol) + (3 \times -286\ kJ/mol)] - [(1 \times -279 kJ/mol)]$$

$$= -1367\ kJ/mole\ of\ ethanol$$

- Molar mass of ethanol = 46 g/mol so:

$$\frac{-1367\ kJ/mol}{46\ g/mol} = -29.7\ kJ\ per\ gram\ of\ ethanol$$

Octane provides ≈1.5 times more energy per gram than does ethanol.

SUMMARY:

Energy:

All mechanical energy can be classified as potential or kinetic:

$$E_{kinetic} + E_{potential} = constant$$
$$\tfrac{1}{2}mv^2 + mgh = constant$$

Units of energy:

SI unit for energy is the Joule (kg m^2/s^2): 4.184 J = 1 cal
1 cal = amount of energy needed to raise the temperature of 1 g of water 1°C

Law of Conservation of Energy:

Energy cannot be created or destroyed
$$\Delta E_{sys} + \Delta E_{surr} = 0$$
where: surr = thermal surr + mechanical surr
$$\Delta E = E_{final} - E_{initial}$$
$$\Delta E = q + w$$
where: $q = -\Delta E_{thermal\ surr}$ and $w = -\Delta E_{mechanical\ surr}$
$$\Delta E = q_p - P\Delta V\ (constant\ pressure)$$
$$\Delta E = q_v\ (constant\ volume)$$

Energy is a state function; it is independent of path

Enthalpy:

$H = E + PV$

$\Delta H = q_p = \Delta E + P\Delta V$ (constant pressure, only $P\Delta V$ work possible)

$\quad (\Delta H = H_{products} - H_{reactants})$

ΔH differs from ΔE by the work term ($P\Delta V$); generally, $\Delta H \approx \Delta E$

Exothermic: $\Delta H < 0$; heat given off to surroundings

Endothermic: $\Delta H > 0$; heat absorbed from surroundings

Enthalpy is a state function; it is independent of path

Calorimetry (Measurement of heat flow):

Heat capacity (C, J/ K): the amount of heat required to raise the temperature of a substance 1 K.

Specific heat capacity (c, J/g K): the amount of heat required to raise the temperature of 1 gram of a substance by 1 K

$$q = c \times m \times \Delta T$$

Molar heat capacity: (C_m, J/mol K) the amount of heat required to raise the temperature of 1 mole of a substance by 1 K.

Coffee cup calorimeter (constant pressure system, $q_p = \Delta H$): allows you to calculate ΔH by measuring the temperature change of water in a Styrofoam cup.

Bomb calorimeter (constant volume system, $q_v = \Delta E$): allows you to calculate ΔE by measuring the temperature change of water in a bomb calorimeter and by knowing the heat capacity of the calorimeter.

Heat of Reaction, ΔH_{rxn}:

- ΔH_{comb} = Heat of Combustion
    1 mole of a substance combines with oxygen in a combustion reaction.
- ΔH_f = Heat of formation
    1 mole of a compound is formed from its elements
- ΔH_{fus} = Heat of fusion
    1 mole of a substance melts
- ΔH_{vap} = Heat of vaporization
    1 mole of a substance vaporizes

- ΔH^o_{rxn} = Heat of reaction measured with all substances in their standard states. The standard states are defined to be:
    - for a gas: 1 atm.
    - for an aqueous solution: 1 M concentration in mol/L
    - for an element or compound: the most stable form at 1 atm. pressure and the temperature of interest, usually 25°C (298 K).

Hess's Law of Constant Heat Summation:
ΔH can be calculated by adding together ΔH's for individual steps

Bond Energies:
Bond energies can be used to calculate ΔH_{rxn}.
    Energy is added to break bonds
    Energy is released when bonds form

Heats of reaction from heats of formation:
$$\Delta H^o_{rxn} = \sum m\Delta H^o_{f(products)} - \sum n\Delta H^o_{f(reactants)}$$

Chapter 6 Exercises:

6.1)   a. If a system absorbs 35 kJ of heat from its surroundings, and does 25 kJ of work on its surroundings, what is the value of ΔE, the change in internal energy?

b. If the surroundings do 20 kJ of work on a system, and the system loses 8 kJ of heat to the surroundings, what is the change in internal energy of the system?

6.2)   a. The relationship between the SI unit of energy, the joule, and the more traditional unit, the calorie, is: 1 calorie = 4.184 J.   If one gram of water requires 1 calorie to raise its temperature by 1°C, how many joules are required to heat 1 kg water by 1°C?

b. Another, larger, unit of energy is the BTU, which is the energy required to heat one pound of water by 1°F. What is the relationship between one joule and one BTU? (1 pound = 453.6 g).

6.3)   How much work is done when 18 grams of water vaporize at 100°C under a constant external pressure of 1 atm?

6.4)   Identify the following reactions or processes as either endothermic or exothermic:
   a.   steam condensing
   b.   natural gas burning
   c.   hydrogen and oxygen gases exploding
   d.   a gas fire radiating heat
   e.   water freezing
   f.   gunpowder exploding
   g.   molten lava cooling down
   h.   magnesium ribbon reacting with oxygen to produce a brilliant white flame
   i.   ether evaporating
   j.   water decomposing into hydrogen and oxygen
   k.   ice melting
   l.   decomposition of calcium carbonate
   m. N_2O_4 decomposing to produce two NO_2 molecules
   n.   iodine crystals subliming

6.5)   If an endothermic reaction takes place in an insulated container, at the end of the reaction, will the temperature of the contents of the container be higher than before the reaction began, lower, or will it remain the same?

6.6)   How much heat is needed to raise 160 g of water from 18.8°C to 37.0°C?

6.7)   10.0 kJ of heat is added to a coil of aluminum wire with a mass of 100 g. What is the final temperature of the wire?

6.8)   The value of ΔH for the reaction shown is -280 kJ

$$C(s) \ + \ 1/2O_2(g) \ \rightarrow \ CO(g) \quad \Delta H = -280 \ kJ$$

a. What is the value of ΔH for the reaction: $2C(s) \ + \ O_2(g) \ \rightarrow \ 2CO(g)$

b. What is the value of ΔH for the reaction: $2CO(g) \ \rightarrow \ 2C(s) \ + \ O_2(g)$

6.9) When natural gas (methane) burns, the quantity of heat released is 890 kJ/mol. The standard molar heat of combustion of carbon (graphite) is -394 kJ/mol. The standard molar heat of combustion of hydrogen is -286 kJ/mol. Calculate the standard molar heat of formation of methane in kJ/mol.

6.10) Determine ΔH_{rxn} for the following reaction

$$2CH_4(g) + 4Cl_2(g) + O_2(g) \rightarrow 2CO(g) + 8HCl(g)$$

given the following set of reaction data:

$$H_2(g) + Cl_2(g) \rightarrow 2HCl(g) \qquad \Delta H° = -92.30 \text{ kJ}$$
$$2H_2(g) + O_2(g) \rightarrow 2H_2O(g) \qquad \Delta H° = -241.8 \text{ kJ}$$
$$2CO(g) + O_2(g) \rightarrow 2CO_2(g) \qquad \Delta H° = -566 \text{ kJ}$$
$$CH_4(g) + 2O_2(g) \rightarrow CO_2(g) + 2H_2O(g) \qquad \Delta H° = -802.3 \text{ kJ}$$

6.11) When benzene is burned in air, considerable heat is released. Calculate the heat released (the enthalpy change) for the following reaction. How much heat is released per mole of benzene?

$$2C_6H_6(l) + 15O_2(g) \rightarrow 12CO_2(g) + 6H_2O(g)$$

$\Delta H_f°$ for $H_2O(g)$ = -242 kJ/mol; $\Delta H_f°$ for $C_6H_6(l)$ = +49 kJ/mol

$\Delta H_f°$ for $CO_2(g)$ = -394 kJ/mol

6.12) Determine ΔH_{rxn} for each of the following reactions using $\Delta H_f°$ values in the text.

a. $POCl_3(l) + 3H_2O(l) \rightarrow H_3PO_4(aq) + 3HCl(aq)$
b. $CuSO_4(s) + 4CO(g) \rightarrow CuS(s) + 4CO_2(g)$
c. $SiO_2(s) + 4HF(g) \rightarrow SiF_4(g) + 2H_2O(g)$

6.13) a. A 30 gram block of lead at 90°C is dropped into water at 20°C in an insulated calorimeter. The final temperature of the water and the lead block is 22°C. How much heat was lost by the lead block? The specific heat of lead is 0.128 J/g K.

b. A 5.0 gram block of ice is dropped into an insulated container of 100 grams of water at 70°C. What is the temperature of the water in the container at the end of the experiment after all the ice has melted? Heat of fusion of ice = 334 J/g.

6.14) A coin appraiser determines the specific heat of some ancient coins to determine if they are pure gold. He heats the coins (113.4 g) to 100°C in a boiling water bath and then transfers them to a calorimeter containing 200 g of water at 20.0°C. The final temperature of the coins and water in the calorimeter is 25.9°C.

a. What is the specific heat of the coins?
b. Are they pure gold?

6.15) When 20 grams of ammonium nitrate is dissolved in 50 grams of water in an insulated container, the temperature drops by 22.4 K. Calculate the heat of solution of ammonium nitrate. Assume that the specific heat capacity of the ammonium nitrate solution is 4.18 J/K g.

Chapter 6 Answers:

6.1) $\Delta E = q + w$; q and w are both positive if heat is gained by the system and work is done on the system, i.e., if both processes increase the internal energy of the system.

  a. q = +35 kJ, w = -25 kJ
    ΔE = +10 kJ, the internal energy of the system increases by 10 kJ

  b. q = -8 kJ, w = +20 kJ
    ΔE = +12 kJ, the internal energy of the system increases by 12 kJ

6.2) a. 4184 J (or 4.184 kJ)

  b. Energy required to heat 1 gram of water by 1°C = 1 cal = 4.184 J
    Energy required to heat 453.6 grams of water by 1°C = 4.184 x 453.6 = 1898 J
    Energy required to heat 453.6 grams of water by 1°F = 1898 x 5/9 = 1054 J

  1 BTU is approximately equal to 1 kJ

6.3) 18 grams of water is one mole. At 100°C and 1 atm, the volume of one mole of an ideal gas is 30.62 L. This equals the change in volume when one mole of liquid is vaporized because the volume of the liquid (18 mL) is negligible compared to   the volume of the vapor.

  w = -PΔV = -1.0 atm x 30.62 L = -30.62 L atm = -3.10 kJ
  The negative sign indicates that the system does work on the surroundings.

  (A useful conversion to remember is that 8.314 J = 0.08206 L atm. These are the values for R in the two different units).

6.4) The first eight are exothermic and heat is liberated; the last six are endothermic, and heat is absorbed.

6.5) An endothermic reaction is a reaction that absorbs energy. The bonds that are broken in the reaction are stronger than the bonds that are formed. If the reaction occurs within an insulated container, where does this energy come from? It must come from the molecules (reactants and products) within the container itself. The kinetic energy of these molecules decreases as energy is withdrawn and the temperature drops.

6.6) 1.22×10^4 J = 12.2 kJ

6.7) T_f = 136°C

6.8) The enthalpy change, ΔH for a reaction is an extensive property, its value depends upon the reaction as it is written. If the reaction is reversed, then the sign of the enthalpy change, ΔH, changes; if the reaction is exothermic in one direction then it must be endothermic in the reverse direction. Likewise, if the coefficients of the equation are changed, then the value of ΔH changes in proportion.

  a. The coefficients are all doubled; the value of ΔH doubles: ΔH = -560 kJ

  b. The equation is reversed (and doubled); the sign of ΔH changes: ΔH = +560 kJ

6.9) It is often useful to write the equations for the reactions described:

1. $CH_4(g) + 2O_2(g) \rightarrow CO_2(g) + 2H_2O(l)$ $\Delta H° = -890$ kJ
2. $C(s) + O_2(g) \rightarrow CO_2(g)$ $\Delta H° = -394$ kJ
3. $H_2(g) + 1/2O_2(g) \rightarrow H_2O(l)$ $\Delta H° = -286$ kJ

The reaction for which the heat of reaction is required is the formation of methane from its constituent elements:

$$C(s) + 2H_2(g) \rightarrow CH_4(g) \qquad \Delta H° = ??\ kJ$$

The ΔH for this reaction can be calculated by summing the ΔH's for a series of reactions which together add up to the same overall reaction (Hess's Law of Heat Summation).

1. $CO_2(g) + 2H_2O(l) \rightarrow CH_4(g) + 2O_2(g)$ $\Delta H° = +890$ kJ (sign changed)
2. $C(s) + O_2(g) \rightarrow CO_2(g)$ $\Delta H° = -394$ kJ
3. $2H_2(g) + O_2(g) \rightarrow 2H_2O(l)$ $\Delta H° = -572$ kJ (value doubled)

These equations add up to the desired equation, so the ΔH values are added:

$$\Delta H_f°(CH_4) = +890 - 394 - 572\ \text{kJ/mol} = -76\ \text{kJ/mol}$$

6.10) $\Delta H_{rxn} = -924.2$ kJ

6.11) The heat of reaction is the difference between the heat of formation of the products and the heat of formation of the reactants in the balanced equation.

$$\Delta H° = 12 \times (-384) + 6 \times (-242) - 2 \times (+49) = -6278\ \text{kJ}.$$

There are two moles of benzene in the equation, so the heat released for one mole is 1/2 this amount, 3139 kJ/mol.

6.12) a. $\Delta H_{rxn} = -335.4$ kJ;   b. $\Delta H_{rxn} = -410.6$ kJ;   c. $\Delta H_{rxn} = -110.2$ kJ

6.13) a. Heat lost = 30 g x 0.128 J/K g x 68°C = 261 J
b. T = 66°C

6.14) a. 0.587 J/g°C
b. No, the specific heat capacity of pure gold is 0.129 J/g°C.

6.15) ΔH = 4.18 J/g K x 22.4 K x 70 g = 6550 J
For one mole of ammonium nitrate, ΔH = 6550 x 80.05/20 = 26.2 kJ/mol

CHAPTER 7
Quantum Theory and Atomic Structure
Energy and matter merge

TOPICS:

- Light

- Quantum Theory

- Atomic line spectra

- Bohr model of the hydrogen atom
    -Energy states of the hydrogen atom

- Wave properties of particles
    -de Broglie
    -Schrodinger: wave mechanical model
    -Heisenberg's uncertainty principle

- Electron arrangements in atoms
    -atomic orbitals
    -quantum numbers

INTRODUCTION:

When atoms and molecules react, they must bump into each other physically. In so doing, their outer electrons interact. Their reactivities depend on the arrangement of these outer (valence) electrons. Chemists like to have a model for atomic structure so they can understand and predict reactivity of elements and compounds. Some models are more successful than others. Experimental evidence can support models or cause them to be discarded. Some models are useful only for certain systems. Bohr's hydrogen atom model is useful for predicting energy levels in the hydrogen atom, but cannot be extended to more complicated systems.

This chapter discusses the theories and models for atomic structure (particularly the arrangement of electrons around an atom) which successfully account for the repetition of properties (periodicity) of elements.

Matter and energy:

In our macroscopic world, we usually think of matter and energy as being distinct from one another: matter has substance that we touch (has mass) and see (is localized); energy has no mass and is delocalized. However, in order to explain experimental observations made about the behavior of atoms, we need to consider that <u>radiant energy and matter both have properties of waves and particles</u>.

It may seem strange to think of matter as having a wavelength associated with it, and light as being composed of particles that have mass. Quantum mechanics blurs the distinction between matter and energy, and challenges us to expand our classical views on the matter.

In the macroscopic world, the wavelengths of baseballs, and dust particles are insignificant compared to the size of the objects, and these objects behave as particles. However, on an atomic level, where electrons weigh 9.11×10^{-31} kg, their wavelengths are about the size of an atom and are significant. The wave-like behavior of electrons can explain why atoms have only certain energy levels.

Light in our macroscopic world behaves as if it were a wave: it bends around the edges of objects (diffracts), and refracts when it passes into a new substance. However, as we will see, light also acts as if it were composed of <u>particles</u> of energy.

SECTION 7.1: Light

Light, the electromagnetic energy that we can see, is only a small part of the electromagnetic spectrum. Light has a wavelength of about 400 nm (violet light) to 750 nm (red light). Radio waves (lower energy than light) have wavelengths of several meters, and x-rays (higher energy than light) have wavelengths of 10^{-10} m (1 Å).

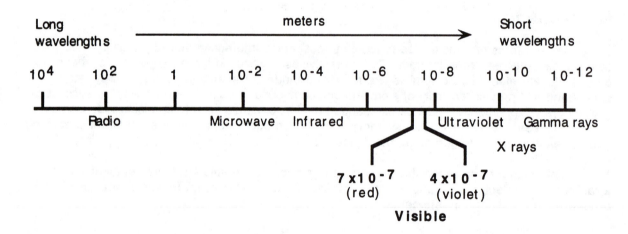

The electromagnetic spectrum

Light that we see is **polychromatic**, made up of many wavelengths. Lasers can generate **monochromatic** light, light at a single wavelength.

The wave nature of light:

The diagram below depicts a wave of light. a) What is its wavelength? b) What is the wave's frequency? c) How fast is the wave traveling?

8.7 x10^{-15} s

650 nm

solutions:

a) The wavelength (λ) is the distance between corresponding points on the wave.
   It is the distance the wave travels during one cycle (m/cycle) = 650 nm

b) The frequency (υ) is the number of cycles per second (cycles/s) is:

$$= \frac{4 \text{ cycles}}{8.7 \times 10^{-15} \text{s}} = 4.6 \times 10^{14} \text{ cycles/s}$$

c) Picture a wave of light one second long. To obtain the distance traveled by the wave
   in one second (its speed), we need to count the waves (frequency) and multiply by
   the length of a wave. (The speed of the wave = wavelength x frequency).

λ

t = 0

t = 1 s

distance = 0

distance = ?

Therefore:

$$\frac{m}{s} = \frac{m}{\text{cycle}} \times \frac{\text{cycles}}{s}$$

$$= \frac{6.5 \times 10^{-7} \text{m}}{\text{cycle}} \times \frac{4.6 \times 10^{14} \text{ cycles}}{s}$$

$$= 3.0 \times 10^8 \text{ m/s}$$

In a vacuum, all electromagnetic radiation travels at 3.00×10^8 m/s, which is the
speed of light:

$$c = \text{frequency} \times \text{wavelength} = 3.00 \times 10^8 \text{ m/s}$$

Exercise:

Choose words from the list below to complete the following paragraph:

        (a) amplitude   (e) intensity
        (b) color   (f) speed
        (c) green   (g) wavelength
        (d) frequency

The wavelength of the light in the visible part of the electromagnetic spectrum determines its _____(1) that we see. Light with a wavelength of 520 nm would have the color _____(2). The _____(3) of the wave, the height of the crest, corresponds to the color's _____(4). All electromagnetic radiation travels at the same _____(5) in a vacuum (the speed of light), but differs in _____, and _____(6&7).

    answers: 1-b, 2-c, 3-a, 4-e, 5-f, 6-d, 7-g.

Classical thinking: Wave vs. particle behavior:

Classify the following as either wave or particle behavior:

    1) Has mass

    2) Has no mass

    3) Is composed of oscillating electric and magnetic fields

    4) velocity = frequency x wavelength

    5) velocity = acceleration x time

    6) Its amplitude is related to its intensity

    7) Bends (**refracts**) when it hits a new medium

    8) Bends around the edge of an object (**diffracts**)

    9) Is either stopped by an object, or continues on its same path

    10) Can be dispersed into component parts by a prism

    11) Causes interference patterns after passing through two adjacent slits

    12) Travels at 3.00×10^8 m/s in a vacuum

answers: wave behavior: 2, 3, 4, 6, 7, 8, 10, 11, 12

          particle behavior: 1, 5, 9

SECTION 7.2: Quantum Theory

We use quantum theory to explain the behavior of electrons and other small particles. Quantum theory challenges us to consider light as a wave or as a stream of particles, and to consider moving objects as particles or as waves. Three principles of the theory which are of chemical interest to us are listed below.

Postulates of Quantum Theory:

1) Energy is quantized. An atom or molecule has only certain states which are characterized by definite amounts of energy. Light exists as a stream of particles, or little packets of quantized energy called photons, which have mass . The energy of a photon is associated with its frequency or wavelength. An atom can change energy states by absorbing or emitting a photon:

$$E_{photon} = h\upsilon = h\frac{c}{\lambda} = \Delta E_{atom}$$

2) Moving objects have wavelengths associated with them which relates to their velocity and mass:

$$\lambda = \frac{h}{mv}$$

3) The allowed energy states of atoms and molecules are described by sets of numbers called quantum numbers. The quantum numbers are associated with individual electrons in the atom. The quantum numbers of all the electrons in an atom designate the energy level, or quantum state, of the atom.

Max Planck, studying blackbody radiation, was the first to suggest that the energy of matter is not continuous, but quantized into discrete units. Each small packet of energy is called a quantum. Energy can be gained or lost in integer multiples of $h\upsilon$:

$$\Delta E = nh\upsilon$$

where n = integer, h = Planck's constant, 6.626×10^{-34} J s, and υ = frequency.

A few years later in 1905, Albert Einstein postulated that light itself exists as a stream of particles, or photons which have mass.

Mass of a photon

1) Use Einstein's equation for the energy of a photon, and his famous equation relating mass and energy ($E = mc^2$) to derive an expression for the mass of a photon.

solution:

$$E_{photon} = \frac{hc}{\lambda} = mc^2$$

$$m_{photon} = \frac{hc}{\lambda c^2} = \frac{h}{\lambda c}$$

2) What is the mass of a photon of blue (w=400 nm) light?

solution:

$$m_{photon} = \frac{h}{\lambda c} = \frac{6.626 \times 10^{-34}\ Js}{400\,nm \times \left(3.00 \times 10^8\ m/s\right)}$$

$$= \frac{6.626 \times 10^{-34}\ kg\,m^2/s}{\left(4 \times 10^{-7}\ m\right) \times \left(3.00 \times 10^8\ m/s\right)}$$

$$= 5.52 \times 10^{-36}\ kg = 5.52 \times 10^{-33}\ g$$

The mass of a photon is so small, it is difficult to think of it as a particle.

The particle behavior of light explains the photoelectric effect. Classical wave mechanics predicted that light of any frequency shining on a metal should cause conduction by freeing an electron once the metal absorbed enough energy. This prediction was not born out by experiments. The photon theory predicts that a metal must absorb a photon of high enough energy in order to release an electron. Increasing the intensity of the light only increases the number of photons, but not their energy. Two Nobel prizes resulted from work on the photoelectric effect: one for studying it (Millikan), and one for explaining it (Einstein).

On reflection, we can see evidence for the particle nature of light all around us. It explains why we tan with visible and UV rays and not with infrared rays (no matter how long we might expose ourselves to them), and why red light is a safelight in photography.

SECTION 7.3: Atomic Line Spectra

Classical physics predicted that a negatively charged particle (an electron), would spiral around a positive particle (a nucleus), and emit a continuous spectrum of light as its energy decreased. However, excited atoms emit light that produce a line spectrum that is characteristic of the element when they are dispersed by a prism. Spectroscopists studied the line spectra of atomic hydrogen, and although they did not know the reason for the discrete spectra, they derived an equation (the Rydberg equation below) that would predict the pattern of the spectral lines:

$$\frac{1}{\lambda} = R\left(\frac{1}{n_1^2} - \frac{1}{n_2^2}\right) \quad n_1, n_2 = \text{integers}$$

where λ = wavelength of a spectral line, n_1 and n_2 are positive integers with $n_2 > n_1$, and R = Rydberg constant = $1.096776 \times 10^7 \text{ m}^{-1}$. Remember, the Rydberg equation applies only to the spectral lines of the hydrogen atom or other one electron systems such as He^+ and Li^{2+}.

Spectral lines in the hydrogen atom:

1) What is the wavelength of the spectral line in the hydrogen atom when $n_1 = 2$ and $n_2 = 3$?

solution:

We plug in the values for n_1 and n_2 into the Rydberg equation and solve for wavelength, λ:

$$\frac{1}{\lambda} = 1.096776 \times 10^7 \text{ m}^{-1}\left(\frac{1}{(2)^2} - \frac{1}{(3)^2}\right)$$

$$\frac{1}{\lambda} = 1.5245 \times 10^6 \text{ m}^{-1}$$

$$\lambda = 6.5595 \times 10^{-7} \text{ m} = 656.5 \text{ nm}$$

2) What is the wavelength of the spectral line in the hydrogen atom when $n_1 = 2$ and $n_2 = 4$?

solution:

$$\frac{1}{\lambda} = 1.096776 \times 10^7 \text{ m}^{-1}\left(\frac{1}{(2)^2} - \frac{1}{(4)^2}\right)$$

$$\frac{1}{\lambda} = 2.056 \times 10^6 \text{ m}^{-1}$$

$$\lambda = 4.864 \times 10^{-7} \text{ m} = 486.3 \text{ nm}$$

3) What is the wavelength of the spectral line in the hydrogen atom when $n_1 = 2$ and $n_2 = 5$?

$$\frac{1}{\lambda} = 1.096776 \times 10^7 \text{ m}^{-1} \left(\frac{1}{(2)^2} - \frac{1}{(5)^2} \right)$$

$$\frac{1}{\lambda} = 2.303 \times 10^6 \text{ m}^{-1}$$

$$\lambda = 4.342 \times 10^{-7} \text{ m} = 434.2 \text{ nm}$$

4) What can we say about the solutions for the Rydberg equation when $n_1 = 2$?

solution:

The spectral lines for n_1=2 (i.e., 656.5, 486.3, and 434.2 nm calculated above) fall in the <u>visible</u> region. of the electromagnetic spectrum. It turns out that if you solve the Rydberg equation for n_1 = 1, it predicts where the spectral lines of a hydrogen atom fall in the <u>ultraviolet</u> region. When n_1 = 3, the equation predicts the lines in the <u>infrared</u> region.

Notice also that the lines become closer together as n_2 increases; the energy levels in an atom become closer together as their distance from the nucleus increases.

SECTION 7.4: The Bohr model of a hydrogen atom

In 1923, Niels Bohr presented a model of the hydrogen atom that predicted the observed line spectra. Bohr assumed a hydrogen atom consisted of a central proton, with an electron moving around it in a circular orbit. He used <u>classical physics</u> to relate the energy of the atom to the radius of the electron's orbit, but incorporated <u>quantum theory</u> into his model by proposing that the angular momentum was quantized. This quantum condition restricted the energy levels in a hydrogen atom to only certain allowable levels which were associated with the fixed orbits:

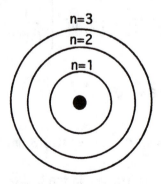

Remember that electrons do not actually circle protons in fixed orbits! This was <u>Bohr's model</u> which was successful for its ability to predict the spectral lines in a <u>one electron system</u>. The model broke down when it was extended to atoms with more than one electron.

Bohr's model for the hydrogen atom:

Choose words from the list below to complete the following paragraph:

        (a) energy   (d) orbits
        (b) excited   (e) photon
        (c) one   (f) radius

Bohr proposed that an atom only emits energy by absorbing or emitting a

_____(1) whose energy equals the difference in energy between the

_____(2). The quantum number n = 1,2,3... is associated with the

_____(3) of the electron's orbit, which in turn, is directly related to the atom's

_____(4). Bohr calculated the energy levels for a hydrogen atom to be:

$$E = -2.18 \times 10^{-18} \text{ J} \left(\frac{1}{n^2} \right)$$

The atom is in its ground state when n = _____(5); all higher levels (n > 1) are

_____(6) states. When a sample of hydrogen is excited, the electrons in different

atoms are excited to different levels. As they fall back to the ground state, they emit energy of

specific wavelengths giving hydrogen its atomic line spectrum.

    <u>answers</u>: 1-e, 2-d, 3-f, 4-a, 5-c, 6-b.

Energy levels of Bohr's orbits:

What is the energy of the first excited state in a hydrogen atom?

<u>solution</u>:

$$E = -2.18 \times 10^{-18} \text{ J} \left(\frac{1}{(2)^2} \right) = -5.45 \times 10^{-19} \text{ J}$$

We <u>define</u> zero energy as the atom's energy when the electron is completely removed from the atom: $E = 0$ when $n = \infty$ (infinity). Energy is therefore negative for any n smaller than ∞. The energy becomes a more negative number as the electron moves from orbits farther from the nucleus to orbits closer to the nucleus. The energy of the most stable orbit (the ground state when $n = 1$) corresponds to the most negative number.

Ionization energy of a hydrogen atom:

1) How much energy is required to remove an electron from a hydrogen atom?

$$H(g) \rightarrow H^+(g) + e^-$$

<u>solution:</u>

The energy difference between any two energy levels in the hydrogen atom is:

$$\Delta E = E_{final} - E_{initial} = -2.18 \times 10^{-18} \, J \left(\frac{1}{n_{final}^2} - \frac{1}{n_{initial}^2} \right)$$

An electron in the ground state of a hydrogen atom is in the n=1 orbit. When it is completely removed from the atom, n=∞.

$$\Delta E = E_{final} - E_{initial} = -2.18 \times 10^{-18} \, J \left(\frac{1}{\infty^2} - \frac{1}{1^2} \right)$$

$$= 2.18 \times 10^{-18} \, J$$

2) How much energy is required to remove an electron from a mole of hydrogen atoms?

<u>solution:</u>

$$\Delta E = \frac{2.18 \times 10^{-18} \, J}{1 \, H \, atom} \times \frac{6.022 \times 10^{23} \, atom}{mol} \times \frac{1 \, kJ}{10^3 \, J} = 1.31 \times 10^3 \, kJ \, / \, mol$$

This is the **ionization** energy of the hydrogen atom, the amount of energy required to form a mole of gaseous H^+ ions from a mole of gaseous H atoms.

Bohr's model of the atom failed to predict the line spectra of any atom other than hydrogen. His model does not take into account electron-electron repulsions, and nucleus-electron attractions of a multiple electron system. In addition, his assumption that electrons circle protons in fixed orbits was incorrect.

SECTION 7.5: The Wave Properties of Particles

de Broglie

Bohr arbitrarily incorporated quantized angular momentum into his calculation for the energy levels of a hydrogen atom because it produced results that agreed with experimental observations. It wasn't until Louis de Broglie showed that particles could have wave properties, that there was a reasonable explanation for <u>why</u> atoms had only discrete energy levels. If electrons exhibit properties of waves, they would have only certain allowed energies since a wave of fixed radius can only have certain frequencies (energies).

The wavelength of particles:

1) Use Einstein's equation relating mass and energy ($E = mc^2$), and Planck's equation for the energy of light ($E = \dfrac{hc}{\lambda}$) to solve for the wavelength of a particle of mass m moving at velocity, v.

   <u>solution:</u>

   $$E = \frac{hc}{\lambda} = mc^2 \quad \text{therefore,} \quad \lambda = \frac{h}{mc}, \quad \text{or}$$

   $$\lambda = \frac{h}{mv} \quad \text{for a particle of any mass moving at velocity v.}$$

2) Determine the wavelength of an electron (mass = 9.11×10^{-31} kg) moving at velocity of 8.0×10^6 m/s.

   <u>solution:</u>

   We'll use the equation $\lambda = \dfrac{h}{mv}$

   ($h = 6.626 \times 10^{-34}$ J s $= 6.626 \times 10^{-34}$ kg m²/s)

   $$\lambda = \frac{6.626 \times 10^{-34} \text{ kg m}^2/\text{s}}{9.11 \times 10^{-31} \text{ kg} \times \left(8 \times 10^6 \text{ m/s}\right)} = 9.09 \times 10^{-11} \text{ m}$$

   In Å, the wavelength of the electron is close to 1: $\lambda = 0.909$ Å

3) Determine the wavelength of a ball (mass = 120 g) moving at a velocity of 45 m/s (about 100 mi/hr).

   <u>solution:</u>

$$\lambda = \frac{6.626 \times 10^{-34} \text{ kg m}^2/\text{s}}{0.120 \text{ kg} \times 45 \text{ m/s}} = 1.23 \times 10^{-34} \text{ m}$$

Notice that although both the wavelengths of the electron and the ball are incredibly short, the wavelength associated with the electron is many orders of magnitude larger than that of the ball, and it is on the same order as the spacing between the atoms in a typical crystal.

A qualitative rule is that if the wavelength of an object moving toward a target is small compared to the size of the target, the interaction will be particle-like. In the above example, the wavelength of the ball, 1.23×10^{-34} m, is much smaller than the size of the bat. If the wavelength of a moving object is about the same magnitude as the size of the target (or larger), the interaction is wave-like. The wavelength of the electron in the example above is approximately the same size as an atom.

Evidence for the wave behavior of electrons:

Waves that pass between adjacent slits cause interference patterns, particularly if the distance between slits is about the same distance as the wavelength. In 1927, Davisson and Germer at Bell Laboratories observed a diffraction pattern when they directed a beam of electrons at a nickel crystal. The atoms in a crystal are about the same distance apart as the wavelength of a moving electron, so the atoms act as "slits" to the beam of electrons which diffract around them and form interference patterns. Since diffraction patterns are usually explained in terms of waves, this result supported de Broglie's idea that moving particles have wavelengths.

Wavelength of particles:

1) How fast would the ball need to be moving in the example above in order for it to cause a diffraction pattern between slits 1 meter apart?

solution:

Efficient interference patterns occur when the distance between slits is approximately the same as the wavelength. In order to have a wavelength of 1 meter, the speed of the ball would need to be:

$$\lambda = 1\text{ m} = \frac{6.626 \times 10^{-34} \text{ kg m}^2/\text{s}}{0.120 \text{ kg} \times v \text{ m/s}}$$

$v = 5.522 \times 10^{-33}$ m/s, not a plausible speed!

2) How close together would the slits need to be to produce effective interference patterns for a particle with a wavelength of 1.23 x 10⁻³⁴ m (the wavelength of the ball in the above example)?

solution:

The slits would need to be approximately the same distance apart as the wavelength of the particle (1.23 x 10⁻³⁴ m, or 1.23 x 10⁻²³ Å) to produce efficient diffraction interference patterns. This distance is many orders of magnitude smaller than the distance between atoms in crystals.

We can see from these examples that it is not possible to determine experimentally if large pieces of matter have wavelengths. However, scientists believe that all matter obeys de Broglie's equation.

Schrodinger: the wave mechanical model of the atom

Schrodinger applied classical equations for waves to the problem of an electron in an atom and developed an equation which allowed him to determine the energy levels and wave properties of the hydrogen atom, and a few other simple systems. The wave equation appears to apply exactly to atoms other than hydrogen, although the solutions are too complex to compute. We want to find functions that satisfy the differential equation:

$$H\psi = E\psi$$

H is a set of mathematical instructions, an operator. E is the energy of the atom, and ψ (psi) is a wave function, a mathematical description of the motion of the electrons as a function of time and position. The wave function, ψ, has no physical meaning, but ψ^2 tells us the probability of finding an electron at a given point.

Main ideas from the Schrodinger wave equation:

• it does not give information on the detailed path of an electron

• for one electron systems, each solution to the equation is associated with a particular wave function, which is also called an atomic orbital. The square of the wave function gives the probability of finding the electron at any point.

• the electron could be anywhere; the probability of finding it at any point never reaches zero. To visualize where the electron spends most of its time, we draw diagrams to show where the electron spends 90% of its time.

Despite the similarity between the terms "orbit" used for Bohr's model, and "orbital" used from Schrodinger's model, they have distinct meanings. Electrons in orbits circle the nucleus at fixed distances; orbitals are regions of space where the electron might be found.

Representations of a 1s orbital:

1) electron density diagram - the density of the dots in a certain region is proportional to the probability of finding an electron there.

2) radial probability plot - a graph of probability plotted as a function of distance from the nucleus. The peak for the ground state H atom appears at the same distance from the nucleus (0.529 Å) as the closest Bohr orbit. The Schrodinger model predicts that the electron spends most of its time at the same distance that the Bohr model predicted it spent all of its time.

3) 90% probability contour - contains 90% of the total probability of finding the electron.

Orbital questions:

1) A hydrogen atom in its ground state has a 90% probability contour diameter of 0.74 Å. What would be the size of the diameter for a 100% probability contour?

<u>answer</u>:

The diameter would be infinite, since there is a <u>chance</u> that the electron could exist anywhere.

2) How can you reconcile the apparent paradox that the electron density diagram for a 1s orbital shows the highest electron density at the nucleus, while the radial probability plot shows the electron density decreasing to 0 at the nucleus?

<u>answer</u>:

The probability of finding a 1s electron decreases as we move away from the nucleus (ψ^2 decreases), but the square of the <u>radius</u> increases:

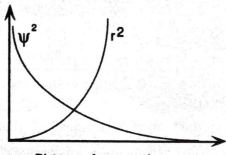

Distance from nucleus

The radial probability curve is a plot of the product of $\left(\psi^2 \times r^2\right)$ vs. distance from the nucleus:

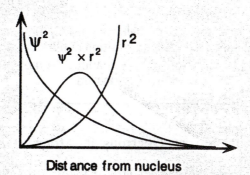

Distance from nucleus

The Heisenberg Uncertainty Principle:

If electrons have wave-like properties, and waves are spread out in space, what can we determine about the position of an electron in the atom? Heisenberg's Uncertainty Principle states that we cannot know simultaneously both the exact position and velocity of a particle. If you try to locate an electron, you must use high energy radiation to "see" it. The high energy (short wavelength) radiation acts like a particle to the electron and knocks it away from where it was.

The uncertainty in the momentum of the particle Δ(mv), multiplied by the uncertainty in its position, Δx, must be equal to, or greater than h/4π (h = Planck's constant):

$$\left(\Delta x\right) \times \Delta\left(mv\right) \geq \frac{h}{4\pi}$$

The Uncertainty Principle:

1) What is the uncertainty in the position of an electron ($m = 9.11 \times 10^{-31}$ kg) moving at 8×10^6 ±1% m/s?

solution:

We calculate the uncertainty in the velocity of the electron, and substitute this value into Heisenberg's equation and solve for the uncertainty in its position (Δx):

$$\Delta v = 8 \times 10^6 \times 0.01 = 8 \times 10^4 \text{ m/s.}$$

$$\Delta x \geq \frac{h}{4\pi m \Delta v} \geq \frac{6.626 \times 10^{-34} \text{ kg m}^2/\text{s}}{4\pi \left(9.11 \times 10^{-31} \text{ kg}\right)\left(8 \times 10^4 \text{ m/s}\right)} \geq 7 \times 10^{-10} \text{ m}$$

2) What is the uncertainty in the position of a ball ($m = 120$ g) moving at 45 ±1% m/s?

solution:
$$\Delta v = 45 \text{ m/s} \times 0.01 = 0.45 \text{ m/s.}$$

$$\Delta x \geq \frac{h}{4\pi m \Delta v} \geq \frac{6.626 \times 10^{-34} \text{ kg m}^2/\text{s}}{4\pi \left(0.120 \text{ kg}\right)\left(0.45 \text{ m/s}\right)} \geq 9.8 \times 10^{-34} \text{ m}$$

The uncertainty in the position of the ball is incredibly tiny compared to the size of a playing field, so you can feel confident that its position can be calculated. The uncertainty in the position of the electron (7×10^{-10} m, or 7 Å) is about 7 times greater than the diameter of an atom, so we do not know exactly where it is located.

SECTION 7.6: The Arrangement of Electrons in Atoms:

We use 3 quantum numbers to characterize an electron's wave function (atomic orbital). The quantum numbers are summarized below

Quantum numbers:			
Name:	letter:	integer:	relates to orbital's:
Principal (energy level*)	n	positive	size
Azimuthal (sublevels)	l	0 to n-1	shape
Magnetic (orbitals)	m_l	$-l,...0,...+l$	orientation

*The energy state of an atom with more than one electron depends on both the n and l values of the occupied orbital.

Azimuthal orbitals:
(sublevels)

number:	letter:	#orientations (-1,...0,...+1)	shape:
$l = 0$	s	1	1s = sphere
$l = 1$	p	3	dumbbell
$l = 2$	d	5	clover leaf, donut dumbbell
$l = 3$	f	7	multilobed

Orientation of orbitals:

s sublevel, $m_l=0$; one orientation

p sublevel, $m_l=-1,0,1$; three orientations (p_x, p_y, p_z)

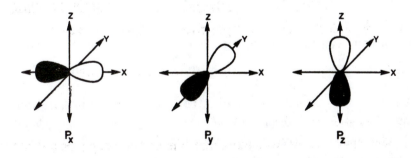

d sublevel, m_ℓ = -2,1,0,1,2; five orientations (d_{xy}, d_{xz}, d_{yz}, $d_{x^2-y^2}$, d_{z^2})

d_{xy} d_{yz} d_{xz}

$d_{x^2-y^2}$ d_{z^2}

Quantum numbers:

1) When n=1, how many values are there for ℓ? How many orbitals are in the first energy level?

   solution:

   ℓ =0 to n-1, so when n=1, ℓ can only have one value (0).

   m_ℓ = ℓ,0,+ℓ, so when ℓ = 0, m_ℓ can only have one value (0).

   Each m_ℓ value corresponds to one atomic orbital, so there is one orbital in the first energy level. The ℓ=0 sublevel is designated by the letter "s", which has only one orientation in space, designated by m_ℓ = 0.

2) When n=2, how many values are there for ℓ? How many orbitals are in the second energy level?

   solution:

   When n=2, ℓ can have the values 0, and n-1=1. The second energy level therefore has 2 sublevels, ℓ=0 (the "s" sublevel) and ℓ=1 (the "p" sublevel).

   For ℓ=0 (s orbital), m_ℓ has only one value (0); for ℓ=1 (p orbital), m_ℓ can have 3 values (-ℓ,0,+ℓ = -1,0,1)

The number of orbitals in the n=2 energy level is 4 (an "s" orbital with one orientation, and "p" orbitals with three orientations in space often called p_x, p_y, and p_z).

3) When n=3, how many values are there for ℓ? How many orbitals are in the third energy level?

solution:

When n=3, ℓ can have the values 0, 1, and 2. The third energy level therefore has 3 sublevels, $\ell=0$ (the "s" sublevel), $\ell=1$ (the "p" sublevel), and $\ell=2$ (the "d" sublevel).

For $\ell=0$ (s orbital), m_ℓ has only one value (0); for $\ell=1$ (p orbital), m_ℓ can have 3 values (-1,0,1); for $\ell=2$, m_ℓ has 4 values (-2,-1,0,1,2).

The "s" orbital has only one orientation ($m_\ell=0$), the "p" orbitals have three orientations ($m_\ell = -1,0,+1$), and the "d" orbitals have five orientations ($m_\ell = -2,-1,0,+1,+2$). The total number of orbitals in the n=3 energy level is therefore: $1 + 3 + 5 = 9$.

4) What is the possible number of ℓ sublevels for a given n?

solution:

When n=1, there is one sublevel (s)

When n=2, there are two sublevels (s and p)

When n=3, there are three sublevels (s, p, and d).

When n=n, there are "n" sublevels.

5) What is the total number of orbitals for a given n?

solution:

When n=1, there is one orbital (1s)

When n=2, there are 5 orbitals (2s, $2p_x$, $2p_y$, $2p_z$)

When n=3, there are 9 orbitals

$$(3s,\ 3p_x,\ 3p_y,\ 3p_z,\ 3d_{xy},\ 3d_{xz},\ 3d_{yz},\ 3d_{x^2-y^2},\ 3d_{z^2})$$

When n=n, there are n^2 orbitals.

Summary Exercise:

Match the appropriate scientists, equations, or concepts from the three lists below with the statements which follow:

Scientists

a) Niels Bohr
b) Louis de Broglie
c) Albert Einstein
d) Werner Heisenberg
e) Max Planck
f) Erwin Schrodinger

Concepts

o) atomic line spectra
p) blackbody radiation
q) Bohr model
r) diffraction patterns
s) photoelectric effect
t) quantum numbers
u) uncertainty principle

Equations

g) $\Delta E = nh\upsilon$

h) $E_{photon} = h\upsilon = h\dfrac{c}{\lambda} = \Delta E_{atom}$

i) $E = mc^2$

j) $\dfrac{1}{\lambda} = R\left(\dfrac{1}{n_1^2} - \dfrac{1}{n_2^2}\right)$ n_2 = integers

k) $E = -2.18 \times 10^{-18}\ J\left(\dfrac{1}{n^2}\right)$

l) $\lambda = \dfrac{h}{mv}$

m) $H\psi = E\psi$

n) $\left(\Delta x\right) \times \Delta\left(mv\right) \geq \dfrac{h}{4\pi}$

1) Experimental results observed by the end of the 1900's which were unexplainable with classical physics. (concepts)

2) Could mathematically predict radiation emitted from blackbodies if he assumed that energy can be gained or lost in <u>integer</u> multiples of $h\upsilon$. (name and equation)

3) His equation was the first to suggest that energy of matter is not continuous but quantized into discrete packets. (name)

4) Postulated that light itself exists as a stream of particles or photons. (name, equation)

5) His idea that mass and energy are alternate forms of the same entity led to a famous equation. (name, equation)

6) Predicted that photons have mass. (name)

7) An equation (named the Rydberg equation) which predicted the wavelengths of emission lines from the hydrogen atom.

8) Developed a model for the hydrogen atom that predicted the observed line spectra for a hydrogen atom; it broke down for more complicated systems. (name, equation)

9) His model for the hydrogen atom restricted the allowed energy levels in a hydrogen atom to only certain allowable levels which were associated with a fixed circular orbit of the electron around the nucleus. (name)

10) Gave the energy levels of a hydrogen atom as:
$$E = -2.17 \times 10^{-18} \text{ J } (1/n^2) \quad \text{(concept)}$$

11) Postulated that electrons exhibit wave properties which could explain why electrons have quantized energy levels in an atom. (name, equation)

12) Experimental result which supported the theory that electrons have wave behavior. (concept)

13) Developed the wave mechanical model of the atom, which allowed the determination of energy levels and wave properties of the hydrogen atom, and a few other simple systems. (name, equation)

14) Stated that it is impossible to know both the exact position and velocity of a particle. (name, equation, concept)

15) Numbers used to specify an electron's energy and wave function (atomic orbital). (concept)

answers: 1-o,p,s; 2-e,g; 3-e; 4-c,h; 5-c,i; 6-c, 7-j; 8-a,k; 9-a; 10-q; 11-b,l; 12-r; 13-f,m; 14-d,n,u; 15-t.

Chapter 7 Exercises:

7.1)   a. If it takes 400 ms for one wavelength of a wave to pass a stationary point, what is its frequency?

   b. What is the wavelength of an infrared wave with velocity of 3.0×10^{10} cm/s and frequency of 8.0×10^{11} s^{-1}?

   c. What is the frequency of a gamma ray wave with velocity of 3.0×10^{10} cm/s and wavelength of 1.2×10^{-12} cm?

7.2)   How many joules of energy are contained in one mole of violet light quanta of frequency 7.0×10^{14} s^{-1}?

7.3)   The energy (in joules) of an electron energy level in the Bohr atom is given by the expression: $E_n = -2.179 \times 10^{-18}/n^2$ J, where n is the quantum number that specifies the energy level. What is the frequency in Hz of the radiation emitted when an electron falls from level n=3 to level n=2?

7.4)   What is the wavelength of a beam of electrons traveling at 2.00×10^6 cm/s? Assume the mass of an electron at this velocity is 9.1094×10^{-31} kg.

7.5)   a. What are the allowed values of the magnetic quantum number m_ℓ associated with the $\ell = 3$ sublevel

   b. If the quantum number n equals 4, what are the allowed values of ℓ associated with this value of n?

   c. What are the allowed values for the quantum number m_s that identifies the property of an electron called "spin"?

7.6)   a. Which orbital quantum number specifies the shape of an atomic orbital?

   b. If an orbital has a single planar node through the nucleus, what kind of orbital is it?

   c. If, in addition to the planar node through the nucleus, the orbital has a single spherical node, to what principal quantum level does the orbital belong?

   d. As the value of the principal quantum number increases, what happens to the energy of the orbital?

7.7)   Describe, in as much detail as possible, a 2p orbital.

7.8)   a. What is the total number of electrons that can occupy the principal quantum level n=4 in one atom?

   b. Derive a general expression to determine the total number of electrons that will completely fill all the orbitals in the principal quantum level n of an atom.

7.9)   Which of the following are not acceptable sets of quantum numbers?

   a. n=4, $\ell = 0$, $m_\ell = 0$ c. n=0, $\ell = 1$, $m_\ell = 0$

   b. n=2, $\ell = 1$, $m_\ell = -2$ d. n=1, $\ell = 0$, $m_\ell = 0$

7.10) Which of the following are not acceptable sets of quantum numbers?

   a. n=3, $\ell = 1$, $m_\ell = -1$ c. n=2, $\ell = 2$, $m_\ell = 0$

   b. n=7, $\ell = 3$, $m_\ell = +3$ d. n=5, $\ell = 1$, $m_\ell = 0$

Chapter 7 Answers:

7.1)   a. The time taken for one wavelength of a traveling wave to pass a stationary point is called the period; it is the reciprocal of the frequency.
   Frequency = 1/400 ms = 2.5 s^{-1}

   b. The velocity of a traveling wave equals the frequency x the wavelength.
   Wavelength = 3.0×10^{10} cm s^{-1}/8.0×10^{11} s^{-1} = 3.8×10^{-2} cm

   c. Frequency = velocity/wavelength = 3.0×10^{10} cm s^{-1}/1.2×10^{-12} cm = 2.5×10^{22} s^{-1}

7.2)   E = $h\upsilon$ = 6.626×10^{-34} J s x 7.0×10^{14} s^{-1} = 4.64×10^{-19} J
   The energy in one mole of photons = 280 kJ.

7.3) 4.6×10^{14} Hz

7.4) $\lambda = \dfrac{h}{mv}$ = 6.626×10^{-34} J s/9.1094×10^{-31} kg x 2.00×10^{4} m/s
   (note the change in units from cm/s to m/s)
   wavelength, λ, = 3.64×10^{-8} m

7.5)   a. The total number of values of m_ℓ is equal to $2\ell + 1$ and is called the orbital degeneracy. If, as in this case, $\ell = 3$, then $2\ell + 1 = 7$, so there are seven different values of m_ℓ for this value of ℓ. m_ℓ = +3, +2, +1, 0, -1, -2, -3

   b. The number of possible values of ℓ is equal to the value of n; in this case there are four possible values of ℓ: 0,1,2, and 3.

   c. There are two possible values of the spin quantum number m_s = ±1/2.

7.6)   a. The secondary or azimuthal quantum number ℓ specifies the shape of an orbital.

   b. p orbital

   c. The total number of nodes is related to the value of the principal quantum number, and therefore to the energy of the orbital. There are two nodes in this orbital-one planar and one spherical. The value of the principal quantum number is therefore 3 (the number of nodes + 1). Since there is a single planar node, $\ell = 1$, so the orbital is a 3p orbital.

   d. As the value of n increases, the energy of the orbital increases. For the hydrogen atom, the energy of each principal quantum level is given by Bohr's equation:

   E_n = $-2.179 \times 10^{-18}/n^2$ J

7.7)   The value of the principal quantum number is 2.
   The value of the azimuthal quantum number is 1.
   The orbital is one of a set of three orbital, all of equal energy, and all at right angles to one another.
   The orbital has a planar node running through the nucleus.

The orbital has no spherical node.
For atoms other than hydrogen, the orbital is the highest in energy within its principal quantum level; the only other type of orbital in this principal quantum level is an s orbital.
The orbital has a capacity for two electrons, each of opposite spin.

7.8)   a. In the n = 4 principal quantum level there are s, p, d, and f orbitals, in sets of 1, 3, 5, and 7 respectively. The total number of orbitals in the n = 4 equals 1 + 3 + 5 + 7 = 16. The total number of electrons that can be accommodated is 32 (2 in each orbital).

b. A general expression for the total number of electrons that can be accommodated by any principal quantum level n of an atom = $2n^2$.

7.9)   b: m_ℓ cannot equal -2; it must be -1, 0, or +1
c: n cannot equal 0, it must be greater than 0.

7.10) c: ℓ cannot equal 2; it must be 0 or 1.

CHAPTER 8
Electron Configuration and Chemical Periodicity

TOPICS:

- Many – electron atoms
  electrostatic interactions
  electron configurations and orbital diagrams

- Periodic trends

- Atomic structure and chemical reactivity

SECTION 8.1: Many-electron atoms

This chapter will discuss the rules for filling orbitals with electrons and how the resulting electron configurations can explain periodic patterns of chemical reactivity.

Approximate solutions to the Schrodinger equation show that the atomic orbitals of many-electron atoms are <u>hydrogen like</u>, so we use the same quantum numbers to describe all atoms. We saw in Chapter 7 that three quantum numbers describe an atom's orbitals:

$$n \text{ (size)}$$

$$l \text{ (shape)}$$

$$m_l \text{ (orientation)}$$

A fourth quantum number, m_S, describes a property of the electron itself. The **spin quantum number**, as it is called, is associated with the spin of the electron about its own axis.

The spin quantum number (m_S):

Choose words from the list immediately below to complete the paragraph that follows the list:

(a) direction	(e) paired
(b) filled	(f) spins
(c) four	(g) two
(d) opposite	

Experiments show that an electron acts as though it _____(1) on its axis. The spin quantum number (m_S) indicates the _____(2) of the electron spin and has the value $+\frac{1}{2}$ or $-\frac{1}{2}$. <u>No two electrons in the same atom can have the same set of</u>

_____(3) <u>quantum numbers</u> (**Pauli exclusion principle**). A consequence of the exclusion principle is that an atomic orbital can hold a maximum of _____(4) electrons and they must have _____(5) spins. When there are two electrons in an orbital, the electrons are said to be _____(6) and the orbital is _____(7).

<u>answers:</u> 1-f, 2-a, 3-c, 4-g, 5-d, 6-e, 7-b.

We saw in Chapter 7 that orbitals closer to the nucleus are more stable (lower energy) than those farther away from the nucleus. The only electrostatic interactions in a hydrogen atom are those between the proton and the electron. For many electron systems, we have to consider both proton-electron attractions and electron-electron repulsions to determine the energy level of an orbital.

Electrostatic interactions:

Determine for each set of charges below, which configuration has the lower energy with respect to infinite separation.

5) Given that the lower the 1 value of an orbital, the more time the electron spends near the nucleus, order the sublevels from lowest energy to highest (for a given "n").

<u>answers</u>:

1-b: A charge of +2 attracts a negative charge more strongly than a charge of +1; stronger attractions make a system lower in energy.

2-a: The energy of a system is lower when opposite charges are close together.

3-a: Attractions make a system lower in energy than repulsions. Like charges repel each other. Energy must be added in (b) to bring the like charges together from infinite separation.

4-a: The two negative charges repel each other in system (b) raising the energy of the system above that of (a). Configuration (b) may be thought of as configuration (a) with an added negative charge, which being closer to the other negative charge than to the positive charge, raises the energy of the system.

5- s < p < d < f. Note: this order does not apply to the hydrogen atom where the energy of sublevels is equal. For many electron atoms, energy of the 2s sublevel is lower than the 2p sublevel due to penetration/screening effects.

Summary of electrostatic effects on orbital's energies:

- Electron-electron repulsions raise energy

- Greater nuclear charge lowers orbital energy

- Electrons in outer orbitals are higher in energy, owing to their greater distance from nuclear charges.

- For a given n value, the energy of the sublevels in many-electron atoms is: s < p < d < f

    We can use our knowledge of energy levels of orbitals to write ground-state electron configurations by adding one electron per element to the lowest energy orbital available (the **aufbau principle**). A useful diagram for remembering the order of orbital energies is:

The order of filling is: 1s 2s 2p 3s 3p 4s 3d 4p 5s 4d 5p 6s 4f 5d 6p 7s, etc.

Order of filling orbitals:

Consider the line connecting 3d, 4p, and 5s. What is the value for n+l in each case?

solution:

Since s, p, d stand for l = 0, 1, 2, respectively,:

$$3d: \quad n=3, \quad l=2, \quad n+l=5$$

$$4p: \quad n=4, \quad l=1, \quad n+l=5$$

$$5s: \quad n=5, \quad l=0, \quad n+l=5$$

We can see that the lines in the above diagram connect orbitals which have n+l = a constant. For a given n+l line, the smallest n-value fills first.

Rules for filling atomic orbitals:

- **Hund's Rule**: when orbitals of equal energy are available, the electron configuration of lowest energy has the maximum number of unpaired electrons with parallel spins

- **Pauli Exclusion Principle**: No two electrons in an atom can have the same set of four quantum numbers. Another, in some ways stronger statement is: Two electrons with the same spin cannot be at the same place at the same time.

- Filled and half-filled sublevels are unusually stable.

Electron configurations show the principal energy level (n value), the sublevel letter (1 value), and the number of electrons in the sublevel. The electron configuration for a helium atom is $1s^2$.

Orbital diagrams give additional information about paired vs. unpaired electrons. An orbital diagram depicts an orbital with a line (or box) grouping the orbitals by sublevel. Arrows indicate an electron and its direction of spin.

Electron configurations and orbital diagrams:

1) Show the electron configurations and orbital diagrams for He, Ne, Ar, and Kr.

   <u>solution</u>:

   He: $1s^2$

   Ne: $1s^2\,2s^2\,2p^6$

   Ar: $1s^2\,2s^2\,2p^6\,3s^2\,3p^6$

   Kr: $1s^2\,2s^2\,2p^6\,3s^2\,3p^6\,4s^2\,3d^{10}\,4p^6$

2) What do the electron configurations for He, Ne, Ar, and Kr have in common?

   <u>solution</u>:

   They all have a filled "p" subshells as their last occupied subshell.

3) Why are noble gases so unreactive (stable)?

solution:

Their outer ("valence") subshells are filled (no vacancies), and the next available subshell has a higher principal quantum number.

similar electron configurations = similar chemical properties

4) What are the electron configurations for C and Si in group 4A?

solution:

C: $1s^2 2s^2 2p^2$

Si: $1s^2 2s^2 2p^6 3s^2 3p^2$

5) Which are the inner shell electrons and which are the valence electrons for Si?

solution:

The inner shell electrons are all those in the 1st and 2nd shells. The valence electrons are those in the 3s and 3p orbitals.

6) What do the electron configurations for C and Si have in common?

solution:

They each have 4 valence electrons.

7) Use a short-hand method to write the electron configurations for Si that eliminates the writing of all the inner electrons.

solution:

$[Ne]3s^2 3p^2$

Electron configurations (special cases):

1) Show the electron configurations orbital diagrams for Cu, Ag, and Au.

Cu: [Ar]4s^1 3d^{10}

$$\underset{4s}{\uparrow} \quad \underset{3d}{\uparrow\downarrow \; \uparrow\downarrow \; \uparrow\downarrow \; \uparrow\downarrow \; \uparrow\downarrow}$$

Ag: [Kr]5s^1 4d^{10}

$$\underset{5s}{\uparrow} \quad \underset{4d}{\uparrow\downarrow \; \uparrow\downarrow \; \uparrow\downarrow \; \uparrow\downarrow \; \uparrow\downarrow}$$

Au: [Xe]6s^1 4f^{14} 5d^{10}

$$\underset{6s}{\uparrow} \quad \underset{4f}{\uparrow\downarrow \; \uparrow\downarrow \; \uparrow\downarrow \; \uparrow\downarrow \; \uparrow\downarrow \; \uparrow\downarrow \; \uparrow\downarrow} \quad \underset{5d}{\uparrow\downarrow \; \uparrow\downarrow \; \uparrow\downarrow \; \uparrow\downarrow \; \uparrow\downarrow}$$

Note: <u>half filled and filled shells are unexpectedly stable</u>.

2) Show the orbital diagrams for Cr, Mo, and W:

Cr: [Ar]4s^1 3d^5

$$\underset{4s}{\uparrow} \quad \underset{3d}{\uparrow \; \uparrow \; \uparrow \; \uparrow \; \uparrow}$$

Mo: [Kr]5s^1 4d^5

$$\underset{5s}{\uparrow} \quad \underset{4d}{\uparrow \; \uparrow \; \uparrow \; \uparrow \; \uparrow}$$

W: [Xe]6s^2 4f^{14} 5d^4

$$\underset{6s}{\uparrow\downarrow} \quad \underset{4f}{\uparrow\downarrow \; \uparrow\downarrow \; \uparrow\downarrow \; \uparrow\downarrow \; \uparrow\downarrow \; \uparrow\downarrow \; \uparrow\downarrow} \quad \underset{5d}{\uparrow \; \uparrow \; \uparrow \; \uparrow \; \_}$$

Note: Generally orbitals fill in the order: ns, (n-1)d, np. However, the energies of the ns and (n-1)d sublevels are so close, deviations in filling patterns occur.

Electron configurations and periodic tables:

We may use our plot of n vs. l (below, left) to construct a periodic table, in two steps. First, flip the diagram, horizontally, to obtain its mirror image (below, right).

Then, second, tilt the diagram to make the lines of constant n+l values horizontal:

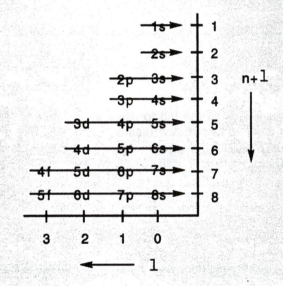

Obtained is a compressed periodic table. Columns in the compressed periodic table correspond to "blocks" of families of elements (called the s, p, d, or f-blocks).
The number of families in the blocks - the blocks' horizontal lengths - are given by the expression $2(2l+1)$, as explained below.

m_l values: - l,...... .,-1,   0,   +1,.........,+1

l values   1 value l values

$2l+1$ values

Each l-value corresponds to an orbital. Each orbital may be occupied by 2 electrons.

<u>Summary of l-values and block lengths:</u>

l symbol	l value	number of orbitals $2l+1$	max. no. of e- in subshell $2(2l+1)$
s	0	1	2
p	1	3	6
d	2	5	10
f	3	7	14

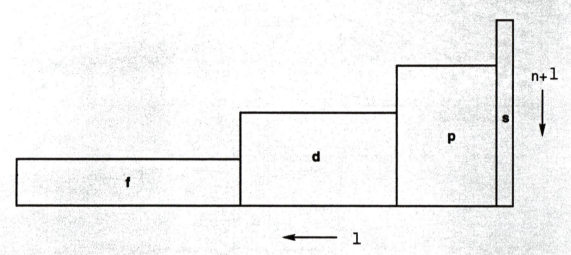

Usually the s-block is moved to the left to place all the metals together, on the left side of the table.   The s-block is shifted down one element, and, He, very different than other members of the ns^2 family, is placed above Ne.

* usually removed and placed below the table, to save space, horizontally.

Electron configuration practice problems:

1) Draw orbital diagrams for atoms with the following electron configurations. How many unpaired electrons does each atom have?

    a) $1s^2 2s^2 2p$

    b) $1s^2 2s^2 2p^6 3s^2$

    c) $1s^2 2s^2 2p^6 3s^2 3p^4$

 solutions:

a) 1s 2s 2p 1 unpaired electron

b) 1s 2s 2p 3s 0 unpaired electrons

c) 1s 2s 2p 3s 3p 2 unpaired electrons

2) Classify the following electron configurations as ground state, excited state, or incorrect:

    a) $1s^2 2s^1 2p^2$ d) $1s^2 2s^1 2d^1$

    b) $1s^2 2s^2 2p^6$ e) $1s^3 2s^2 2p^6$

    c) $1s^2 2s^2 2p^6 3s^1 4f$

 solutions:

 a) excited state, b) ground state, c) excited state

 d) wrong – there is no d subshell in the 2nd energy level

 e) wrong – there cannot be 3 electrons in an "s" orbital (or any orbital).

3) What is the maximum number of electrons in an atom that can have the following quantum numbers:

    a) $n = 3$

    b) $n = 3$, $m_s = +\frac{1}{2}$

    c) $n = 3$, $l = 2$

    d) $n = 3$, $l = 2$, $m_l = -2$

    e) $n = 3$, $l = 2$, $m_l = -2$, $m_s = +\frac{1}{2}$

    f) $n = 4$, $m_l = 1$

solutions:

a) 18 electrons:

The 3rd shell contains the following orbitals:
$3s$, $3p_x$, $3p_y$, $3p_z$, $3d_{xy}$, $3d_{xz}$, $3d_{yz}$, $3d_{x^2-y^2}$, $3d_{z^2}$ each of which can hold 2 electrons.

b) 9 electrons:

Half of the electrons in the 3rd shell orbitals have a spin of $+\frac{1}{2}$, and half have a spin of $m_s = -\frac{1}{2}$

c) 10 electrons:

The five orbitals with $l = 2$ (the 5 "d" sublevels), can hold 10 electrons.

d) 2 electrons:

The m_l value (–2) defines one orbital (a specific orientation of a "d" orbital), and 2 electrons may exist in one orbital.

e) 1 electron:

The four quantum numbers define a unique electron state in an atom.

f) 6 electrons:

When $n = 4$, l may equal 0,1,2, or 3. For the three sublevels $l = 1,2$, or 3, m_l may have the value 1, so there are 3 orbitals (which can hold 6 electrons) in the 4th shell that contain an m_l value of 1.

4) What is wrong with the following sets of quantum numbers?

a) $n = 1$, $l = 1$, $m_l = 0$, $m_s = +\frac{1}{2}$

b) $n = 1$, $l = 0$, $m_l = 1$, $m_s = +\frac{1}{2}$

c) $n = 5$, $l = 2$, $m_l = -3$, $m_s = -\frac{1}{2}$

d) $n = 3$, $l = 0$, $m_l = -1$, $m_s = -\frac{1}{2}$

solutions:

a) For the first shell, l has only one value (0 = "s" orbital)

b) $m_l = -l$, ...0,...+l, so when $l = 0$, m_l can only equal 0. (This one value for m_l indicates the s orbital's one orientation in space).

c) For $l = 2$, m_l can equal –2,–1,0,1,2 (not –3).

d) For $l = 0$, m_l can equal only 0.

Experimental support for electron configurations:

1) What is the electron configuration for the sodium atom, Na?

answer: $1s^2 2s^2 2p^6 3s^1$

2) Which electron(s) would you <u>predict</u> from theory (its electron configuration) would be most easily removed from a sodium atom?

answer:

One would expect that electrons farthest away from the positively charged nucleus would be the easiest to remove. The one electron in the outermost 3rd energy level of a sodium atom should be substantially easier to remove than the electrons closer to the nucleus in the 1st and 2nd energy levels.

In fact, experiments show that the ionization energies (the energy required to remove electrons from a gaseous atom) for a sodium atom fall into three groups with one very low value (the 3s electron), eight which gradually increase in magnitude (the 8 electrons for which n=2), and two very large value (the 2 electrons in the 1s shell which are bound very tightly to the nucleus).

SECTION 8.2: Periodic trends:

Atomic size and electron configuration influence ionization energy (the energy required to remove an electron from a gaseous atom), and electron affinity (the energy involved when an electron adds to an atom). The size of an atom is the main factor that determines the lengths of covalent bonds; bond lengths influence bond <u>strength</u>, and bond strength determines chemical reactivity (the breaking and forming of bonds).

How do we determine an atom's size? We assume that atoms closest to each other in elementary substances are touching.

Atomic size:

What is the atomic radius of a cesium atom and a fluorine atom based on the following data?

5.24 Å

1.42 Å

F F

solution:

radius of Cs = 1/2 diameter = 5.24/2 = 2.62 Å

radius of F = 1/2 diameter = 1.42/2 = 0.71 Å

Keep in mind that because an electron cloud is not rigid, the size of an atom will depend to some extent on the nature of the neighboring atom(s).

Periodic Trends in Atomic Size:

To determine the size of atoms, we must consider both the principal quantum number of its valence electrons, and the effective nuclear charge on those electrons.

Atomic size trends:

1) Down a group:

Consider the first three elements of group 1A (the alkali metals), H, Li and Na:

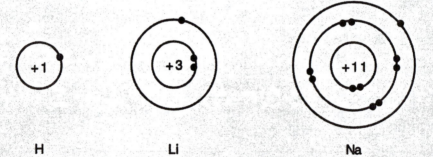

| H | Li | Na |

What is the effective nuclear charge (Z_{eff}) on the valence electron for each atom, and how does atomic radius change as we move down the group?

solution:

Electrons in inner shells "shield" outer shell electrons from the positive charge of the nucleus. To a very rough first approximation, each inner shell electron cancels the charge of one proton in the nucleus, so the outer "s" electron for all elements in group 1A is attracted by the same nuclear charge ($\approx$ +1).

The average distance of an electron from a nucleus of constant charge increases with the principal quantum number, so **atomic radius increases as we move down a group in the periodic table.**

2) Across a period:

Consider the first three elements of the 3rd period (Na, Mg, and Al):

What is the approximate effective nuclear charge (Z_{eff}) on the outer electrons for each atom, and how does atomic radius change as we move across the period?

<u>solution</u>:

Electrons which are approximately the same distance from the nucleus are not effective at shielding one another from the nuclear charge. Therefore, as we move across a period, as each additional proton is added to the nucleus, the effective charge on the outer electrons increases by approximately one. The increasing effective nuclear charge pulls outer electrons in more tightly so **atomic radius decreases as we move across a period**.

Note: The size trends hold most regularly for the main-group elements; minor variations occur in the transition elements.

Atomic size exercises:

1) Arrange the following groups of atoms in order of increasing size:

a) Mg, Ba, Ca	c) C, H, F
b) Ca, Br, Kr	d) B, Al, Ga

<u>solutions</u>:

a) Mg < Ca < Ba (size increases down a group)

b) Kr < Br < Ca (size decreases across a period)

c) H < F < O

c) B < Ga < Al

   (size normally increases down a group, so you might expect gallium to be larger than aluminum. However, the effective nuclear charge is large for gallium, see below.)

2) Write the electron configuration for Ga. Use the configuration to explain the high Z_{eff} felt by gallium's 4p electron.

solution: $1s^2 2s^2 2p^6 3s^2 3p^6 4s^2 3d^{10} 4p^1$

The 3d electrons are not effective at shielding the nuclear charge from the 4p electrons, so the 4p electron in gallium feels an increased Z_{eff} from all the protons added during the transition series.

Ionization energy and electron affinity:

It requires energy to remove electrons from atoms ($IE_1 > 0$), and energy is usually released when an electron adds to an atom (EA_1 usually < 0). It always requires energy to add an electron to an anion ($EA_2 > 0$).

$Na(g) \rightarrow Na^+(g) + e^-$ ΔE = first ionization energy = $IE_1 > 0$

$Na^+(g) \rightarrow Na^{2+}(g) + e^-$ ΔE = second ionization energy = $IE_2 > IE_1$

$Na(g) + e^- \rightarrow Na^-(g)$ ΔE = first electron affinity = EA_1 usually < 0

$Na^-(g) + e^- \rightarrow Na^{2-}(g)$ ΔE = second electron affinity = $EA_2 > 0$

Trends in ionization energy:

In general, the closer an electron is to the nucleus the more difficult it is to remove, so **ionization energy decreases down a group and increases across a period.**

1) Arrange the following groups of atoms in order of increasing first ionization energy:

    a) Mg, Ba, Ca   c) C, O, F
    b) Ca, Br, Kr   d) B, Al, Ga

solutions: (compare these to the solutions for atomic radii above)

    a) Ba < Ca < Mg   c) C < O < F
    b) Ca < Br < Kr   c) Al < Ga < B

2) What element in period three would have the following series of ionization energies (kJ/mol)?

580	1815	2740	11,600
IE_1	IE_2	IE_3	IE_4

solution:

The large jump in ionization energy occurs when the all the valence electrons have been removed from a subshell. For this atom, the jump occurs after three electrons have been removed, so the element is aluminum with 3 valence electrons:
$1s^2 2s^2 2p^6 3s^2 3p^1$.

Trends in electron affinity:

1) Elements in groups 6A and 7A (the halogens) attract electrons strongly and have highly negative (exothermic) electron affinities. In ionic compounds, they form negative ions.

2) Elements in groups 1A and 2A lose electrons readily and do not attract them, and have positive (endothermic), or slightly negative, electron affinities. In ionic compounds, they form positive ions.

3) The noble gases (group 8A) in general neither lose nor gain electrons and have positive (endothermic) electron affinities.

SECTION 8.3: Atomic Structure and Chemical Reactivity

Large atoms with low ionization energies exhibit more metallic behavior than small atoms with high ionization energies. Therefore, <u>metallic behavior increases as you go down a group (as the atoms get larger), and decreases across a period (as the atoms get smaller)</u>.

Atomic size, ionization energy, and metallic behavior:

1) Indicate whether each of the trends listed increases or decreases as you move down, and from right to left across the periodic table:

a) metallic behavior   b) atomic size   c) ionization energy

<u>answers</u>:

a) metallic behavior increases, b) atomic size increases, c) ionization energy decreases

2) Given that metallic character increases down a groups and decreases across a period, draw a line along which metallic character is approximately constant.

solution:

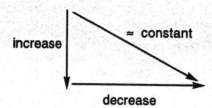

Notice the similar diagonal line on the periodic table which separates metals from nonmetals.

Element oxides:

Consider the reactions of a metal oxide, CaO, and a nonmetal oxide, NO_2 with water:

$$CaO(s) + H_2O \rightarrow Ca(OH)_2(aq)$$

$$3NO_2(g) + H_2O \rightarrow 2HNO_3(aq) + NO(g)$$

Which of the oxides formed an acid and which formed a base?

answer:

CaO acts as a base by removing protons from water to produce OH^- ions. NO_2 reacts with water to form nitric acid (HNO_3). As metallic behavior increases down a group, element oxides become more basic. As metallic behavior decreases across a period, element oxides become more acidic.

Monatomic Ions:

Main group atoms gain or lose electrons to attain noble gas configurations (filled outer p subshells). It often requires too much energy to remove all the electrons necessary for transition metals to attain noble gas configurations, so they often attain a **pseudo-noble gas configuration**, in which valence "s" electrons are lost and a filled d shell remains.

Although the 4s orbitals are lower in energy than the 3d orbitals when they are filled by the aufbau process, once the orbitals are filled with electrons, the 3d orbital becomes lower in energy than the 4s. The 4s electrons are lost first in the formation of Period 4 transition metal ions. Except for the lanthanides, electrons with the highest n value are always lost first.

Electron configuration of ions:

1) Predict the ions that Na, Mg, and Al will form and give their electron configurations.

     answers: Na^{1+}: $1s^2 2s^2 2p^6$, Mg^{2+}: $1s^2 2s^2 2p^6$, Al^{3+}: $1s^2 2s^2 2p^6$

2) Which noble gas is **isoelectronic** (the same electron configuration) with the above ions?

     answer: Ne

3) Arrange the above ions and Ne in order of increasing atomic size:

     Answer: $Al^{3+} < Mg^{2+} < Na^{1+} < Ne$

Since the electron configurations are the same for all four species, it is the nuclear charge that determines the atomic size. The highest nuclear charge of aluminum (+13), pulls the electrons in the most tightly followed by magnesium (+12,) sodium (+11), and neon (+10).

4) Write the electron configuration for the transition metal, Zn. What would you predict would be the ion it forms in compounds?

     answer: $[Ar]4s^2 3d^{10}$

Zinc loses its two 4s electrons to form Zn^{2+} with an electron configuration of $[Ar]3d^{10}$. This configuration is called a pseudo-noble gas configuration.

5) Write the electron configurations thallium (Tl), lead (Pb), and bismuth (Bi) and predict the ions they would likely form in compounds.

     answers:

        Tl: $[Xe]6s^2 4f^{14} 5d^{10} 6p$ Bi: $[Xe]6s^2 4f^{14} 5d^{10} 6p^3$

        Pb: $[Xe]6s^2 4f^{14} 5d^{10} 6p^2$

To obtain pseudo-noble gas configurations, Tl, Pb, and Bi would need to lose their 6s and 6p electrons forming: Tl^{3+}, Pb^{4+}, and Bi^{5+}. These ions do in fact form, but the most common ions formed by these elements are the ones which retain the two "s" electrons (sometimes called an **inert pair**) and lose the "p" electrons: Tl^{1+}, Pb^{2+}, Bi^{3+}.

Magnetic Properties:

The magnetic properties of transition metals and their compounds can support or refute a given electron configuration. Elements or compounds with all paired electrons are **diamagnetic** and are not attracted by a magnetic field. Elements or compounds with unpaired electrons are **paramagnetic**, and are attracted by an external magnetic field.

Predicting magnetic properties:

Write the electron configuration for each of the following transition metals and predict if it would be paramagnetic or diamagnetic:

a) Fe^{3+} (Z=26)   b) Fe^{2+}

c) Cu^{2+} (Z=29)   d) Cu^{1+}

<u>answers</u>:

a) Fe: $[Ar]4s^2 3d^6$

Fe^{3+}: $[Ar]3d^5$

There is one electron in each of the five d orbitals, so we would predict that Fe^{3+} is paramagnetic with five unpaired electrons.

b) Fe: $[Ar]4s^2 3d^6$

Fe^{2+}: $[Ar]3d^6$

Four of the d orbitals contain one unpaired electron; the remaining d orbital contains two paired electrons. Fe^{2+} is paramagnetic with four unpaired electrons

c) Cu: $[Ar]4s3d^{10}$

Cu^{2+}: $[Ar]3d^9$

Four of the d orbitals contain paired electrons; the remaining d orbital contains an unpaired electron. Cu^{2+} is paramagnetic with 1 unpaired electron

d) Cu^{1+}: $[Ar]3d^{10}$

All five d orbitals contain paired electrons; Cu^{1+} is diamagnetic.

A word of caution: as you will see in Chapter 22, the magnetic properties of transition metal complexes depend both on the electron configuration of the metal, and the attached ligands. For example, Fe^{2+} with attached water molecules is paramagnetic with four unpaired electrons, but Fe^{2+} with attached CN⁻ molecules is diamagnetic with all electrons paired.

Ionic size:

Choose words from the list below to complete the following paragraph:

a) anion
b) decrease
c) increase(s)
d) larger

e) more
f) proton
g) smaller

Cations, since they have fewer electrons than their parent atoms, are

_____(1) than their parent atoms; anions which have more electrons than their

parent atoms are _____(2) than their parent atoms. Consider the hydrogen atom. If

it loses its electron to form the cation, H^+, all its orbitals are empty and it is only as large as a

_____(3). If it gains an electron into its 1s orbital to form the anion, H^-, the

electron-electron repulsions cause the electrons to occupy _____(4) space. Ionic

size _____(5) down a group just as does atomic size. Across a period, we move from

cations of metals to anions of nonmetals. There is a small _____(6) in ionic size

among the cations, followed by a large _____(7) in size when we reach the anions, and

then another small then another small _____(8) in size among the anions.

answers: 1-g, 2-g, 3-f, 4-e, 5-c, 6-b, 7-c, 8-b.

Chapter 8 Exercises:

8.1) a. Electrons in which type of atomic orbital are best at shielding other electrons from the nuclear charge?

b. Electrons in which type of atomic orbital (s, p, d, or f) are the most shielded by electrons in other orbitals from the nuclear charge?

c. Electrons in which type of orbital penetrate nearest to the nucleus?

d. In a many-electron atom, electrons in which type of orbital within a single principal quantum level experience the lowest effective nuclear charge?

8.2) The energy required to excite the 2s electron of a lithium atom to a 2p orbital is far greater than the energy required to move an electron in the 2s orbital of the Li^{2+} ion to a 2p orbital. Explain why.

8.3) What are the electron configurations of

    a. germanium
    b. krypton
    c. zinc
    d. iridium

8.4) State, in your own words, the meaning of the Aufbau Principle.

8.5) Write out the electron configuration and give the set of four quantum numbers for a valence electron for each of the following elements or ions.

    a. Sb   b. Br⁻   c. Al   d. Fe^{2+} e. Sm

8.6) According to Hund's Rule of maximum multiplicity, how many singly-occupied orbitals are there in the valence shell of the ground state of

    a. a carbon atom
    b. nitrogen atom
    c. an iron atom
    d. a sulfur atom

8.7) Pair each element with its classification:

    a. Chromium   1. noble gas
    b. Strontium   2. representative metal
    c. Bromine   3. transition metal
    d. Neon   4. alkali metal
    e. Sulfur   5. halogen
    f. Lithium   6. nonmetal
    g. Lead   7. lanthanide
    h. Gadolinium   8. alkaline earth metal

8.8)   How many valence electrons do the following elements have in their valence shells?

        a. sulfur   d. silicon
        b. argon   e. chlorine
        c. magnesium   f. aluminum

8.9)   Which of the sixteen representative elements in Periods 2 and 3 of the Periodic Table have:

        a. the smallest atomic radius
        b. the highest ionization energy
        c. the highest electron affinity
        d. the most metallic character

8.10) Which of the elements in the third Period of the Periodic Table has a relatively low 1st ionization energy, a higher second ionization energy, but a very high third ionization energy?

8.11) The first ionization energy for the second period elements are given below.

Li	Be	B	C	N	O	F	Ne
0.52	0.90	0.80	1.09	1.40	1.31	1.68	2.08

a. Explain why the first ionization energy generally increases from left to right.
b. Explain the variations of the trend at the positions of B and O.

8.12) Which is the most probable ion produced by the removal of valence electron(s) from, or the addition of electrons to,

        a. an aluminum atom   d. a calcium atom
        b. a phosphorus atom   e. a sulfur atom
        c. a chlorine atom

8.13) Which is the largest ion?

        sodium Na^+ chlorine Cl^-
        rubidium Rb^+ iodine I^-

Chapter 8 Answers:

8.1) a. s orbital; b. f orbital; c. s orbital; d. f orbital

8.2) In a lithium atom the 2s and 2p orbitals are separated in energy because of the shielding effect. The 2s orbital is pulled down in energy because it experiences a greater effective nuclear charge. Therefore it requires considerable energy to excite an electron from the 2s to the 2p level.

In the Li^{2+} ion, there is only one electron and there cannot be any shielding. As a result the 2s and 2p orbitals are approximately equal in energy and it is easy to move an electron between them.

8.3) a. Ge: $1s^2 2s^2 2p^6 3s^2 3p^6 4s^2 3d^{10} 4p^2$
      b. Kr: $1s^2 2s^2 2p^6 3s^2 3p^6 4s^2 3d^{10} 4p^6$
      c. Zn: $1s^2 2s^2 2p^6 3s^2 3p^6 4s^2 3d^{10}$
      d. Ir: $1s^2 2s^2 2p^6 3s^2 3p^6 4s^2 3d^{10} 4p^6 5s^2 4d^{10} 5p^6 6s^2 4f^{14} 5d^7$

8.4) The Aufbau Principle states that the atomic orbitals of an atom fill in order of increasing energy. The lower energy orbitals fill first. The Aufbau Principle allows the easy determination of the most probable ground-state electron configuration of an atom. There are a few instances where Hund's rule of maximum multiplicity takes precedence. For example, in copper, where the ground-state configuration is $[Ar]4s^1 3d^{10}$ not $[Ar]4s^2 3d^9$.

One way to determine the order of filling orbitals is to use the Periodic Table, horizontally row by row. Another method is to use the "n+1 rule" which states that the atomic orbitals of an atom fill in order of increasing "n+1" value, and if "n+1" is the same for two orbital, the orbital with the lowest n value fills first. For example, 3d is filled before 4p.

8.5) a. $1s^2 2s^2 2p^6 3s^2 3p^6 4s^2 3d^{10} 4p^6 5s^2 4d^{10} 5p^3$ $n=5$ $l=1$ $m_l=+1$ $m_s=+1/2$
      b. $1s^2 2s^2 2p^6 3s^2 3p^6 4s^2 3d^{10} 4p^6$ $n=4$ $l=1$ $m_l=+1$ $m_s=-1/2$
      c. $1s^2 2s^2 2p^6 3s^2 3p^1$ $n=3$ $l=1$ $m_l=-1$ $m_s=+1/2$
      d. $1s^2 2s^2 2p^6 3s^2 3p^6 3d^6$ $n=3$ $l=2$ $m_l=+2$ $m_s=+1/2$
      e. $1s^2 2s^2 2p^6 3s^2 3p^6 4s^2 3d^{10} 4p^6 5s^2 4d^{10} 5p^6 6s^2 4f^6$ $n=4$ $l=3$ $m_l=+2$ $m_s=+1/2$

8.6) a. two; valence shell configuration is $2s^2 2p^2$
      b. three; valence shell configuration is $2s^2 2p^3$
      c. four; valence shell configuration is $4s^2 3d^6$
      d. two; valence shell configuration is $3s^2 3p^4$

8.7) a-3; b-8; c-5; d-1; e-6; f-4; g-2; h-7
      Some elements fit into more than one category.

8.8) a. 6 valence electrons   d. 4 valence electrons
      b. 8 valence electrons   e. 7 valence electrons
      c. 2 valence electrons   f. 3 valence electrons

8.9) a. Atomic size decreases left to right, due to the increase in the nuclear charge, and increases top to bottom, due to the accumulation of more and more principal quantum levels. Neon is the smallest of the sixteen elements.

b. Neon also has the highest ionization energy; it is difficult to remove a tightly bound electron from a small atom.

c. Chorine has the highest electron affinity. Fluorine would have been the logical choice, but the fluorine atom is very small and addition of another electron is inhibited by a large electron repulsion in the crowded space around the nucleus.

d. The most metallic element is the one furthest to the left and lowest in the Periodic Table: sodium in this example.

8.10) The fact that it is the third ionization that is very high indicates that this ionization breaks up a stable "noble gas configuration". In other words, the atom originally has two valence electrons. The element is magnesium.

8.11) a. The increase in nuclear charge moving from left to right across a period results in smaller and smaller atoms and therefore an increase in the energy needed to remove an electron from the valence shell.

b. Electrons gain additional energy stabilization when an energy shell or subshell is filled or half-filled. Be has a higher ionization energy because an electron must be removed from a filled s subshell. N has a higher ionization energy because an electron must be removed from a half-filled p subshell.

8.12) Sufficient electrons should be added or removed to achieve a valence shell configuration equivalent to that of a noble gas.

a. Removal of three electrons: Al^{3+}
b. Addition of three electrons: P^{3-}
c. Addition of one electron: Cl^-
d. Removal of two electrons: Ca^{2+}
e. Addition of two electrons: S^{2-}

8.13) Although for atoms there is a general decrease in size from left to right across the representative elements in the Periodic Table, there is no such trend for ions because the charges invariably differ. Monatomic ions on the left side typically have positive charges whereas monatomic ions on the right have negative charges.

For example, which is larger, rubidium Rb^+ or iodine I^-? It is sometimes convenient to consider isoelectronic series around the end of one period to the beginning of the next. In this case, consider:

$$Se^{2-} \quad Br^- \quad Kr \quad Rb^+ \quad Sr^{2+}$$

The nuclear charge increases, but the number of electrons remains the same. So the ions progressively get smaller and smaller through the series. Rb^+ is smaller than Br^-. Since Br^- is smaller than I^-, then Rb^+ must be smaller than I^-.
I^- is the largest ion, Na^+ is the smallest.

CHAPTER 9

Models of Chemical Bonding

TOPICS:

- Chemical bonds
    - -Ionic, covalent, and metallic
- Ionic bonding model
    - -Lattice energy and the Born-Haber cycle
- Covalent bonding model
- Electronegativity and bond polarity
- Lewis structures
    - -Resonance
    - -Formal charge
- Heats of reactions from bond energies

SECTION 9.1: Chemical bonds

Atoms bond together to obtain stable (filled shell) electron configurations. Electrical conductivity is the best guide to determine the type of bonding. We usually classify chemical bonds in one of three ways:

"Metallic"   Conducts in the solid state by electron flow

"Ionic"   Conducts in the molten state by motion of ions

"Covalent"   Non-conductor

We can picture bonds as valence electrons attracted by atomic "cores". A "core" or "kernel" is an atom without its valence electrons.

1) Metallic (loosely held electrons by large cores):

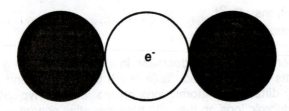

examples: Ca^{2+}, Na^+

2) Covalent (tightly held electrons by small cores):

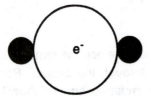

examples: F^{7+}, C^{4+}

3) Ionic (electrons held tightly by small core):

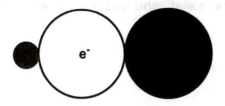

example: F^{7+} and Na^{1+}

Ionic, covalent, and metallic bonding:

Classify the following as typical of ionic bonding (i), covalent bonding (c), or metallic bonding (m).

1) Nonmetals share a localized pair of electrons between atoms.

2) Valence electrons from many atoms are pooled and delocalized to form a sea of electrons.

3) Reactive metals in group 1A or 2A transfer electrons to reactive nonmetals in group 7A (or the top of group 6A).

4) The electrostatic attraction between positive and negatively charged particles causes a rigid three dimensional array of atoms to form.

5) Exist as separate molecules with relatively weak intermolecular forces.

6) The chemical formula represents the cation to anion ratio in the compound (empirical formula)

7) The chemical formula reflects the number of atoms in the molecule (molecular formula).

8) The bonding in H_2O

9) The bonding in NaCl

10) The bonding in diamond

11) The bonding in solid silver

    <u>answers</u>: 1-c, 2-m, 3-i, 4-i, 5-c, 6-i, 7-c, 8-c, 9-i, 10-c, 11-m.

The properties of an ionic solid, a metal, and a covalent compound:

Warm up exercises:

1) What is the electron configuration for sodium?

   <u>answer</u>: [Ne]3s^1

2) What is the electron configuration for chlorine?

   <u>answer</u>: [Ne]3s^{2}sp^5

3) Describe the easiest way for sodium and chlorine to obtain noble gas configurations.

   <u>answer</u>:

   Sodium loses one electron to obtain the electron configuration of the noble gas neon. Chlorine gains one electron to obtain the electron configuration of the noble gas argon. Na$^+$ and Cl$^-$ both have noble gas electron configurations.

4) Consider the familiar ionic solid, sodium chloride (NaCl), and the metal and covalent nonmetal from which it is composed, sodium (Na), and chlorine (Cl$_2$). Identify each of the following as a property of Na, NaCl, or Cl$_2$:

   a) BP = 883°C, MP = 97.8°C

   b) BP = -35°, MP = -101°C

   c) BP = 1413°C, MP = 801°C

   d) shatters under stress

   e) deforms under stress

   conducts electricity:

         f) when dissolved in water

         g) when melted

         h) as a solid

         i) does not conduct electricity

   j) a pale green gas at room temp.

   k) a soft, silvery-white metal at room temperature

   l) a white solid at room temperature

   m) violently decomposes water producing a basic solution and hydrogen gas.

   n) burns with a yellow flame

   <u>answers</u>:

   sodium, Na (metallic): a, e, g, h, k, m, n

Metals do not usually crack, but deform with stress because metal ions can slide past each other through the electrons to move to new lattice positions.

Metals are good conductors of heat and electricity as both solids and liquids due to their mobile electrons.

chlorine, Cl_2 (covalent): b, i, j

Chorine, like most covalent substances, consists of individual molecules which have relatively weak attractions for each other (intermolecular attractions). These compounds have low melting and boiling points because the individual molecules stay intact (no covalent bonds are broken); it is only the relatively weak intermolecular forces <u>between</u> molecules which must be overcome.

> <u>Note</u>: Remember that covalent compounds with low melting and boiling points are not breaking covalent bonds. Covalent bonds are extremely strong (240 kJ/mol for the bond in chlorine, 347 kJ/mol for a typical single bond between carbon atoms), and covalent substances which form a network of covalent bonds throughout the sample have extremely high melting points and boiling points. Diamond, which consists of a three dimensional array of carbon atoms covalently bonded together, melts at 3550°C and is used to test the hardness of other substances because it is so hard.

Covalent compounds are poor conductors of electricity because the electrons are localized (not free to move), and no ions are present.

sodium chloride, NaCl (ionic): c, d, f, g, I

NaCl (ionic): Ionic compounds usually have high melting points and boiling points because it requires large amounts of energy to free ions from lattice positions in ionic compounds .

Ionic compounds are hard and brittle because strong electrostatic forces hold the ions in place throughout the crystal. If enough pressure is applied to overcome the attractions, ions of like charge are brought close together and their repulsions crack the sample.

The ions of an ionic compound are generally immobilized in a solid, especially at ordinary temperatures, and do not conduct electricity. When an ionic solid melts or dissolves, the ions can move and carry an electric current.

The octet rule:

Atoms lose, gain, or share electrons to attain a filled outer level (a noble gas configuration). An outer level contains eight (an octet of) electrons: two from the "s" orbital, and six from the three "p" orbitals.

SECTION 9.2: Ionic bonding

Ionic compounds form when metals transfer electrons to nonmetals to form ions with noble gas electron configurations; the electrostatic attraction between the cations and anions draws them into a three dimensional array of ions.

Formation of magnesium oxide:

1) What is the electron configuration for magnesium?

    <u>answer</u>: $[Ne]3s^2$

2) How does magnesium obtain a noble gas configuration in an ionic compound?

    <u>answer</u>:

    Mg loses its two "s" electrons to become Mg^{2+}, and attain the same electron configuration as neon. $([Ne]3s^2 - 2e^- = [Ne])$

3) What is the electron configuration for oxygen?

    <u>answer</u>: $[He]2s^22p^4$

4) How does oxygen obtain a noble gas configuration in its compounds?

    <u>answer</u>:

    Oxygen gains two electrons to become O^{2-} and attain the same electron configuration as Ne. $([He]2s^22p^4 + 2e^- = [He]2s^22p^6 = [Ne])$.

    Magnesium transfers two of its electrons to oxygen to form Mg^{2+} and O^{2-}, both with the electron configuration of neon. These ions form an ionic solid with an empirical formula of MgO.

5) Is an electron transfer process endothermic or exothermic?

    <u>answer</u>:

    It requires energy to remove two electrons from a gaseous magnesium atom (2180 kJ/mol). It also requires energy to add two electrons to a gaseous oxygen atom. Addition of the first electron is exothermic (-141 kJ/mol), but addition of the second electron is endothermic (878 kJ/mol). Therefore, the electron transfer process requires 2917 kJ/mol.

    Why does MgO form at all if it requires so much energy to transfer the electrons? As you will see in the next problem, the large amount of energy released when separated gaseous ions are packed together to form an ionic solid, the **lattice energy**, makes the entire process energetically favorable.

Lattice energy and the Born-Haber cycle:

Lattice energy is defined as the change in energy for the following reaction:

$$Mg^{2+}(g) + O^{2-}(g) \rightarrow MgO(s)$$

Notice that the ions are in their <u>gaseous</u> state. We can calculate the lattice energy by breaking the reaction down into steps, the sum of which gives the overall reaction. (Review Hess's Law, Chapter 6: $\Delta H_{total} = \Delta H_1 + \Delta H_2 + \Delta H_3...$).

(ΔH values in kJ/mol)

1) Use Hess's law to determine the lattice energy of MgO.

<u>solution</u>:

Using Hess's Law: $\Delta H_f^o = \Delta H_1 + \Delta H_2 + \Delta H_3 + \Delta H_4 + \Delta H_{lattice\ energy}$

The heat of formation of magnesium oxide is an exothermic process: heat is released ($\Delta H_f^o = -602$ kJ/mol). Notice that all the steps taking elemental magnesium and oxygen to gaseous ions are endothermic (positive ΔH values). The lattice energy releases a large amount of energy to make the overall process energetically favorable.

$\Delta H_{lattice\ energy} = \Delta H_f^o - \Delta H_1 - \Delta H_2 - \Delta H_3 - \Delta H_4$

$$= -602 - 150 - 2180 - 247 - 737$$

$$= -3920 \text{ kJ/mol}$$

The lattice energy is greater as the charges on the ions increases, and the radii of the cation and anion become smaller:

$$\Delta H_{lattice}^o \propto \frac{cation\ charge \times anion\ charge}{cation\ radius + anion\ radius}$$

Lattice energy prediction:

Would you expect the lattice energy to be greater for MgO or for NaCl?

solution:

MgO with +2 and -2 charges on its ions makes the numerator in the above equation larger than NaCl with +1 and -1 cation and anion charges.

The smaller ionic radii of Mg^{2+} and O^{2-} make the denominator smaller than for Na^+ and Cl^-, also making $\Delta H^\circ_{lattice}$ larger, so the lattice energy is greater for MgO.

SECTION 9.3: Covalent bonding model

The H_2 molecule:

Why is a hydrogen molecule more stable than two hydrogen atoms? Consider hydrogen atoms 1 and 2 separated by a large distance (no interactions, 0 energy state):

1   2

As the two atoms become closer together, what are the forces of attraction between the atoms? (Answer: Attractive forces occur between particles of opposite charge: proton 1 attracts electron 2; proton 2 attracts electron 1). What are the forces of repulsion between the atoms? (Answer: Repulsive forces occur between particles of like charge: electron 1 repulses electron 2; proton 1 repulses proton 2).

A graph of the energy of a H_2 molecule as a function of the distance between the two nuclei looks like this:

(The calculation is based on an accurate, if approximate, solution of the Schrodinger equation).

The H₂ molecule:

1) Based on the previous drawing, what is the energy of the two separated hydrogen atoms?

   <u>answer</u>:

   When there is no interaction between hydrogen atoms, we define the energy to be 0.

2) How long is a bond in a H₂ molecule?

   <u>answer</u>:

   0.74 Å: the distance between bonded hydrogen atoms represents an energy minimum.

3) What can you say about the attractive and repulsive forces when the hydrogen atoms are 0.74 Å apart?

   <u>answer</u>:

   At a distance of 0.74 Å (the bond length of a hydrogen molecule), attractive forces balance repulsive forces. If we attempt to bring the atoms any closer together, the repulsive forces exceed the attractive forces, and the system becomes less stable.

Covalent bonds between atoms may be single or multiple. In general, single bonds are longer and weaker than multiple bonds.

Bond type	Bond order	Bond energy (kJ/mol)	Bond length (Å)
single C–C	1	347	1.54
double C=C	2	614	1.34
triple C≡C	3	839	1.20

SECTION 9.4:   Electronegativity and bond polarity:

Most bonds do not fit neatly into either ionic or covalent bonding models; they have a little of each character. We saw that two identical hydrogen atoms share electrons between them to form a hydrogen molecule, and that magnesium transfers two electrons to oxygen to form magnesium oxide. Between these extremes, we have atoms that do not transfer electrons, but they are different enough so that they do not share them equally. In these cases a **polar covalent bond** forms. In a polar covalent bond, the shared electrons spend more time around one atom than the other, giving the atoms partial positive and negative charges.

Polar covalent bonds:

1) Predict which of the following molecules contain polar covalent bonds, and show the partial charges.

        a) HCl   b) O_2 c) HF   d) H_2O

answers:

The symbol "∂" means "partial"; ∂+ means that atom has a partial positive charge, ∂- means that atom has a partial negative charge.

a) polar: $\overset{\partial+\ \ \partial-}{H-Cl}$ b) nonpolar   c) polar: $\overset{\partial+\ \ \partial-}{H-F}$ d) polar: $\underset{\partial+\quad\quad\partial+}{\overset{\partial-}{H\diagdown O\diagup H}}$

2) Can you predict the bond energy for a H – Cl bond given that the bond energy for H – H is 432 kJ/mol, and the bond energy for Cl – Cl is 239 kJ/mol?

answer:

You might expect that the bond energy for H – Cl would be the average of the H – H and Cl – Cl bond energies:

$$\text{Expected H – Cl bond energy} = \frac{432\text{ kJ/mol} + 239\text{ kJ/mol}}{2} = 336\text{ kJ/mol}$$

However, the measured bond energy for H – Cl is 427 kJ/mol. We attribute this discrepancy to the ionic character in the polar H – Cl bond. The electrostatic attraction between the partially charged atoms leads to greater bond strength, and the actual bond energy is higher than expected.

Linus Pauling developed a scale of **Electronegativity** of atoms based on the difference between the expected and actual bond energies,

$$\Delta = (\text{H-Cl})_{actual} - (\text{H-Cl})_{expected}$$

Electronegativity:

Choose words form the list below to complete the following paragraph:

    a) attract   e) increases
    b) covalent   f) ionic
    c) decreases   g) polar covalent
    d) fluorine   h) same

Electronegativity is the relative ability of bonded atoms to _____(1) the shared electrons. Atoms with very different electronegativities form _____(2) bonds, those with intermediate electronegativity differences form _____(3) bonds, and those with no electronegativity difference form _____(4) bonds. A purely covalent bond exists only between two of the _____(5) atoms. Electronegativity, in general, increases as an atom's size _____(6), so electronegativity _____(7) as you go up a group in the periodic table and _____(8) from left to right across the periodic table. The most electronegative element is _____(9) to which Pauling assigned an electronegativity of 4.0.

Answers: 1-a, 2-f, 3-g, 4-b, 5-h, 6-c, 7-e, 8-e, 9-d.

Polarity and boiling points:

Match the following compounds with their values for ΔEN (the difference in electronegativity between their two atoms), and their expected boiling points.

	ΔEN	bp. °C
1) PCl_3	a) 0	i) 1413
2) Cl_2	b) 0.5	ii) 1412
3) NaCl	c) 0.9	iii) 183
4) $SiCl_4$	d) 1.2	iv) 76
5) $MgCl_2$	e) 1.5	v) 59
6) $AlCl_3$	f) 1.8	vi) 56
7) SCl_2	g) 2.1	vii) -34.6

Answers:

	Cl_2	SCl_2	PCl_3	$SiCl_4$	$AlCl_3$	$MgCl_2$	NaCl
Expected:	a	b	c	d	e	f	g
	vii	vi	v	iv	iii	ii	i
Observed:	vii	v	iv	vi	iii	ii	i

All of the elements in the above compounds are in the third period of the periodic table. Sodium and chlorine have the greatest difference in electronegativity and they combine to form NaCl, an ionic solid with high melting and boiling points. As we move from sodium across the table towards chlorine, the electronegativity of the elements increases bringing them closer in value to the electronegativity of chlorine, and the resulting compounds show less ionic character as the difference in electronegativity of their atoms decreases.

As bonds become less polar, boiling points tend to decrease. You probably could not predict exactly the order of the boiling points, because the boiling point for $SiCl_4$ is unusually low (56°C) due to weak forces between molecules and it falls out of order.

SECTION 9.5: Lewis structures

The **Lewis structure** (after G.N. Lewis) of a molecule shows atoms with their valence electrons. The Lewis structure does not necessarily show the correct geometry of the atoms, but their sequence (or connectivity) must be correct:

$$H-\overset{\displaystyle ..}{\underset{\displaystyle ..}{O}}-H$$

is a correct Lewis structure for water even though water is actually a bent molecule. In Lewis structures, lines indicate electron pairs in covalent (shared electron-pair) bonds.

The Lewis structure of water:

1) Show an oxygen atom with its valence electrons.

    <u>answer</u>:

    Oxygen in group 6A has 6 valence electrons. We place these around four sides oxygen one at a time, and then pair up electrons as needed:

    The number of unpaired electrons shows the number of bonds the nonmetal usually forms. Oxygen usually forms two bonds which gives it the two electrons it needs to attain a noble gas configuration.

2) Show two Hydrogen atoms with their valence electrons.

    <u>answer</u>:

    Each hydrogen atom needs one electron to fill its 1s shell.

            H·   **·H**

3) Show, using dots for electrons, how oxygen and hydrogen combine to form water, by sharing electrons

$$H\cdot\cdot\overset{\displaystyle ..}{\underset{\displaystyle ..}{O}}\cdot\cdot H$$

Method for drawing Lewis structures:

1) Decide on the sequence of atoms: usually place the atom with the lowest group number in the center, except for hydrogen, which never goes in the center.

H   O   H

2) Add up the total number of valence electrons

(for main group atoms, # valence e- = group number)

valence electrons in H_2O= valence electrons from O + 2 x valence electrons from H

= 6 + 2(1) = 8

3) Draw single bonds between atoms and subtract two valence electrons for each bond.

H−O−H

It requires two electrons for each single bond, so electrons left to place are:
8 - 4 = 4 .

4) Distribute remaining electrons to give octets to all atoms except hydrogen (central atom last)

H−Ö−H

5) If there are not enough electrons for the central atom, make multiple bonds.

Oxygen has an octet, so multiple bonds are not needed.

Lewis octet structure practice problems:

1) Draw the Lewis electron dot symbol for the atoms C, N, O, and F.

solutions:

·Ċ·   ·N̈·   ·Ö·   ·F̈:

2) How many hydrogen atoms would be needed for each atom to give it an octet of electrons? Draw each structure.

H
|
H–C–H
|
H

H–N̈–H
|
H

H–Ö–H

H–F̈:

4-H
(methane)

3-H
(ammonia)

2-H
(water)

1-H
(hydrofluoric acid)

3) Draw Lewis structures for the following compounds:

a) CO_2

b) CO

c) CN^-

d) NO^+

e) NOF

f) H_2CO

g) C_2N_2 (atom order is NCCN)

h) N_2

solutions:

a) CO_2 valence electrons = 4 + (2×6) = 16

Carbon is placed in the middle because it is in the lowest group. If we use four electrons to make single bonds between carbon and oxygen, we would need 16 more electrons to complete octets around the carbon and oxygen atoms.

:Ö–C̈–Ö:

CO_2 has only 16 valence electrons, so we must make multiple bonds. Double bonds between carbon and oxygen use 8 electrons and leave 8 to place around the two oxygen atoms as lone pairs to complete their octets:

Ö=C=Ö

b) CO valence electrons = 4 + 6 = 10

The only way to give carbon and oxygen octets with 10 electrons is to use 6 electrons to make a triple bond, and use the other four electrons to give each atom a lone pair.

:C≡O:

c) CN^- valence electrons = 4 + 5 + 1 = 10

:C≡N:⁻

Notice that CN^- is **isoelectronic** (same number of electrons) with CO and has essentially the same structure.

d) NO^+ valence electrons = 5 + 6 - 1 = 10

$$:N{\equiv}O:^+$$

NO^+ is isoelectronic with both CO and CN^-. If we start with CO, and replace carbon with nitrogen, nitrogen brings with it an extra valence electron. The +1 charge of NO^+ "cancels" the extra electron, and we can predict, without counting all valence electrons, that CO and NO^+ are isoelectronic.

e) NOF valence electrons = 5 + 6 + 7 = 18

$$\ddot{O}{=}\ddot{N}{-}\ddot{F}:$$

f) H_2CO valence electrons = (2x1) + 4 + 6 = 12

$$H{-}\overset{\ddot{O}}{\underset{|}{C}}{-}H$$

g) N_2 valence electrons = (2x5) = 10

$$:N{\equiv}N:$$

A nitrogen molecule contains a triple bond.

h) C_2N_2 valence electrons = (2x4) + (2x5) = 18

The only way to obtain octets on all the atoms is to use 12 electrons to make triple bonds between the carbon and nitrogen atoms, and use the last four electrons to give each nitrogen a lone pair.

$$:N{\equiv}C{-}C{\equiv}N:$$

Exceptions to the octet rule:

Be and B are often have fewer than 1 electron per valence-shell orbital. Atoms with low-lying (valence-shell) d orbitals available can <u>expand</u> their valence shells and accept more than eight electrons. Phosphorus and sulfur can accommodate 12 electrons around them; iodine can accommodate 14. Molecules with an odd number of electrons form a <u>free radical</u> with an unpaired electron and do not obey the octet rule.

Lewis structures:

Show Lewis structures for the following compounds:

a) SO_2 c) PCl_5

b) SF_6

solutions:

a) SO_2 valence electrons = 6 + (2x6) = 18

$$:\ddot{O}=\overset{\displaystyle ..}{S}=\ddot{O}:$$

Sulfur expands its octet to hold 10 electrons.

b) SF_6 valence electrons = 6 + (6x7) = 48

$$\begin{array}{c} :\ddot{F}: \\ :\ddot{F}\underset{:\ddot{F}:}{\overset{|}{\underset{|}{S}}}\ddot{F}: \\ :\ddot{F}: \end{array}$$

Sulfur expands its octet to hold 12 electrons.

c) PCl_5 valence electrons = 5 + (5x7) = 40

$$\begin{array}{c} :\ddot{Cl}: \\ :\ddot{Cl}-\overset{|}{\underset{|}{P}}\overset{\textstyle ..}{\underset{\textstyle :\ddot{Cl}:}{\ddot{Cl}:}} \\ :\ddot{Cl}: \end{array}$$

Phosphorus expands its octet to hold 10 electrons.

Resonance (delocalized bonding):

What is the Lewis structure for the nitrite anion, NO_2^-? NO_2^- has 24 valence electrons. In order to achieve octets around each atom, a double bond is needed between the nitrogen and one of the oxygen atoms:

$$\cdot \overset{\cdot\cdot}{\underset{\cdot\cdot}{O}} \overset{\overset{\cdot\cdot}{N}}{} \overset{\cdot\cdot}{\underset{\cdot\cdot}{O}} :^{-}$$

Experiments show that NO_2^- has only one bond type, not one shorter (and stronger) double bond and two longer single bonds. In order to represent this idea, we use **resonance** structures, with a two headed resonance arrow between them.

$$^{-}:\overset{\cdot\cdot}{\underset{\cdot\cdot}{O}} = \overset{\overset{\cdot\cdot}{N}}{} \overset{\cdot\cdot}{\underset{\cdot\cdot}{O}}\cdot \quad \longleftrightarrow \quad \cdot\overset{\cdot\cdot}{\underset{\cdot\cdot}{O}} = \overset{\overset{\cdot\cdot}{N}}{} \overset{\cdot\cdot}{\underset{\cdot\cdot}{O}}:^{-}$$

Both the drawings and the term _resonance_ can be misleading because they seem to imply that the molecule switches back and forth, or resonates between structures. The actual structure is a hybrid of the two resonance structures. The bonds in the nitrite anion can be thought of as containing (on average) 1.5 electrons.

Resonance structure:

Show resonance structures for CO_3^{2-}. How many electrons, on average, are in each bond of CO_3^{2-}?

solution:

Valence electrons = 4 + (3x6) + 2 = 24

Each bond can be thought of as containing 1.3 electrons.

Formal charge:

We use "formal charges" as a bookkeeping device to determine, for example, which resonance structures contribute the most to a molecule's electron configuration. An atom has a formal charge if the number of electrons it "controls" does not equal its number of valence electrons. We consider an atom to control its lone pair electrons (or any single, not shared, electrons), and to control one of the electrons in any shared electron-pair bond to other atoms. For example, consider the Lewis structure for ammonia:

$$H-\underset{\underset{H}{|}}{\overset{\cdot\cdot}{N}}-H$$

The nitrogen atom "controls" its lone pair electrons plus one electron from each of the three bonds to hydrogen for a total of five electrons. Nitrogen, in group 5A has five valence electrons and so has no formal charge in an ammonia molecule.

Now consider the Lewis structure for one resonance form of CO_3^{2-}:

$$\left[\underset{:\overset{\cdot\cdot}{O}:\quad\quad:\overset{\cdot\cdot}{O}:}{\overset{\overset{\cdot\cdot}{\overset{\cdot\cdot}{O}\cdot}}{\underset{\|}{C}}}\right]^{2-}$$

The carbon atom "controls" four electrons (one electron from each of the 4 bonds to oxygen). Carbon, in group 4A, has four valence electrons and so has no formal charge in CO_3^{2-}.

The doubly bonded oxygen atom controls two lone pairs, and one electron from each double bond for a total of six electrons. Since oxygen has six valence electrons, it has no formal charge.

The singly bonded oxygen atoms each control three lone pairs, and one electron from the single bond for a total of 7 electrons. Since oxygen has six valence electrons, each singly bonded oxygen has a formal charge of -1. Carbon, with four single bonds, has a formal charge of 0. The formal charges in a molecule must add up to the charge on the molecule (in this case, -2).

Formal charge practice problem:

1) Draw resonance forms for the molecule SCN⁻ and assign formal charges to each atom.

$$:\overset{\cdot\cdot}{S}=C=\overset{\cdot\cdot}{N}:^{-}\quad\longleftrightarrow\quad:\overset{\cdot\cdot}{\underset{\cdot\cdot}{S}}-C\equiv N:^{-}\quad\longleftrightarrow\quad:S\equiv C-\overset{\cdot\cdot}{\underset{\cdot\cdot}{N}}:^{-}$$

(a)   (b)   (c)

2) Given that:

- smaller formal charges are preferable to larger ones
- a more negative formal charge should reside on a more electronegative atom,

Which resonance form above is the most important contributor to the resonance hybrid of the thiocyanate ion?

<u>solution:</u>

Resonance structure "a" formal charges:
    S = 0; C = 0; N = -1

Resonance structure "b" formal charges:

$$S = -1; C = 0; N = 0$$

Resonance structure "c" formal charges:

$$S = +1; C = 0; N = -2$$

We eliminate resonance structure "c" because of the +2 formal charge on nitrogen, and the positive formal charge on sulfur. We pick "a" over "b" because nitrogen is more electronegative than sulfur, and therefore should be given the negative charge.

SECTION 9.6: Heats of Reaction

We saw in chapter 6 that we can use bond energy data and Hess's law to calculate heats of reaction, ΔH_{rxn}:

$$\Delta H^\circ_{rxn} = \Delta H^\circ_{reactant\ bonds\ broken} + \Delta H^\circ_{product\ bonds\ formed}$$

Heat of reaction problem:

For the following reaction, use bond energy data to obtain ΔH_{rxn}.

solution:

	ΔH_{sys} (kJ/mol bonds)	
One H-C bond is broken	414.2	ΔH_1
One Cl-Cl bond is broken	242.7	ΔH_2
One C-Cl bond is formed	-330.5	ΔH_3
One H-Cl bond is formed	-431.0	ΔH_4

Using Hess's law we write:

$$\Delta H_{rxn} = \Delta H_1 + \Delta H_2 + \Delta H_3 + \Delta H_4$$
$$\Delta H_{rxn} = 414.2 + 242.7 - 330.5 - 431.0 = -104.6\ kJ.$$

The reaction is exothermic because more energy is released forming the C-Cl and H-Cl bonds than is needed to break the H-C and Cl-Cl bonds. Note that this bond energy value probably would not match the calorimetric value exactly because bond energies are average values obtained from many different compounds. The bond energy in a particular substance would be close, but probably not equal to the average.

Chapter 9 Exercises:

9.1) Explain, in terms of the relative electronegativities of the participants, the difference between covalent, ionic, and metallic bonding.

9.2) a. Although molecular compounds are always covalent, some covalent substances are not described as molecular. Explain with examples.

b. Is it possible for a compound to contain both covalent and ionic bonds?

c. What is the difference between a metallic bond and an ionic bond?

9.3) In the Lewis electron-dot symbols for atoms of the following elements, how many electron dots are placed around the symbol for the element?

        a. germanium (Ge)
        b. silicon (Si)
        c. phosphorus (P)

9.4) Calculate the lattice energy of sodium chloride if:

the sublimation energy of sodium metal is 108 kJ/mol
the dissociation energy for chlorine is 242 kJ/mol of molecules
the first ionization energy for sodium is 495 kJ/mol
the electron affinity of a chlorine atoms is -348 kJ/mol
the standard enthalpy of formation of sodium chloride is -411 kJ/mol

9.5) Which salt has the highest lattice energy?

   RbCl   RbBr   KI $MgCl_2$ $MgBr_2$ CaI_2

9.6) a. In which one of the following diatomic molecules is the bond the strongest? In which molecule is the bond the longest?

      H_2 F_2 Cl_2 Br_2 I_2

b. Which carbon-carbon bond is the longest?

      C-C   C=C   C≡C

c. Which triple bond is the strongest?

      N≡N   C≡C   C≡N   C≡O

9.7) Calculate the enthalpy of formation of ammonia $\Delta H_f^\circ (NH_3)$ if
the bond energy of an N-H bond = 391 kJ/mol
the bond energy of an H-H bond = 436 kJ/mol
the bond energy of an N≡N bond = 945 kJ/mol

9.8) Classify the following compounds as covalent, polar covalent, ionic but with some covalent character, or ionic.

HF   KCl   MgS Br_2 CCl_4 $CaCl_2$ O_2 SO_2

9.9) Which of the following elements, arranged in the order in which they appear in the Periodic Table, is the most electronegative? Which is the least electronegative?

B   N   F
Al   P   Cl
Ga   As   Br

9.10) a. What is the formal charge on the nitrogen atom in the Lewis electron-dot structure for the nitrite ion NO_2^-?

b. What is the formal charge on the carbon atom in the Lewis electron-dot structure for the carbonate ion CO_3^{2-}?

c. What is the formal charge on the sulfur atom in the electron-dot structure for the molecule sulfur dioxide in which one of the bonds is written as a single bond and the other is written as a double bond?

9.11) For which of the following molecules or polyatomic ions is the drawing of resonance structures desirable in order to achieve a more realistic picture of the bonding?

N_2 SO_4^{2-} C_2H_2 CCl_4 CO_3^{2-} H_2O CH_4 NaCl NO_2^- BF_4^-

9.12) Which of the following twelve molecules do not obey the octet rule?

HF PF_3 OF_2 XeF_2 $CHCl_3$ PCl_5
CH_4 BCl_3 H_2S NH_3 CO_2 SF_6

Chapter 9 Answers:

9.1) Electronegativity is a measure of the tendency of an element in a compound to attract electrons to itself. It is therefore a predictor of how an element will behave when involved in bonding with other elements.

If two elements involved in a bond are both electronegative, they both tend to attract the bonding electrons and the compromise is to share them - a covalent bond.

If one element is electronegative and the other electropositive, the one that is electropositive tends to give electrons to the element that is electronegative. The result is the formation of ions and an ionic bond.

If both elements are electropositive, then neither wants the electrons. The result is a delocalization of the electrons from their respective atoms and the formation of a metallic bond.

9.2)   a. Substances that exist with their atoms bonded covalently in a three dimensional lattice are not molecules in the accepted sense and are often referred to as covalent networks. Examples are: diamond - C_n, silicon carbide $(SiC)_n$, quartz $(SiO_2)_n$, boron nitride $(BN)_n$.

b. Yes - polyatomic ions contain covalent bonds, but are involved in ionic bonding with other ions. An example is ammonium nitrate. This is an ionic compound containing ammonium ions and nitrate ions. However, both the ammonium ion NH_4^+ and the nitrate ion NO_3^- contain polar covalent bonds.

c. Metallic bonding occurs in a metallic lattice and involves the delocalization of valence electrons throughout the lattice. This global sharing of electrons holds the lattice together. Ionic bonding involves the transfer of electron(s) from one species to another and the formation of oppositely charged ions. Attraction between the opposite charges holds the ions together in the solid ionic lattice.

9.3) A Lewis symbol for an element indicates the number of electrons an atom of the element has in its valence shell: a. 4; b. 4; c. 5

9.4) Lattice energy = 787 kJ/mol NaCl

9.5) The lattice energy of a salt depends upon the charges on the ions, the sizes of the ions, and the crystal symmetry. The charge on the ion is usually the predominant factor. Since magnesium and calcium form ions with a 2+ charge, these salts will have a higher lattice energy than the salts of rubidium and potassium. Magnesium is smaller than calcium, and chlorine is smaller than bromine or iodine, so, of the salts listed, magnesium chloride will have the highest lattice energy.

9.6)   a. H_2 has the strongest bond; I_2 has the longest bond

b. The C-C single bond is the longest bond. As more pairs of electrons are shared, the closer the two nuclei are drawn together.

c. The difference in electronegativity contributes to the bond strength. A triple bond between C and O is the strongest.

9.7) $\Delta H_f^\circ(NH_3)$ = bonds broken - bonds made =

945 + 3 x 436 - 6 x 391 kJ = -93 kJ for 2 mol ammonia = -46 kJ/mol

9.8) There is a continuum in the character of bonding from pure covalent (equal sharing of electrons) to ionic (transfer of electrons). The description of a particular bond is based upon the difference in the electronegativities of the participants.
Note that there is no such thing as a pure ionic bond, some sharing always takes place. However, in cases where the electronegativity difference is large (e.g. NaCl, the bond is usually said to be ionic.

HF	polar covalent
KCl	ionic
MgS	ionic but with some covalent character
Br_2	covalent
CCl_4	polar covalent (although the molecule is nonpolar)
$CaCl_2$	ionic
O_2	covalent
SO_2	polar covalent

9.9) Fluorine is the most electronegative element in the Periodic Table. Of the elements listed above, you might have expected gallium to be the least electronegative.
In fact, compared to gallium, aluminum has a slightly lower electronegativity, a slightly lower first ionization energy, and a larger atomic radius-all contrary to the expected trends. This anomaly is due to the position of gallium immediately following the first transition series. Filling the first set of d orbitals leads to an increase in nuclear charge of 10+ without a compensating increase in the shielding of the nuclear charge. The 4s and 4p electrons are therefore held more tightly than might otherwise have been expected.

9.10) a. 0; b. 0
c. +1; the presence of this formal charge on the sulfur in this structure indicates that a better structure, minimizing the formal charge might be drawn.

9.11) Resonance structures are drawn when different arrangements of single and double bonds are possible. The actual structure, and the actual distribution of electrons in the molecule, is somewhere between the extremes represented by the resonance structures.

The sulfate ion, the carbonate ion, and the nitrite ion are all molecules for which different arrangements of bonds can be drawn. For example, the sulfate ion is usually drawn with two double S=O bonds and two S-O single bonds. These four bonds can be arranged in six different ways-six different resonance structures. In reality, the electrons are distributed so that each bond has a bond order of 1.5, and each bond is the same.

9.12) The octet rule is often not obeyed. Although terminal atoms in a molecule almost always have a complete set of 8 electrons in their valence shells (except H), elements other than C,N,O and F often disobey the rule when they are the central atom in a molecule or polyatomic ion.
XeF_2 has 10 electrons in the valence shell of Xe
PCl_5 has 10 electrons in the valence shell of P
BCl_3 has only 6 electrons in the valence shell of B
SF_6 has 12 electrons in the valence shell of S

CHAPTER 10

Shape of Molecules and Theories of Covalent Bonding

TOPICS:

- VSEPR and molecular shape
  - Ionic, covalent, and metallic
- Molecular polarity
- Valence bond theory
- Molecular orbital theory

SECTION 10.1: Valence-shell electron-pair repulsion (VSEPR) and molecular shape

Lewis structures give us a method for drawing molecules with their bonds and lone pairs, but they do not predict the geometry of a molecule. We can predict the shapes of molecules by drawing them in a way which <u>minimizes repulsions between groups of electrons while allowing them to approach the nucleus as close as possible</u>. (We define a "group" of electrons as a single, double, or triple bond, a lone pair, or a single electron).

Valence-shell electron-pair repulsion (VSEPR)

What geometry minimizes valence-shell electron-pair repulsions while allowing them to approach the nucleus as close as possible for the following compounds?

      a) CH_4 b) NH_3 c) H_2O d) HF

<u>solution</u>:

a) tetrahedral   b) trigonal pyramid   c) bent   d) linear

   (109.5°)   (107.3°)   (104.5°)

(The terms describing the geometry refer to the nuclei, <u>not</u> the electron clouds).

In each case, the bond angles are close to <u>tetrahedral</u>.

In the drawings, the shaded balls represent hydrogen atoms, and the lines connecting the balls to the central atom represents two-electron covalent bonds. The lines connecting hydrogen atoms are only there to emphasize the tetrahedral geometry of the electron clouds (<u>they do not represent electrons</u>). We can draw the molecules without these lines, and use "rabbit ears" for the lone pairs, to come up with the following structures:

The above example demonstrates several things:

1) The decrease in bond angles from methane to ammonia to water shows that <u>lone pairs take up more room than bonded electron pairs</u>. It may be useful to draw a lone pair as a large loop to remind yourself of the extra space needed for a lone pair.

2) A <u>tetrahedral</u> arrangement allows four electron pairs to be as close to the nucleus as possible.

3) We can talk about the geometry of the <u>electrons</u> (tetrahedral for all cases above), or the structure of the <u>molecules</u>. In order to predict the structure of a molecule, we need to know the arrangement of the electrons.

Trigonal planar vs. trigonal pyramid:

The atoms in an ammonia molecule form a trigonal pyramid because the lone pair on nitrogen takes up one of four tetrahedral sites. Imagine (hypothetically) removing the lone pair from nitrogen: the three hydrogen atoms would spread out into a trigonal planar arrangement to minimize electron repulsion. Which element has two fewer electrons than nitrogen? (Answer: boron).

3 groups around a central atom:

1) What is the geometry of a boron trichloride molecule(BCl_3)?

<u>solution</u>:

The number of valence electrons in BCl_3 = 3 + (3x7) = 24 (= 12 pairs)

:Ċl:

B

:Ċl: :Ċl:

Boron is electron deficient with six electrons. The geometry of boron trichloride is trigonal planar. The bond angles are 120°.

2) What is the geometry of a carbonate molecule (CO_3^{2-})?

solution:

The number of valence electrons in CO_3^{2-} = 4 + (3x6) + 2 = 24 (= 12 pairs)

$$\cdot\ddot{O}\cdot \quad \rceil^{2-}$$

C

:Ö: :Ö:

A double bond counts as one "domain". Three domains around carbon make its arrangement of atoms (and electron pairs) trigonal planar. The bond angles are 120°.

3) What is the geometry of a molecule of formaldehyde (CH_2O)?

solution:

The number of valence electrons in CH_2O = 4 + (2x1) + 6 = 12 (= 6 pairs)

$\cdot\ddot{O}\cdot$

C

H   H

The arrangement of atoms is trigonal planar, but the bond angles are distorted from 120° because of the different atoms around carbon.

Bent vs. linear:

Water is a bent molecule because its arrangement of electron pairs is tetrahedral: the two lone pairs on oxygen take up tetrahedral sites and the two bonds to hydrogen take up the other two tetrahedral sites. Imagine (hypothetically) removing the two lone pairs from the oxygen. The two hydrogen atoms would spread out into a linear arrangement to minimize electron repulsion. What element has four fewer electrons than oxygen? (Answer: Beryllium).

2 groups around a central atom:

1) What is the geometry of gaseous beryllium chloride ($BeCl_2$)?

solution:

The number of valence electrons in $BeCl_2$ is 2 + (2x7) = 16 (= 8 pairs)

$$:\overset{..}{\underset{..}{Cl}}-Be-\overset{..}{\underset{..}{Cl}}:$$

Beryllium is electron deficient with four electrons around it. The shape of $BeCl_2$ is linear.

2) What is the shape of carbon dioxide (CO_2)?

solution:

The number of valence electrons in CO_2 is 4 + (2x6) = 16 (=8 pairs)

$$:\overset{..}{O}=C=\overset{..}{O}:$$

Carbon dioxide is linear

3) What is the shape of ozone (O_3)?

solution:

O_3 has 6 x 3 = 18 valence electrons (= 9 pairs)

$$:\overset{..}{O}=\overset{..}{O}-\overset{..}{\underset{..}{O}}: \longleftrightarrow :\overset{..}{\underset{..}{O}}-\overset{..}{O}=\overset{..}{O}:$$

Ozone is a bent molecule.

Five groups around a central atom produce an interesting situation where not all the five sites are geometrically equivalent. Phosphorus pentachloride forms a **trigonal bipyramid** with three chlorine atoms in axial positions, and two chlorine atoms in equatorial positions:

5 groups around a central atom:

Given the bond angles in a trigonal bipyramid:

$$\begin{array}{c} Cl \\ 90° \quad \underset{|}{\overset{|}{}} \quad Cl \\ Cl - P \cdots\cdots\overset{}{\smile} - 120° \\ \blacktriangleleft \\ Cl \\ | \\ Cl \end{array}$$

1) Where would a lone pair exist to minimize electron-electron repulsions?

solution:

A lone pair takes an equatorial site which has only two neighbors 90° from it; An axial site has three neighbors 90° away.

2) What is the shape of sulfur tetrachloride (SCl_4)?

solution:

The number of valence electrons in SCl_4 is 6 + (4x7) = 34 (=17 pairs)

$$\begin{array}{c} Cl \\ | \\ \bigcirc S \cdots\cdots Cl \\ | \blacktriangleright Cl \\ Cl \end{array}$$

The lone pair on sulfur takes an equatorial site of the trigonal bipyramid; the geometry of the atoms is something like a seesaw. Lone pairs distort bond angles from the ideal 90° and 120°.

3) What is the shape of bromine trifluoride?

solution:

The number of valence electrons in BrF_3 is 7 + (3x7) = 28

$$\begin{array}{c} F \\ | \\ \bigcirc Br \, \bigcirc \\ | \blacktriangleright F \\ F \end{array}$$

The two lone pairs on bromine take two equatorial sites. The atoms form a distorted "T" shape.

4) What is the shape of the triiodide ion, I_3^-?

solution:

The number of valence electrons in I_3^- is $(3 \times 7) + 1 = 22$

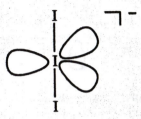

The three lone pairs all take equatorial positions. The ion is linear.

Six groups around a central atom form an octahedral arrangement. Don't get confused by the term octahedral for six groups: if we filled in the sides, there would be eight to form an octahedron. Notice that all the positions are equivalent. Once an electron pair takes a site, however, the second electron pair minimizes its interactions with the first electron pair by taking the site opposite (180°) from the first lone pair.

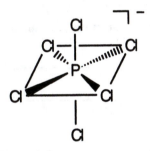

6 groups around a central atom:

1) What is the shape of sulfur hexafluoride (SF_6)?

solution:

The number of valence electrons in $SF_6 = 6 + (6 \times 7) = 48$ (=24 pairs, of which $3 \times 6 = 18$ are lone pairs on F, not shown below).

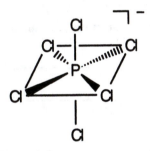

2) What is the shape of iodine pentafluoride (IF_5)?

The number of valence electrons in $IF_5 = 7 + (5 \times 7) = 42$

Iodine pentafluoride is a **square pyramid**.

3) What is the shape of xenon tetrafluoride (XeF_4)?

solution:

The number of valence electrons in $XeF_4 = 8 + (4 \times 7) = 36$

XeF_4 is **square planar**. The electron pairs orient opposite each other to minimize lone-pair/lone-pair repulsions.

Additional exercises:

1) Describe the environment of the carbon atoms in acetone (CH_3COCH_3):

solution:

• The number of valence electrons in acetone is: carbons: $(3 \times 4) = 12$; hydrogens: $(6 \times 1) = 6$; oxygens: 6; Total: 24 (= 12 pairs)

• The Lewis structure of acetone is:

• The structure of the carbon atoms bonded to 3 hydrogen atoms and the central carbon atom is tetrahedral.

• The structure of the central carbon atom with three groups (a double bond and two single bonds) is trigonal planar.

2) What would you predict the geometry of phosphine (PH_3) to be?

<u>solution</u>:

The number of valence electrons in PH_3 = 5 + (3x1) = 8.
The Lewis structure of phosphine is the same as ammonia:

We would predict that the structure of phosphine with four groups around it (three single bonds and a lone pair), like ammonia, would be <u>tetrahedral</u> and therefore would have bond angles close to 109°. [Remember the bond angles in ammonia (107°) are slightly less than 109° because the lone pair takes up more room than a bonded electron pair, and pushes the bonded atoms together]. However, unexpectedly, the bond angles in phosphine are 94°, almost right angles! In general, as atoms get larger, they contain a more "spread out" lone pair.

Summary of molecular geometry:

Number of electron groups (domains)	Arrangement of electron groups	Bond angle(s)
2	linear	180°
	$O=C=O$ $Cl-Be-Cl$ $H-C\equiv N$	
3	trigonal planar	120°

| 4 | tetrahedral | 109° |

| 5 | trigonal pyramid | 90° & 120° |

| 6 | octahedral | 90° |

The common cases of 4 electron pair arrangements:

2-domains: CO_2, HCN

$$O=C=O \qquad H-C\equiv N$$

3-domains: CO_3^{2-} (trigonal planar), O_3 (bent)

4-domains: CH_4, NH_3, H_2O, HF

The common cases of 5 electron pair arrangements:

3-domains: FSN, SO_2

4-domains: SOF_2

5-domains: PF_5, SF_4, ClF_3, XeF_2

The common cases of 6 electron pair arrangements:

3-domains: SO_3

4-domains: SO_2F, NSF_3

5-domains: SOF_4

6-domains: SF_6, ClF_5, XeF_4

0 Formal Charge Environments for atoms that satisfy the octet rule:

Determine the formal charge on each atom.

1) CARBON:

      a)   b)   c)   d)

2) NITROGEN:

      a)   b)   c)

3) OXYGEN:

      a)   b)

4) FLUORINE:

solutions:

All atoms are in 0 formal charge environments. Can you think of specific molecules that would fit each electron arrangements?

      Some answers: 1. a) CH_4, b) CO_2, c) H_2CO, d) HCN

                     2. a) NH_3, b)$CH_3NHNHCH_3$, c) N_2

                     3. a) H_2O, b) O_2

                     4. a) HF

SECTION 10.2: bond polarity

The geometry of some molecules causes the molecules to be polar. We saw in chapter 9 that atoms of different electronegativity share electrons unequally. The polar covalent bond in H–F causes that molecule to have a **dipole moment:**

$$H\text{---}F$$
$$\longmapsto$$

The polar bonds in a linear molecule such as BeF_2 cancel each other out so the molecule has no net dipole:

$$F\text{---}Be\text{---}F$$
$$\longleftarrow\!\!\!+\;\;+\!\!\!\longrightarrow$$

If water were a linear molecule, it would have no net dipole. However, the bent geometry of water causes it to have polarity, regions of positive and negative charge:

Polar molecules have stronger intermolecular forces than nonpolar molecules and have higher melting and boiling points as a result.

Polar molecules:

Which of the following molecules would you expect to be polar?

a) CO_2 c) SCl_4
b) $SiCl_4$ d) XeF_4

solutions:

a) The structure of CO_2 is: $O\!\!=\!\!C\!\!=\!\!O$

Since it is a linear molecule, its polar bonds cancel each other out, and there is no net dipole.

b) The structure of $SiCl_4$ is tetrahedral:

There is no net dipole.

c) The structure of SCl_4 is a seesaw:

This molecule has a dipole moment.

d) The structure of XeF₄ is square planar:

The symmetry of this molecule causes it to have no net dipole.

SECTION 10.3: Valence bond theory and orbital hybridization

We have used valence shell electron pair repulsion theory to help us predict the shape of molecules. Now we will look at another model that describes the <u>bonds</u> in molecules as the overlap of a half-filled atomic orbital of one atom with a half-filled atomic orbital from another atom. The space formed by the overlapping orbitals can hold two electrons with opposite spins.

Valence bonds in simple molecules:

1) Consider a hydrogen molecule, H₂. Describe the orbitals which overlap to form its bond.

<u>solution</u>:

First we will write the electron configuration for the two hydrogen atoms:

H: ↿   H: ↿
        1s   1s

The two atomic orbitals overlap to form a bond:

H ( ) H

This bond, which is cylindrically symmetric, is called a sigma (σ) bond. Its cross section resembles an s orbital.

P orbitals can form two types of bonds. Two p orbitals can overlap end-to-end to form a sigma (σ) bond, which has its highest electron density along the bond axis:

p orbitals:
sigma bond

P orbitals can also overlap side-to-side to form a "pi" (π) bond. In cross section the pi orbital resemble a p orbital.

p orbitals:
pi bond

A double bond in this picture contains one σ bond and one π bond. For example, a carbon-carbon double bond (represented below) consists of a σ bond, in which the shared electrons exist between the atoms, and a π bond, in which the shared electrons spend time above and below the σ bond. The π bond forms from p orbitals on the two carbon atoms. The σ bond forms from hybridized "sp^2" orbitals on the carbon atoms (discussed later).

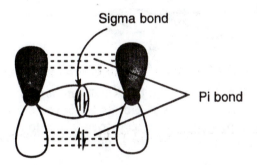

A carbon-carbon double bond
consisting of a sigma bond
and a pi bond.

A triple bond contains one σ bond and two π bonds.

Valence bonds in more complex molecules (hybridization):

We saw that VSEPR theory predicts that methane, CH_4, has a tetrahedral geometry. Lets consider the orbitals which overlap to form the bonds in methane (CH_4).

The electron configurations for the carbon and hydrogen atoms are:

C: ⇅ ⇅ ↑ ↑ __
   1s 2s 2p

H's: ↑
     1s

It is not obvious how to make four bonds to carbon since there are only <u>two</u> half filled orbitals. If we excite one electron into the empty 2p orbital from the 2s orbital, we would have four half filled orbitals:

⇅ ⇅ ↑ ↑ __ ⟶ ⇅ ↑ ↑ ↑ ↑
1s 2s 2p   1s 2s 2p

What would you expect the bond angles to be of bonds formed between hydrogen's 1s orbital, and the 2p orbitals from carbon? (Answer: Since the carbon 2p orbitals are perpendicular to one another, the C-H bonds formed with these orbitals would be oriented at 90° angles from one another):

— hydrogen 1s orbitals

In the drawing, the black ball represents a carbon core with 2 of its 2p orbitals. The hydrogen 1s orbitals overlap to form bonds oriented at 90° angles.

However, we know from experiment that a methane molecule is tetrahedral with bond angles of 109.5°.

Linus Pauling works us around this problem with his proposal that <u>orbitals in a molecule are different from the orbitals in the isolated atoms</u>. We can combine the 2s orbital and the three 2p orbitals mathematically to give new **hybrid** atomic orbitals which have 109.5° bond angles between them. We call these hybrid orbitals **sp³**.

⇅ ⇅ ↑ ↑ __ ⟶ ⇅ | ↑ ↑ ↑ ↑ |
1s 2s 2p   1s | sp³ |

four equivalent hybrid sp³ orbitals

Now four hydrogen atoms can combine with the four <u>equivalent</u> half-filled sp³ atomic orbitals on carbon to form a tetrahedral methane molecule:

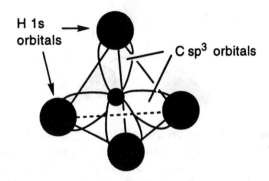

Valence bond theory and hybridization:

1) How would the valence bond theory describe the atomic orbitals of nitrogen in ammonia (NH₃) and the atomic orbitals of oxygen in water (H₂O)?

    <u>solution</u>:

    Since the bond angles in ammonia and water are close to tetrahedral, we use sp³ hybrids to describe the bonding in these molecules. The lone pair on nitrogen in ammonia, and the two lone pairs on oxygen in water reside in nonbonding sp³ orbitals.

2) Describe the orbitals which overlap to form the bonds in boron trifluoride (BF₃).

    <u>solution</u>:

    We know from VSEPR theory that boron trifluoride is trigonal planar with 120° bond angles. To make three equivalent orbitals around boron, we mix the 2s orbital and two of the 2p orbitals to obtain three hybrid **sp²** orbitals with 120° bond angles.

    A half filled 3p³ orbital from fluorine overlaps with a half filled hybrid sp² orbital on boron to form a sigma bond:

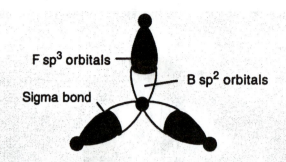

F sp³ orbitals —
B sp² orbitals
Sigma bond

Although it is the hybrid orbitals on the <u>central atom</u> (boron) which describe the 120° bond angles of a trigonal planar molecule, we assume that minimizing electron pair repulsion is also important for peripheral atoms. In order for the four groups around each fluorine (a single bond and three lone pairs) to minimize electron repulsions, the fluorine atoms must be sp^3 hybridized.

Notice that two of boron's p orbitals are hybridized, but the third p orbital remains unhybridized and lies perpendicular to the plane made by the hybridized bonds. Interaction between this empty p orbital and a lone pair of electrons on fluorine cause the B-F bond distance to be shorter than expected for a purely single B-F bond.

3) Describe the bonding in carbon dioxide (CO_2).

<u>solution</u>:

Carbon dioxide is a linear molecule containing double bonds between carbon and the oxygen atoms.

$$O = C = O$$

Carbon needs a pair of atomic orbitals oriented in opposite directions. This requires a new type of hybridization which mixes one s orbital and one p orbital to form 2 **sp** hybrid orbitals oriented 180° from each other.

C:

1s   2s   2p   1s   sp   2p

Oxygen has three groups around it (a double bond and two lone pairs) and so requires sp^2 hybridization to obtain a trigonal planar structure.

O:

1s   2s   2p   1s   sp²   2p

The sigma bonds in carbon dioxide form from overlap of the carbon half filled sp hybrid orbitals with the half filled oxygen sp^2 orbital on each oxygen (see drawing below).

The pi bonds form from the side-to-side overlap of the electrons in the unhybridized p orbitals from carbon and oxygen:

sp hybridized orbitals
for a carbon atom

sp^2 hybridized orbitals
for an oxygen atom

4) The shapes of molecules with expanded octets have atoms with atomic d orbitals available to mix with s and p orbitals. A trigonal bipyramid shape requires hybrid orbitals mixed from a "s", three "p", and one "d" orbital. An octahedral shape requires hybrid orbitals mixed from one "s", three "p", and two "d" orbitals. What would be the name for these two hybrids?

Solution:

Trigonal bipyramid: five dsp^3 hybrid orbitals

Octahedral: six d^{2}sp^3 hybrid orbitals

Summary of Hybrid orbitals:

Hybrid orbital	number of groups	shape	bond angles	example
sp	2	linear	180°	O=C=O
sp^2	3	trigonal planar	120°	SO$_3$
sp^3	4	tetrahedral	109.5°	CH$_4$

dsp^3	5	trigonal pyramid	120° and 90°	PCl_5
d^2sp^3	6	octahedral	90°	SF_6

SECTION 10.4: Molecular orbital (MO) Theory and Electron delocalization

In valence bond theory, an <u>atom</u> forms hybrid orbitals from combinations of its atomic orbitals. These orbitals overlap with orbitals from other atoms to form bonds containing localized electrons. In MO theory, as two atoms come together to form a molecule, the <u>molecule</u> forms new orbitals which are delocalized over the entire molecule. Each molecular orbital can be occupied by two electrons and has a particular energy (and shape). We can write electron configurations for the molecule just as we write electron configurations for atoms. MO theory is useful for explaining the magnetic and spectral properties of molecules, but it is more difficult to picture molecular orbitals than the hybrid orbitals of valence bond theory, which are sometimes called Localized Molecular Orbitals (LMO).

The MO description of the bond in a hydrogen molecule:

When two hydrogen atoms bond, their 1s orbitals combine to form two molecular orbitals:

The σ1s orbital (the bonding orbital) is lower in energy than the atomic orbitals from which it is composed; the σ*1s orbital (the antibonding orbital) is higher in energy than the original atomic orbitals. The bonding MO is spread mostly between the nuclei of the hydrogen atoms; the antibonding MO has a node between the nuclei and when it is occupied, the molecule is less stable than the separate atoms. Atomic orbitals must have similar energy and orientation to overlap effectively.

We fill molecular orbitals with electrons the same way we fill atomic orbitals:

- Orbitals fill in order of increasing energy
- An orbital can hold a maximum of two electrons with opposite spins
- Electrons do not pair up until all orbitals of equal energy contain one electron.

Molecular Orbitals:

Choose words from the list below to complete the following paragraphs:

a) antibonding	e) energy
b) atomic	f) higher
c) between	g) lower
d) beyond	h) valence

When a molecule forms, atomic orbitals of similar _____(1), and hence of similar size, and the right symmetry overlap to form molecular orbitals. Inner atomic orbitals are too low in energy (and therefore too small) to overlap sufficiently to form molecular orbitals; it is the _____(2) atomic orbitals which combine to form molecular orbitals. Some combinations of atomic orbitals form bonding molecular orbitals, orbitals that are _____(3) in energy than the separate atomic orbitals. Some combinations of atomic orbitals form antibonding molecular orbitals, orbitals that are _____(4) in energy than the separate atomic orbitals. Every bonding molecular orbital has a corresponding _____(5) orbital. Electrons in bonding orbitals have a high probability of being found _____(6) the nuclei; electrons in antibonding orbitals have a high probability of being found _____(7) the nuclei. The number of molecular orbitals formed always equals the number of _____(8) orbitals combined to form them.

<u>answers</u>: 1-e, 2-h, 3-g, 4-f, 5-a, 6-c, 7-d, 8-b.

Hydrogen's MO's:

Consider the drawing at the beginning of this section for the molecular orbitals of a hydrogen molecule.

1) What is the electron configuration for the H_2 molecule?

<u>solution</u>: $(\sigma 1s)^2$

The σ indicates the orbital's symmetry. The subscript "1s" indicates the atomic orbitals used to form the molecular orbital. (An asterisk (*) indicates an antibonding orbital).

2) What is the bond order for this molecule?

solution:

bond order = 1/2[(# of e⁻ in bonding MO's) - (# of e⁻ in antibonding MO's)]

$$= 1/2 \ (2) = 1$$

Note: A bond order > 0 implies the molecule is more stable than the separate atoms. In general, the higher the bond order, the stronger the bond.

3) Would you predict that H_2^+ would exist? H_2^-? H_2^{2-}? What are their electron configurations?

solution:

• If we remove an electron from the MO diagram of H_2 to form H_2^+, the bond order is:

1/2 (1) = 1/2;   electron configuration = $(\sigma 1s)^1$

• If we add an electron to the MO diagram of H_2 to form H_2^-, the bond order is:

1/2 (2-1) = 1/2;   electron configuration = $(\sigma 1s)^2 (\sigma^* 1s)^1$

• If we add two electrons to the MO diagram of H_2 to form H_2^{2-}, the bond order is:

1/2 (2-2) = 0;   electron configuration = $(\sigma 1s)^2 (\sigma^* 1s)^2$

H_2^+ and H_2^- have bond orders > 0, so they should exist (in fact, they have both been detected spectroscopically). The bond order of H_2^{2-} = 0, so this molecule is predicted not to exist.

The MO description of the bonding in an oxygen molecule:

An oxygen atom contains valence electrons in its 2s and 2p orbitals. When two oxygen atoms bond, their 2s orbitals overlap to form σ bonding and antibonding orbitals. One 2p orbital from each oxygen atom overlaps end-to-end to form σ bonding and antibonding orbitals; the other two 2p orbitals from each oxygen atom overlap side-to-side to form two sets of π bonding and antibonding orbitals. We fill in the orbitals with the valence electrons from the oxygen atoms using the filling rules. There are 12 valence electrons (6 from each oxygen atom).

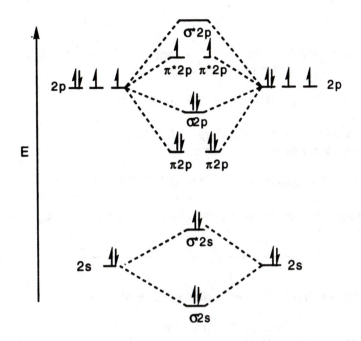

Oxygen's MO's:

Consider the drawing above for the molecular orbitals of an oxygen molecule.

1) What is the electron configuration for the O_2 molecule?

   <u>solution</u>:

   $$(\sigma 2s)^2(\sigma 2s^*)^2(\pi 2p)^4(\sigma 2p)^2(\sigma 2p^*)^2$$

2) What is the bond order for this molecule?

   <u>solution</u>:

   bond order = 1/2[(# of e⁻ in bonding MO's) - (# of e⁻ in antibonding MO's)]

   = 1/2 (8-4) = 2

   Note: A bond order of 2 implies a bond which is approximately twice as strong as a bond order of 1 (a double bond).

3) Is oxygen paramagnetic or diamagnetic?

   <u>solution</u>:

   Oxygen contains two unpaired electrons, so it is paramagnetic. One of the strengths of MO theory is its ability to explain the paramagnetism in molecules such as oxygen.

4) Would O_2^+ have a stronger or weaker bond than O_2? What about O_2^-?

- If we remove an electron from the MO diagram, we remove it from an antibonding orbital, so the bond order becomes larger:

$$\text{bond order} = 1/2 \ (8-3) = 2.5.$$

- If we add an electron to the MO diagram for O_2^-, we add it to an antibonding orbital, and the bond order becomes smaller:

$$\text{bond order} = 1/2 \ (8-5) = 1.5.$$

We would predict O_2^+ would have a stronger bond than O_2, and O_2^- would have a weaker bond than O_2.

The bonding in heteronuclear diatomic molecules:

In polar covalent molecules, bonding MOs are closer in energy to the atomic orbitals of the more electronegative atoms. When two atoms of a diatomic molecule are very different, we can not use the energy level diagram for homonuclear molecules. We devise a new diagram for each molecule.

The main strengths of the molecular orbital model are that it predicts relative bond strengths and magnetism of simple diatomic molecules. It is difficult to apply qualitatively to polyatomic molecules.

In practice, we often combine the localized electron and molecular orbital models. Lets consider the molecule benzene (C_6H_6), which requires resonance structures since all C-C bonds are known to be equivalent:

We consider the sigma bonds between carbon atoms to be localized electrons in sp^2 hybrid orbitals giving the benzene ring its planar structure (valence bond theory). The remaining p orbital on each carbon atom is perpendicular to the plane of the benzene ring and contain the remaining electrons. These 6 p orbitals can combine to form π molecular orbitals (MO theory). The electrons in these π molecular orbitals are delocalized above and below the ring. You will often see the structure of benzene written with a circle inside the carbon six-membered ring to indicate its delocalized π bonding.

Chapter 10 Exercises:

10.1) How many nonbonding pairs of electrons do the following compounds have in valence orbitals on the central atom?

NH_3 H_2O BF_3 $BeCl_2$ HCN

H_2S CH_4 OF_2 PCl_3 $CHCl_3$

10.2) Which of the following species have a trigonal bipyramidal arrangement of electron pairs around the central atom?

PCl_5 SF_4 BrF_3 SOF_4 I_3^-

TeF_5^- ICl_4^- IOF_5 BrF_5 NH_3

10.3) What is the arrangement of electron pairs, bonding and nonbonding, around the central atoms, and what are the shapes of these molecules or ions?

ICl_4^- SOF_2 NO_3^-

10.4) With which one of the five VSEPR electron pair arrangements are the "square-planar" and "square-pyramidal" molecular geometries associated?

10.5) Which two of the following molecules have a dipole moment?

CH_4 XeF_4 SO_3 SF_6

H_2S PCl_3 BF_3 XeF_2

10.6) a. If one 2s orbital and two 2p orbitals combine to form a set of hybrid orbitals, which set is formed?

b. If a set of hybrid orbitals is formed from one d orbital, one s orbital, and two p orbitals from the same principal quantum level, how many orbitals would be created and what would their designation be?

10.7) The number of hybrid orbitals within a set around the central atom in a molecule always equals:

a. the number of electrons that need to be accommodated around the atom

b. the number of sigma bonds that the atom will form

c. the number of atomic orbitals used to make the set

d. the number of π bonds that need to be catered for

e. the total number of sigma and pi bonds that will be formed by the atom

10.8) What orbital hybridization is necessary to describe the molecular geometry of

a. XeF_2 b. BrF_5 c. PCl_5

10.9) When the valence orbitals of carbon are sp^2 hybridized, in what type of bonds will the carbon atom participate?

10.10) If the internuclear axis between two bonded atoms is the z-axis, which orbitals on the two atoms could be used to form a π bond between the atoms?

10.11) What are the bond orders in the diatomic ions OF^+, F_2^+, and N_2^{2-}?

10.12) a. How many orbitals are involved in the delocalized π system of the carbonate ion?

b. What is the advantage of considering the delocalization of π electrons in molecules and ions like the carbonate ion?

Chapter 10 Answers:

10.1) Use the VSEPR approach to determine the number of bonding and nonbonding pairs of electrons around the central atom in each molecule:

NH_3	one lone pair on the N	H_2S	two lone pairs on the S
H_2O	two lone pairs on the O	CH_4	no lone pairs on the C
BF_3	no lone pairs on the B	OF_2	two lone pairs on the O
$BeCl_2$	no lone pairs on the Be	PCl_3	one lone pair on the P
HCN	no lone pairs on the C	$CHCl_3$	no lone pairs on the C

10.2) Do not be misled by the number of terminal atoms in the molecule or ion. Use the VSEPR approach to determine the total number of bonding *and* nonbonding pairs of electrons around the central atom. It is the *total* number of electron pairs that ultimately determines the geometry of the molecule or polyatomic ion.

PCl_5	trigonal bipyramid	TeF_5^-	octahedral
SF_4	trigonal bipyramid	ICl_4^-	octahedral
BrF_3	trigonal bipyramid	IOF_5	octahedral
SOF_4	trigonal bipyramid	BrF_5	octahedral
I_3^-	trigonal bipyramid	NH_3	tetrahedral

10.3)

ICl_4^-	octahedral arrangement	two lone pairs	square planar molecule
SOF_2	tetrahedral arrangement	one lone pair	trigonal pyramidal molecule
NO_3^-	trigonal arrangement	no lone pairs	trigonal planar molecule

10.4) Both the square-planar and the square-pyramidal geometries arise from an octahedral arrangement of electron pairs around the central atom. In the square-planar geometry there are two lone pairs opposite on another on two sides of the plane. In the square-pyramidal geometry there is one lone pair.

10.5)

CH_4	tetrahedral	nonpolar
SO_3	trigonal planar	nonpolar
BF_3	trigonal planar	nonpolar
H_2S	V-shaped	*polar* (dipole moment)
XeF_4	square-planar	nonpolar
SF_6	octahedral	nonpolar
PCl_3	trigonal pyramid	*polar* (dipole moment)
XeF_2	linear	nonpolar

10.6) The number of hybrid orbitals in a set equals the number of atomic orbitals used to create the set. The designation incorporates letters corresponding to the orbitals used, with superscripts denoting the number of each orbital used.

     a. Three sp^2 hybrid orbitals
     b. Four sp^2d hybrid orbitals

10.7) Hybridization of orbitals in valence bond theory is a procedure usually reserved for the orbitals involved in sigma bonding and the orbitals containing lone pairs of electrons around the central atom (b is incorrect). In other words, hybrid orbitals contain sigma bonding or nonbonding electrons. There is no relation between the hybridization and the number of electrons that need to be accommodated-some electrons might be π bonding (a is incorrect). It is the number of sigma bonds and lone pairs that determines the number of hybrid orbitals required (d and e are incorrect). The correct answer is c. The number of hybrid orbitals in a set equals the number of atomic orbitals used to generate the set.

10.8)   a. XeF_2 has a trigonal bipyramidal arrangement of electron pairs around the xenon; it is a linear molecule. The hybridization is that required for a trigonal bipyramidal arrangement, sp^3d.

b. BrF_5 has an octahedral arrangement of electron pairs around the bromine; it is a square-pyramidal molecule. The hybridization is that required for an octahedral arrangement, sp^3d^2.

c. PCl_5 has a trigonal bipyramidal arrangement of electron pairs around the phosphorus; it is a trigonal bipyramidal molecule; there are no lone pairs. The hybridization is that required for a trigonal bipyramidal arrangement, sp^3d.

10.9)   Carbon is in Group 4, it has four valence electrons and it participates in four bonds. The sp^2 hybrid orbitals participate in three sigma bonds; the remaining p orbital is used for one π bond.

10.10) For a π bond, the orientation is correct if the two orbitals coincide with an overlapping p orbital shape when viewed down the internuclear axis (the z axis in this case). For example, an s orbital can never participate in π bond formation because there is no orientation of an s orbital that gives it a double-lobed p orbital shape.

Possible combinations are:

p_x and p_x, p_y and p_y, d_{xz} and p_x, d_{yz} and p_y, d_{xz} and d_{xz}
(not p_x and p_y and not p_z and p_z, which would lead to a sigma bond)

10.11) Use the molecular orbital energy level diagram for a diatomic molecule. Fill up the energy levels according to the Aufbau Principle, Hund's Rule of Maximum Spin Multiplicity, and the Pauli Exclusion Principle. The bond order is the number of bonding electrons minus the number of antibonding electrons divided by two. It is sometimes useful to consider isoelectronic species. For example, OF^+ is isoelectronic with O_2 and has a bond order of 2.0. F_2^+ has one more antibonding electron and therefore must have a bond order of 1.5 (0.5 less), and N_2^{2-} is isoelectronic with O_2 and must also have a bond order of 2.0.

10.12) a. The delocalized π system of the carbonate ion, CO_3^{2-}, involves four p_π orbitals, each perpendicular to the plane of the molecule and one on each of the four atoms. Since there are four atomic orbitals involved, four molecular orbitals result. The pair of π bonding electrons is in the lowest molecular orbital. This is an orbital that extends over all four atoms and electrons in this orbital are therefore delocalized over the entire molecule.

b. Considering the delocalization of the π electrons eliminates the need to draw the resonance structures required in valence bond theory.

TOPICS:

- Physical states and phase changes
- Intermolecular forces
- The liquid state
- The solid state
    - metallic solids
    - network covalent solids
    - ionic solids
    - molecular solids
- Changes in state
- Vapor pressure

INTRODUCTION:

In the last two chapters we saw how chemical bonds form from electrostatic attractions between particles.

Chemical bonds:

Determine if the following statements describe 1) covalent, 2) ionic, or 3) metallic bonds.

a) electrostatic attraction between anions and cations

b) electrostatic attraction between cations and delocalized valence electrons

c) electrostatic attraction between positive nuclei and electron pairs

d) non-conductor

e) conducts in the molten state by motion of ions

f) conducts in the solid state by electron flow

g) no large cores

h) both small and large cores

i) no small cores

<u>answers</u>: 1 (covalent)-c, d, g; 2 (ionic)-a, e, h; 3 (metallic)-b, f, i.

The strength of bonds between atoms in a molecule determines chemical reactivity since bonds break and form during chemical reactions. In this chapter, we discuss the electrostatic attractions <u>between</u> molecules (intermolecular attractions) which determine physical properties of substances.

Intermolecular attractions:

Which electrostatic attraction must be overcome (a or b in the drawing below) in order for a water molecule to pass from a liquid state into the vapor state?

solution:

When water evaporates, no hydrogen and oxygen covalent bonds (a) break; the water molecules must overcome <u>intermolecular</u> attractions (b) to one another. These intermolecular attractions are much smaller (usually 5% or less) than the attractions which form a typical covalent bond because there is more distance between the charged particles, and the charges are generally smaller.

SECTION 11.1: Physical states and phase changes

We saw in Chapter 1 (and again in Chapter 5), that, broadly speaking, there are three ways which we classify matter:

Touching & ordered (solid)	Touching & disordered (liquid)	Not touching (gas)

States of matter:

Use the above drawings as a guide to classify the following as characteristics of solids, liquids or gases:

   a) molecules are far apart
   b) molecules touch and are fixed in position
   c) molecules touch but have no long range order
   d) essentially no interparticle interactions
   e) interparticle forces are strong
   f) condensed states of matter
   g) easily compressed

<u>answers</u>: a-gas, b-solid, c-liquid (could also be amorphous solids), d-gas, e-solid & liquid, f- solid & liquid, g-gas

Particles in a substance change their distance from each another and their freedom of motion when the substance changes its state of matter. It always takes more energy to completely separate particles (vaporize a liquid) than to overcome the forces holding the molecules in their fixed positions (melt a solid).

Phase changes:

Match the words with the phase change it describes and indicate whether it is an exothermic or endothermic process:

1) condensation	a) solid → liquid
2) deposition	b) solid → gas
3) freezing	c) liquid → gas
4) melting	d) liquid → solid
5) sublimation	e) gas → liquid
6) vaporization	f) gas → solid

answers: 1-e, 2-f, 3-d, 4-a, 5-b, 6-c.
1,2,3 - exothermic; 4,5,6 - endothermic

Heat of deposition:

What is the heat of deposition for one mole of water given:

$$H_2O\ (g) \rightarrow H_2O\ (l) \qquad \Delta H = -\Delta H^{o}_{vap} = -40.7 \text{ kJ/mol } (100°C)$$

$$H_2O\ (l) \rightarrow H_2O\ (s) \qquad \Delta H = -\Delta H^{o}_{fus} = -6.02 \text{ kJ/mol } (0°C)$$

solution:

If we add the above two equations we obtain:

$$H_2O\ (g) \rightarrow H_2O\ (s) \qquad \Delta H = \text{heat of deposition}$$

$$\Delta H^{o}_{dep} = -\Delta H^{o}_{vap} - \Delta H^{o}_{fus}$$

$$\Delta H^{o}_{dep} = -40.7 \text{ kJ/mol} - 6.02 \text{ kJ/mol} = -46.7 \text{ kJ/mol}$$

Ice crystals form on a cold window from the deposition of water vapor.

SECTION 11.2: Intermolecular forces

London dispersion forces

Bromine is a liquid at room temperature and ordinary pressure. What causes the the nonpolar bromine molecules to attract one another? The boiling point of bromine is fairly low (58.8°), so the attractive forces must not be all that strong. **London dispersion forces** are weak intermolecular attractions between nonpolar molecules caused by momentary oscillations of electron density which create transient dipoles. These dipoles <u>induce</u> dipoles in neighboring molecules, and the induced dipoles attract each other. Dispersion forces exist in all molecules, but are overpowered by stronger dipole-dipole interactions in polar molecules. Dispersion forces generally increase with molar mass because electrons farther from the nucleus are easier to distort (larger atoms are more **polarizable**).

Dipole-dipole attractions

Dipole-dipole attractions exist when polar molecules (molecules with dipole moments) line up so their partial positive and negative charges attract each other. In a collection of polar molecules, the dipoles arrange themselves to maximize the attractions between molecules.

The **hydrogen bond** is a special type of particularly strong dipole-dipole attraction, and it stands out as the most important intermolecular force. Hydrogen bonds occur between molecules that have a hydrogen atom bound to a small, highly electronegative atom with lone pairs (such as oxygen, nitrogen, or fluorine). The hydrogen bond is particularly strong due to the great polarity of the bond, and the small size of hydrogen which allows close approach of the dipoles. Water has a much higher boiling point than expected from trends for hydrides due to hydrogen bonding.

Intermolecular forces:

Rationalize the difference in boiling points for each of the following pairs of substances:

a) Cl_2 (-34.6°C) and Br_2 (58.8°C)

b) H_2O (100°C)and H_2S (-60.7°C)

c) pentane (36.1°C) and cyclopentane (49.2°C)

d) HF (20°C) and HCl (-85°C)

e) CH_3OCH_3 (35°C) and CH_3CH_2OH (79°C)

<u>solutions</u>:

a) Both chlorine and bromine are nonpolar molecules with weak dispersion intermolecular forces. At room temperature, chlorine is a gas and bromine is a dark red liquid due to its higher molar mass and greater number of electrons.

b) Hydrogen sulfide has a higher molar mass than does water, but hydrogen bonding in water molecules makes it have a much higher boiling point. Sulfur is not small enough (or electronegative enough) to allow hydrogen bonding.

c) The molar mass of pentane is 72, and the molar mass of cyclopentane is 70 g/mole. They are both nonpolar molecules. Cyclopentane has the higher boiling point because its disc shape makes more intermolecular contact causing it to favor the liquid state.

d) Both HCl and HF are polar molecules with dipole-dipole interactions. The higher boiling point of HF is due to hydrogen bonding.

e) The higher boiling point of ethanol over diethyl ether is due to hydrogen bonding between the hydrogen from an O-H group on one molecule and the oxygen on another molecule.

Hydrogen bonding:

Hydrogen-bonding is the most important intermolecular force.

1) Give explanations for the boiling points of the two sets of compounds:

A		B	
CH_4	-164°C	CH_3CH_3	-88.6
NH_3	-33.35°C	CH_3NH_2	-6.3
H_2O	100°C	CH_3OH	65
HF	19.54	CH_3F	-78.4

answer:

Hydrogen bonding in ammonia, water, and hydrogen fluoride cause these compounds to have higher boiling points than methane which has no hydrogen bonding.

Methyl amine (CH_3NH_2), and methanol (CH_3OH) have higher boiling points than ethane (CH_3CH_3) and fluoromethane (CH_3F) because of hydrogen bonding. (Fluoromethane has no hydrogen bonding because there are no H-F bonds).

2) Acetic acid (CH_3CO_2H, structure below) dimerizes in non-polar solvents.

What would you predict the structure of its dimer to be?

<u>solution</u>:

Hydrogen bonding causes the oxygen atom on one molecule to hydrogen bond with the hydrogen atom of the OH bond on another molecule.

$$H_3C-C\cdots\ O-H\cdots O=C-CH_3$$

3) What organic compounds are soluble in water?

<u>answer</u>:

In general, organic compounds that contain a group capable of hydrogen bonding with water are soluble in water.

4) What intermolecular force is responsible for the pairing of bases in DNA?

<u>answer</u>:

The hydrogen bond.

SECTION 11.3: The liquid state

There are three properties of liquids that we can explain by their intermolecular forces: surface tension, capillarity, and viscosity.

Surface tension

Surface tension is the energy required to increase the surface area by a given amount. The stronger the intermolecular forces, the stronger the surface tension. Molecules in the interior of the liquid are attracted by molecules in all directions, while the molecules at the surface are attracted only by molecules below it and on each side, pulling it into the interior of the liquid.

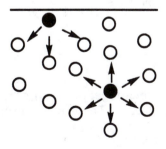

Capillarity

The ability of a liquid to rise in a tube is due to the competition between internal forces within a liquid (cohesive forces) and those between the liquid and the wall of a container (adhesive forces).

Exercise:

Which example shows a liquid with stronger cohesive forces than adhesive forces ?

(A)   (B)

solution:

Liquid "a" has stronger attractions to itself (cohesive forces) than to the glass walls (adhesive forces) causing the liquid to form a convex meniscus. Liquid "b" has stronger attractions to the glass walls (adhesive forces) than to itself (cohesive forces) causing the liquid to move up the tube and form a concave meniscus. Glass is a polar substance. Liquid "a" is a nonpolar liquid such as mercury. Liquid "b" is a polar substance such as water.

Viscosity

Both gases and liquids exhibit viscosity, the resistance to flow, which depends both on the strength of intermolecular interactions, and the shape of the molecule. Viscosity decreases with an increase in temperature. We warm up chocolate fudge sauce to make it become less viscous and flow more easily. Large molecules which can become entangled make liquids more viscous than smaller, less tangled molecules.

SECTION 11.4:   The Solid State

In **crystalline** solids, atoms pack together in an orderly fashion. The smallest repeating unit of the lattice is its **unit cell**. Three common unit cells are shown below:

| Simple cubic | Body-centered cubic | Face-centered cubic |

If the unit cell is repeated in three directions, it forms a crystal lattice. Crystal structures can be determined by **X-ray diffraction**, a technique which allows us to reconstruct the structure of a crystalline solid by analyzing the diffraction patterns made by x-rays scattering off the atoms in the crystal.

The balls in the above drawings can represent ions, atoms, or molecules.

• When balls = ions, we have an <u>ionic solid</u> such as sodium chloride, NaCl.

• When balls = molecules we have a <u>molecular solid</u> such as ice.

• When balls = atoms, there are three types of solids possible:

    - a <u>network covalent solid</u> , a network of covalently bonded atoms.
      (example: diamond)

    - a <u>metallic solid</u> , an array of metal atoms bonded by metallic bonding forces.
      (example: gold)

    - an <u>atomic solid</u>, a network of atoms held together by dispersion forces.
      (example: a solidified noble gas).

Properties of crystalline solids:

For each type of solid listed, determine whether it must overcome intermolecular forces or intramolecular forces (break chemical bonds) to pass into the gas phase. What would you predict would be the boiling points (high or low) for each type of solid?

        1) Ionic solid   4) Metallic solid

        2) Molecular solid   5) Atomic solid

        3) Network covalent solid

<u>solution</u>:

• 1 (<u>ionic solids</u>), 3 (<u>network covalent solids</u>), and 4) (<u>metallic solids</u>) must break chemical bonds in order to pass into the vapor phase. They have high boiling points reflecting the strength of the bonds that must be broken.

• 2 (<u>molecular solids</u>), and 5 (<u>atomic solids</u>) must overcome weak intermolecular (or interatomic) forces to pass from a liquid to a vapor.

Molecular solids have lattice points occupied by individual molecules. The molecules stay intact, but separate as they pass into the gas phase. These solids have a wide range of boiling points reflecting the many types of van der Waals forces that exist between molecules. Since van der Waals forces are not all that strong, the boiling points for molecular solids are generally fairly low.

Atomic solids have lattice points occupied by atoms held together by dispersion forces. They have low boiling points due to the weakness of London dispersion forces.

Determining the number of atoms per unit cell:

Determine the number of atoms per unit cell for the three common types of unit cells drawn at the beginning of this section. What is the coordination number for each atom in the solids formed by these unit cells?

solution:

1) simple cubic: There is one atom at each corner of a cube making 8 atoms. Each of the atoms is shared by eight other unit cells, so only 1/8th of it belongs to the unit cell in question. 8 atoms x 1/8/unit cell = 1 atom/unit cell.

Each atom in the solid has 6 closest neighbors, so the coordination number = 6.

2) body-centered: Just as with the simple cubic unit cell, the atoms on the corners of the cube each contribute 1/8th of an atom making 1 atom/unit cell. The atom in the center of the cube belongs only to that unit cell, so there are 1+1 = 2 atoms/unit cell.

Each atom in the solid has 8 nearest neighbors so the coordination number = 8.

3) face-centered: There is one atom/unit cell from the 8 atoms on the corners. The 6 atoms on the faces of the cube are each shared by two unit cells making 6 x 1/2 = 3 more atoms/unit cell. The total atom per unit cell is therefore 3+1 = 4 .

Each atom in the solid has 12 nearest neighbors so the coordination number = 12.

The face-centered cube with 4 atoms/unit cell is the most efficiently packed.

There are two packing arrangements that pack spheres as efficiently as possible. To see why this is so, imagine making three layers of marbles. We make the first layer so all the marbles are as close together as possible. We place the second layer of marbles in the triangular spaces made by the first layer. We can then stack the third layer so the marbles are directly over marbles in the first layer, or we can stack the third layer so the marbles lie over spaces in the first layer. The first arrangement in which the layers are stacked in an ABABAB pattern is called **hexagonal closest packing**. It is based on the hexagonal unit cell. The second arrangement has the layers stacked in an ABCABCABC pattern and is called **cubic closest packing**. It is based on the face-centered cubic cell.

Atomic radius from unit cell measurements:

Metallic gold crystallizes in a face-centered cubic lattice with its atoms touching along the diagonal of the face of the unit cell. If the unit cell measures 4.07×10^{-10} m on a side, what is the apparent radius of the gold atom?

<u>solution:</u>

We can use the Pythagorean theorem to calculate the diagonal across the unit cell from the edge length of the unit cell:

$\leftarrow$ 4.07×10^{-10} m

$$(edge)^2 + (edge)^2 = (diagonal)^2$$

$$2 \times (4.07 \times 10^{-10}m)^2 = (diagonal)^2$$

$$5.75 \times 10^{-10} \text{ m} = diagonal$$

We can see from the drawing that the diagonal equals the radius of a gold atom at each corner plus the diameter of the gold atom in the center of the face, or 4r.

$$radius = \frac{diagonal}{4} = \frac{5.75 \times 10^{-10} \text{ m}}{4} = 1.44 \times 10^{-10} \text{ m}$$

Metallic solids:

We can picture metals as a regular array of metal cations in a "sea" of valence electrons. The mobile electrons conduct heat and electricity, and the cations move easily as the metals is shaped into a sheet or a wire. The band model describes the sea of electrons as a continuous band of molecular orbitals formed from the valence atomic orbitals of the many metal atoms in a metal crystal.

Energy

conduction band

valence band

Electrons in filled molecular orbitals (**valence band**) can be excited into empty orbitals (**conduction band**) where they are mobile and are free to conduct electricity or transmit thermal energy.

Semiconductors have a small energy gap between the valence and conduction bands. Thermally excited electrons can pass the gap and allow a small current to flow. **Insulators** have a large gap between the bands which is too large for electrons to jump even when heated.

Network covalent solids:

Strong covalent bonds link atoms together in a three dimensional network. Diamond is a network covalent solid of tetrahedrally linked carbon atoms. It is an insulator because there is a large gap between the filled and empty levels. The silicates are an important class of network covalent solids with an extended array of covalently bonded silicon and oxygen atoms. Silicates form the structure of clays, rocks, and many minerals.

Molecular solids:

Molecular solids have strong covalent bonds within the molecules, but weak forces between molecules. Ice is the most common example of a molecular solid. It takes 6 kJ of energy to melt 18 grams (1 mole) of ice, however it takes ≈47 kJ of energy to break a mole of covalent O-H bonds. Some molecular solids have only London dispersion forces holding nonpolar molecules together, and so we would expect it would take even less than 6 kJ of energy to melt a mole of these solids.

Ionic solids:

Ionic solids are held together by the electrostatic attraction between anions and cations. Usually one ion lies in the spaces formed by the packing of the other ion. From a ball and stick model of a sodium chloride crystal lattice it may look as though the unit cell is simple cubic unless you look at <u>one type of ion</u> only. The sodium ions form a face-centered unit cell. The chloride ions form an interlaced face-centered cubic cell. The drawing below shows the front face of the face-centered cube. For clarity, not all atoms are shown on the back face of the crystal lattice.

Amorphous solids:

Amorphous solids are noncrystalline. Their structure can be compared to a liquid (which has no long-range molecular order), whose atoms are frozen in place.

SECTION 11.5: Changes in State

If we add energy at a constant rate to a substance, such as ice, and plot the temperature vs. time, we obtain a heating curve.

Heating curve for water:

A heating curve for water looks like:

1) How much energy is required to take 18 grams of water (1 mole) from -10°C to its melting point?

   solution:

$$q = n \times C_{ice} \times \Delta T$$
$$= (1 \text{ mol}) \times (37.6 \text{ J/mol°C}) \times (0°C -(-10°C)) = 376 \text{ J} \quad (0.376 \text{ kJ})$$

2) How much energy is required to melt 18 grams of water at 0°C?

   solution:

$$q = n \times \Delta H°_{fus}$$
$$= (1 \text{ mol}) \times (6.02 \text{ kJ/mol}) = 6.02 \text{ kJ}$$

3) How much energy is required to take 18 grams of water from its melting point to its boiling point?

   solution:

$q = n \times C_{water} \times \Delta T$

$\quad = (1 \text{ mol}) \times (75.4 \text{ J/mol°C}) \times (100°C - 0°C) = 7540 \text{ J} \quad (7.54 \text{ kJ})$

4) How much energy is required to vaporize 18 grams of water at 100°C?

    solution:

$\quad q = n \times \Delta H°_{vap}$

$\quad\quad = (1 \text{ mol}) \times (40.7 \text{ kJ/mol}) = 40.7 \text{ kJ}$

5) How much energy is required to raise the temperature of 18 grams of water vapor from 100°C to 110°C?

    solution:

$\quad q = n \times C_{water\ vapor} \times \Delta T$

$\quad\quad = (1 \text{ mol}) \times (33.1 \text{ J/mol°C}) \times (110°C - 100°C) = 331 \text{ J} \quad (0.331 \text{ kJ})$

6) What is the total heat required to take 18 grams of solid water from -10° to water vapor at 110°C?

    solution:

The total heat required equals the sum of the heat required for each step of the process (Hess's Law):

$\quad$ Total heat = 0.376 kJ + 6.02 kJ + 7.54 kJ + 40.7 kJ + 0.331 kJ

$\quad\quad$ = 55.0 kJ

Notice that of all the steps, it takes by far the most heat to separate molecules from each other as they change from the liquid to the vapor phase.

It takes less time for steam to increase 10° in temperature than it takes for water to increase 10° in temperature. The heat capacity of water vapor is smaller than the heat capacity of water, so it takes less energy to raise 1 mole of water vapor 10°C than to raise 1 mole of liquid water 10°C.

Within a given state of matter (solid, liquid, or gas), added energy increases the temperature of the system and so increases the average kinetic energy of the molecules. During a change in state, added energy increases the potential energy of the particles by breaking some hydrogen bonds (solid to liquid), or by breaking all hydrogen bonds so that the molecules can move apart (liquid to gas).

SECTION 11.6: Vapor pressure

Liquid-gas equilibria

Choose words from the list below to complete the following paragraph:

a) condense   e) increases
b) constant   f) pressure
c) equilibrium   g) kinetic energy
d) higher   h) vapor pressure

Water evaporates as surface water molecules with enough _____(1) overcome attractions to other water molecules and escape from the liquid. If water is in a closed container, molecules that enter the vapor phase collide with the walls and create _____(2) inside the container. At the same time, some of the gaseous water molecules that collide with the liquid surface are attracted strongly enough to remain there (they _____(3)). When the rate of evaporation = the rate of condensation, the pressure of the vapor is _____(4). The pressure exerted by the vapor in the flask at equilibrium is called the _____(5). If we disturb the system, it adjusts itself to regain an _____(6) state. The vapor pressure of water inside a large container is the same as the vapor pressure of water inside a small container. If we increase the temperature, the vapor pressure _____(7). The vapor pressure of a liquid is affected by intermolecular forces. Molecules with weak intermolecular forces vaporize more easily than molecules with strong intermolecular forces and therefore have _____(8) vapor pressures.

answers: 1-g, 2-f, 3-a, 4-b, 5-h, 6-c, 7-e, 8-d.

Vapor pressure increases as temperature increases, but the relationship is not linear. We can plot ln P vs. 1/T to obtain a straight line.

The Clausius-Clapeyron equation:

$$\ln P = \frac{-\Delta H_{vap}}{R} \times \frac{1}{T} + C$$

If you plot the natural log of the vapor pressure (ln P) vs. 1/T (Kelvin), what is the slope of the line? What is the y intercept?

solution:

The Clausius-Clapeyron equation is the equation of a straight line (y = mx + b). y = ln P, and x = 1/T.

The slope (m) = $\dfrac{-\Delta H_{vap}}{R}$

The y-intercept (b) = C, a constant characteristic of the liquid in question.

Vapor pressure of water at different temperatures:

The vapor pressure of water at room temperature (25°C) is 23.8 torr, and the heat of vaporization of water is 41.2 kJ/mol. What is the vapor pressure of water at 80°C?

solution:

The constant, C, does not depend on temperature, so we can solve the Clausius-Clapeyron equation for C at both temperatures and set them equal to one another:

$$\ln P_{T_1} + \frac{\Delta H_{vap}}{RT_1} = C = \ln P_{T_2} + \frac{\Delta H_{vap}}{RT_2}$$

$$\ln P_{T_1} - \ln P_{T_2} = \frac{\Delta H_{vap}}{R}\left(\frac{1}{T_2} - \frac{1}{T_1}\right)$$

$$\ln \frac{P_{T_1}}{P_{T_2}} = \frac{\Delta H_{vap}}{R}\left(\frac{1}{T_2} - \frac{1}{T_1}\right)$$

Now we can substitute in our values for water:

$$P_{T_1} = 23.8 \text{ torr}$$
$$T_1 = 25 + 273 = 298 \text{ K}$$
$$T_2 = 80 + 273 = 353 \text{ K}$$
$$\Delta H_{vap} = 41.2 \text{ kJ/mol} = 41,200 \text{ J/mol}$$
$$R = 8.3148 \text{ J/K mol}$$

$$\ln \frac{23.8 \text{ torr}}{P_{T_2}} = \frac{41,200 \text{ J/mol}}{8.3148 \text{ J/K mol}}\left(\frac{1}{353 \text{ K}} - \frac{1}{298 \text{ K}}\right)$$

$$P_{T_2} = 317 \text{ torr}$$

Boiling:

Choose words from the list below to complete the following paragraph:

a) atmosphere
b) lower
c) 1 atm.
d) 100°C
e) 94°C

A liquid boils when the vapor pressure equals the pressure of the _____(1). No bubbles can form in the interior of the liquid if the vapor pressure inside the bubbles is _____(2) than the atmospheric pressure. The **normal boiling point** is the temperature at which the vapor pressure of the liquid is _____(3). The normal boiling point of water is _____(4); water boils at _____(5) in New York City. In Boulder, Colorado (5430 ft. above sea level), water boils at _____(6). It takes longer to boil an egg in Boulder than in New York because atmospheric pressure is less at 5,430 ft. than at sea level, and the water boils at a _____(7) temperature.

   answers: 1-a, 2-b, 3-c, 4-d, 5-d, 6-e, 7-b.

Melting:

Choose words from the list below to complete the following paragraph:

a) compress
b) denser
c) higher
d) larger
e) less
f) pressure
g) volume

Atoms or molecules in solids and liquids are close together, and it is therefore difficult to _____(1) them. Since there is not a large volume change between liquid and solid states, the melting point of solids is only slightly affected by _____(2). For most substances, it takes slightly _____(3) temperatures to reach the melting point when pressure is applied. Water is an exception: large intermolecular spaces in ice make it _____(4) dense than liquid water. Ice takes up a _____(5) volume than liquid water (anyone who has had their pipes burst during a cold snap can attest to this). When we apply pressure, ice can reduce its _____(6) by changing to liquid water. We can glide over ice on skates because the _____(7) from our blades melts the ice and provides lubrication. We are so used to seeing ice float in water, that it doesn't seem strange, but in almost every other substance, the solid state is _____(8) than the liquid.

   answers: 1-a, 2-f, 3-c, 4-e, 5-d, 6-g, 7-f, 8-b.

Sublimation:

Up to their melting points, solids always have a lower vapor pressure than the corresponding liquids. Substances which have high enough vapor pressures to sublime under normal conditions are not too common: dry ice, moth balls, and solid room deodorizers are familiar examples.

Phase diagrams:

Phase diagrams show the phase changes of a substance at different combinations of temperature and pressure in a closed container.

Phase diagram of carbon dioxide:

Consider the phase diagram for carbon dioxide, CO_2:

1) Identify regions A, B, and C as solid, liquid, or gas.

    answers: A = solid, B = liquid, C = gas.

2) Identify the critical point (the point where vapor can no longer be condensed to a liquid regardless of the applied pressure).

    answer: D

3) Identify the triple point (the point at which solid, liquid, and gas are all at equilibrium).

    answer: E

4) What would the atmospheric pressure need to be in order for CO_2 to exist as a liquid?

    answer: at least 5.1 atm.; fire exists as a liquid in fire extinguishers.

5) What is the temperature of dry ice?

    answer: -78°C

A unique substance:

What common substance has the following properties?

It is the chief constituent of our bodies.

It reacts violently with sodium metal to produce aqueous sodium hydroxide and
   hydrogen gas.

It boils at about 90°C at an elevation of 10,000 ft.

It expands when it freezes.

It has a high specific heat capacity, higher than almost all liquids.

It has a high heat of vaporization.

It has high surface tension and high capillarity.

It's solid state is less dense than its liquid state.

It is a good solvent for table salt and for table sugar.

A cupful of it's liquid occupies a volume of approximately 100 gallons at 100°C and
      atmospheric pressure, as a gas.

It does not mix with oil.

It is one of the products when natural gas burns.

It is the only product of combustion of hydrogen.

It has a density of 1.0 g/mL at 25°C and 1 atm.

It can be electrolytically decomposed into the elements hydrogen and oxygen.

<u>answer</u>: water

Chapter 11 Exercises:

11.1) a. The heat of vaporization of water at 100°C plus the heat of fusion of ice at 0°C does not equal the heat of sublimation of ice. Why?

b. Does $\Delta H^{\circ}_{fus} + \Delta H^{\circ}_{vap} = \Delta H^{\circ}_{sub}$?

11.2) a. What are the three principal types of intermolecular force?

b. What is the difference between an intermolecular force and an intramolecular force? Is a hydrogen bond in water an intramolecular or an intermolecular bond?

11.3) What is the enthalpy change when 20.0 grams of steam initially at 110°C and 760 torr is cooled to ice at -10.0°C?

Heat capacity of ice = 2.00 J/g °C
Heat of fusion of ice = 334 J/g
Heat capacity of water = 4.18 J/g °C
Heat of vaporization of water = 2.26 kJ/g
Heat capacity of steam = 2.00 J/g °C

Is heat liberated or absorbed when this change takes place?

11.4) What amount of heat is released when 75.0 g of steam at 115°C cools to 37°C, body temperature? Why does a burn due to steam tend to be more severe than one caused by boiling water?

11.5) How much heat is needed to melt 160 g of snow at -12°C and warm the resulting water to 37°C? Why might it be dangerous to eat large quantities of snow in the winter?

11.6) A system of ice and water at 0.0°C has a mass of 5.0 grams. The system is converted completely to water at 0.0°C by supplying 1.34 kJ of heat. How much water was initially present?

The heat of fusion of ice = 334 J/g

11.7) If 40.0 g of ice at -8°C is added to 100 g of water at 25°C, what is the final temperature of the mixture?

11.8) Ammonia boils under atmospheric pressure at -33°C. If the molar heat of vaporization of ammonia is 24 kJ/mol, at what temperature would ammonia boil under a pressure of 6.0 atm?

11.9) What do the following represent on a temperature-pressure phase diagram for a single substance?

a. the areas between lines on the diagram
b. a line on the diagram
c. a triple point where lines meet on the diagram
d. the critical point on the diagram

11.10) Calculate how much of the space within a body-centered unit cell is actually occupied by the atoms, i.e. what is the occupancy of available space within the cube.

The edge-length of a body-centered cubic unit cell is $4r/\sqrt{3}$, where r = radius of the atom involved. The volume of a sphere=$4\pi r^3/3$.

11.11) An unknown element crystallizes in a face-centered cubic lattice with a unit cell edge length of 352 pm. The density of the metal is 8.92 g/cm^3. What is the element?

11.12) X-ray crystallography reveals that strontium has a face-centered cubic lattice with a unit cell length of 608 pm. What is the atomic radius of Sr?

11.13) Tungsten, W, is used in the filaments of light bulbs. It forms a body-centered metallic lattice and has an atomic radius of 141 pm. What is the length of the W unit cell?

11.14) Thorium, Th, has the largest molar mass of any element with a face-centered cubic lattice. X-ray crystallography studies indicate the atomic radius of the element is 180 pm. The molar mass of Thorium is 232 g/mol. What is the volume of the Th unit cell? What is the density?

11.15) Europium, Eu, is the only rare-earth element that has a body-centered cubic lattice. The density of Eu is 5.26 g/cm^3. The molar mass of Europium is 152 g/mol. What is the volume of the Eu unit cell? What is the atomic radius of Eu?

Chapter 11 Answers:

11.1) Hess's Law of Constant Heat Summation states that the enthalpy change for a reaction is independent of the route taken between the reactants and products. It is a specialized statement of the First Law of Thermodynamics. To apply Hess's Law, the starting point and ending point for each route must be the same and the routes must be complete (no missed steps)

a. The heat of vaporization of water at 100°C plus the heat of fusion of ice at 0°C does not equal the heat of sublimation of ice because the heat required to heat the water from 0°C to 100°C has not been included. The temperature of the sublimation is also not specified.

b.) Under conditions of 1 bar pressure and 25°C for all the enthalpy changes,

$$\Delta H^o_{fus} + \Delta H^o_{vap} = \Delta H^o_{sub}$$

11.2) a. The strongest intermolecular force is hydrogen bonding. The strength of a hydrogen bond lies between 10 and 40 kJ/mol.

The second strongest intermolecular force is the attraction between molecules possessing a dipole moment (dipole-dipole attraction). The strength of this type of bond lies between 5 and 25 kJ/mol.

The third type of intermolecular force is London dispersion. The strength of this bond varies from very weak (about 0.05 kJ/mol) to quite strong (about 40 kJ/mol) depending upon the size of the molecule.

b. An intermolecular bond is a bond between molecules; an intramolecular bond is a bond within a molecule. A hydrogen bond is an intermolecular bond between water molecules. The oxygen atom of one water molecule hydrogen-bonds to the hydrogen atom of another water molecule.

11.3) Heat liberated when the steam is cooled to 100°C = 20 g x 2.00 J/g °C x10°C = 400 J
Heat liberated when the steam is condensed to water = 20 g x 2260 J/g = 45200 J
Heat liberated when the water is cooled to 0°C = 20 g x 4.18 J/g °C x100°C = 8360 J
Heat liberated when the water is frozen = 20 g x 334 J/g = 6680 J
Heat liberated when the ice is cooled to -10°C = 20 g x 2.00 J/g °C x 10°C = 400 J

Adding all the steps involved, the total enthalpy change = 61.0 kJ

Heat is liberated in each step of the process; energy is released when a substance cools and energy is also released when bonds are formed.

11.4) 1.88×10^5 J; The heat released by the condensation of the steam represents the major fraction of the heat transferred.

11.5) 8.20×10^4 J; The heat required to melt and warm the ice comes from the body core. This heat loss could cause hypothermia.

11.6) If the heat of fusion if ice = 334 J/g, then 1.34 kJ is sufficient heat to melt 4 grams of ice. If the total mass of the system is 5 grams, there must have been 1 gram of water originally present.

11.7) 0°C; 14,000 J of heat are needed to melt the 40.0 g of ice, but only 10,400 J of heat is available in the 100 g of H_2O as it cools from 25°C to 0°C.

11.8) The relationship between the vapor pressure and the temperature is provided by the Clausius-Clapeyron equation. Recall that a liquid will boil when its vapor pressure equals the external (atmospheric) pressure.

Remember to convert the temperatures into degrees Kelvin.

$$\ln(p_2/p_1) = -\Delta H_{vap}/R(1/T_2 - 1/T_1)$$

$$\ln(6/1) = -24000/8.314(1/T_2 - 1/240)$$

The new boiling point T_2, under an increased external pressure of 6 atm, is 9.0°C

11.9) a. the conditions under which only one phase is the thermodynamically stable phase
b. the conditions under which two phases can coexist in equilibrium together.
c. the conditions under which three phases can coexist in equilibrium with one another.
d. the temperature above which it is impossible to liquefy a gas no matter how high the pressure.

11.10) The number of atoms within a body-centered unit cell is 2; one at the center and one on the corners (eight corners each contributing one-eighth). The atoms touch along the diagonal of the cube so that the length of the body diagonal is equal to 4r, where r is the radius of the atom involved. Thus, as stated in the question, the edge length of the unit cell is $4r/\sqrt{3}$.

The volume of the unit cell is $(4r/\sqrt{3})^3$

The volume of the two atoms occupying the unit cell is $2 \times 4\pi r^3/3$.

The fraction of the cell occupied by the atoms is $8\pi r^3/3 \div (4r/\sqrt{3})^3 = 0.68$ or 68%

11.11) There are four atoms in a face-centered cubic unit cell. The volume of this unit cell is $(352)^3$ pm^3. The density of the metal is 8.91 g/cm^3. Since the density equals the mass/volume, the mass of four atoms must be equal to the density x volume:

The mass of four atoms = 8.92 g/cm^3 x $(352)^3$ x $(10^{-10})^3 cm^3$ = 3.89 x 10^{-22}
The mass of one atom = 3.89 x $10^{-22}/4$ = 9.73 x 10^{-23} g
The mass of one mole of atoms = 9.73 x 10^{-23} g/atom x 6.022 x 10^{23} atom/mol
= 58.6 g/mol; The element is probably nickel.

11.12) 215 pm

11.13) 326 pm

11.14) 1.32 x 10^{-22} cm^3; 11.7 g/cm^3

11.15) 9.60 x 10^{-23} cm^3; 198 pm

CHAPTER 12

The Properties of Mixtures: Solutions and Colloids

TOPICS:

- Types of solutions
    - Concentration units

- Energy changes during solution

- Equilibrium process
    - Temperature effects
    - Pressure effects

- Colligative properties of solutions
    - vapor pressure lowering
    - boiling point elevation
    - freezing point depression
    - osmotic pressure

- Colloids

SECTION 12.1: Types of Solutions

In Chapter 2 we saw that elements and compounds are pure substances, and that mixtures are impure substances. This chapter will concentrate on homogeneous mixtures (solutions).

Review exercise:

Choose words from the list below to complete the following paragraph:

(a) aqueous	(e) homogenous
(b) colloid	(f) mixture
(c) gaseous	(g) solid
(d) heterogeneous	(h) solution

A combination of pure substances is called a _____(1). There are two kinds of mixtures: a mixture that has visible boundaries is a _____(2) mixture; a mixture with no visible boundaries is said to be a _____(3) mixture. A homogeneous mixture is also called a _____(4). A substance dissolved in water is an _____(5) solution. Air is a _____(6) solution consisting of mainly oxygen and nitrogen molecules. Brass is a _____(7) solution consisting of mainly copper and zinc. A heterogeneous mixture consisting of particles in suspension which do not settle out is called a _____(8).

answers: 1-f, 2-d, 3-e, 4-h, 5-a, 6-c, 7-g, 8-b.

Liquid-liquid solutions:

Liquids with similar structures and therefore similar intermolecular forces of about the same type and magnitude are **miscible** (soluble in all proportions).

Like dissolves like:

Which of the following pairs of liquids would you predict would be miscible with each other?

1)

    a) carbon tetrachloride (CCl₄)   b) chloroform (CHCl₃)

2)

    a) ethyl alcohol (CH₃CH₂OH)   b) methyl alcohol (CH₃OH)

3)

    a) hexane (C₆H₁₄)   b) pentane (C₅H₁₂)

4)

    a) octyl alcohol (C₈H₁₇OH)   b) n-butyl alcohol (C₄H₉OH)

__answers__:

Each of the above pairs contains liquids which are miscible with each other.

1) Chloroform and carbon tetrachloride have similar intermolecular forces despite the fact that chloroform is nonpolar, and chloroform is polar. Dipole forces make only a minor contribution to the intermolecular forces in chloroform.

2) Ethyl alcohol and methyl alcohol each contain an -OH group which causes the molecules to hydrogen bond and gives them similar intermolecular forces.

3) Molecules of pentane and hexane are held together by nonpolar dispersion forces of about the same magnitude.

4) Octyl alcohol and n-butyl alcohol each contain a nonpolar hydrocarbon region and a polar -OH group. They have similar intermolecular forces and are miscible.

5) Which of the above pairs would you predict would be miscible with water?

answers

Only pair "2", ethyl alcohol and methyl alcohol, are miscible with water. The -OH group in these alcohols allows them to hydrogen bond, and therefore the intermolecular forces between the alcohols and water in a solution are approximately the same as those of the pure liquids.

Note that octyl alcohol and n-butyl alcohol also each contain an -OH group, but the longer carbon-hydrogen chains in these molecules require many hydrogen bonds in water to break in order for them to insert themselves into the water structure. In general, solubility in water decreases with longer hydrocarbon chain length.

Solid-liquid solutions:

Solids always have limited solubilities in liquids. The intermolecular forces in solids are much stronger than those of liquids at the same temperature. Compare the solubility of iodine (a bluish black solid with a violet vapor) to bromine (Br_2, a dark red liquid) in carbon tetrachloride at 25°C. Bromine, a liquid, is infinitely soluble in carbon tetrachloride; iodine, a solid, forms a saturated solution when the mole fraction of I_2 is only 0.011. The dispersion forces in $I_2(s)$ are an order of magnitude greater than those in $CCl_4(l)$. Low-melting solids are more soluble than high-melting solids of similar structure.

Solid-liquid solution:

Choose words from the list below to complete the following paragraph:

a) saturated   d) solvent
b) solubility   e) temperature
c) solute

When we dissolve table salt (sodium chloride) in water, we say that sodium chloride is the _____(1), and water is the _____(2). The **solute** is the part of the solution which we consider to be dissolved in the other, the **solvent**. The _____(3) is usually present in larger amount than the _____(4).

In 100 mL of hot water (100°C), we could add 39 g of sodium chloride before salt would no longer go into solution. We have a _____(5) solution, one which contains (at a given temperature) as much of a solute as it can retain in the presence of excess _____(6). The maximum amount of a substance that dissolves in a fixed amount of a solvent is its _____(7). At 0°C, only 36 grams of sodium chloride dissolves. Solubility depends on the system's _____(8).

answers: 1-c, 2-d, 3-d, 4-c, 5-a, 6-c, 7-b, 8-e.

Solubilities:

The structures of water, isopropyl alcohol (rubbing alcohol), n-hexane (a major component of gasoline), and carbon tetrachloride are:

water   isopropyl alcohol   n-hexane   carbon tetrachloride

Using the general rule that **LIKE DISSOLVES LIKE**, which solvent would you choose to dissolve the following?

a) bike grease   d) soap

b) oil-based paint   e) candle wax

c) latex paint   f) DDT:

answers:

a) Grease is a mixture of nonpolar hydrocarbon molecules.
   Both gasoline (n-hexane) and carbon tetrachloride would dissolve it.

b) An oil-based paint would require a nonpolar solvent such as n-hexane or carbon tetrachloride.

c) A latex (water-based) paint dissolves in water, or an alcohol such as isopropyl alcohol.

d) Soap is a molecule with a long nonpolar hydrocarbon chain, and a polar, ionic head. The hydrocarbon tails surround grease molecules, and the polar ionic heads hydrogen bond and dissolve in water.

e) Waxes dissolve in nonpolar solvents such as hexane and carbon tetrachloride.

f) DDT has a structure similar to carbon tetrachloride, and therefore is soluble in CCl_4. It tends to concentrate in the fatty tissue of animals. Its insolubility in water prevents it from being washed out of soil by rainfall.

Gas-liquid solutions:

In general, the higher the boiling point of a gas, the closer its intermolecular forces will be to the magnitude of intermolecular forces of a liquid, and the more soluble it will be in a liquid. The best solvent for a gas will be the one whose intermolecular forces are the most similar to those of the gas.

Solubility of noble gases:

Consider the noble gases He, Ne, Ar, Kr, Xe, and Rn.

1) Which noble gas is the most soluble in water?

2) Are the noble gases more soluble in water or in benzene?

answers:

1) Rn, with the highest boiling point, is the most soluble noble gas. The boiling points of the noble gases increase with increasing molar mass from He (-269°C) to radon (-62°C).

2) The solubility of the noble gases is one to two orders of magnitude greater in benzene than in water because the attractive dispersion forces between nonpolar benzene molecules is similar to the dispersion forces between atoms in the noble gases.

Oxygen dissolves in water and allows fish and other aquatic life to live. Tap water is slightly acidic (it has a pH of about 6) because carbon dioxide dissolves in the water and forms carbonic acid:

$$CO_2(g) + 2H_2O(l) \rightarrow H_2CO_3(aq) \rightleftharpoons H_3O^+(aq) + HCO_3^-(aq)$$

Gases are typically not very soluble in water because gas molecules interact weakly with highly hydrogen bonded water molecules. In a carbonated beverage, more CO_2 is dissolved by bottling the drink under pressure.

Other solutions:

 Gases are infinitely soluble in each other to form gas-gas solutions. Solids diffuse to only a small extent, and therefore must be heated to high temperatures and cooled to form solutions. Brass, sterling silver, and steel are examples of metal mixtures called alloys. Gases can dissolve into solids by filling spaces between the closely packed particles. Oxygen reacts with iron to form iron oxide, or rust.

 Ways to express concentration:

 In Chapter 3 we saw that concentration (c) is the amount per unit volume (n/V). For the specific case when n = moles, and V = liters, concentration is given the special name molarity (M).

$$M = \frac{\text{moles solute}}{\text{liter solution}}$$

Molarity is not always a convenient unit to use since volume (liters of solution) depends on temperature, and volumes are not necessarily additive. Molality (m) expresses the concentration of a solution by moles of solute per kg of solvent:

$$m = \frac{\text{moles solute}}{\text{kg solvent}}$$

Mass percent, %(w/w), is a common way to record impurities in a solid. Volume percent, %(v/v), is often used for liquids and gases. Mole percent gives us a picture of the actual number of solute particles present in the solution.

$$\text{mass \%} = \frac{\text{mass solute}}{\text{mass solution}} \times 100\%$$

$$\text{volume \%} = \frac{\text{volume solute}}{\text{volume solution}} \times 100\%$$

$$\text{mole \%} = \frac{\text{moles solute}}{\text{moles solution}} \times 100\%$$

Lead poisoning:

 Lead and lead compounds inhibit development of cognitive processes in children. Recent guidelines have lowered the acceptable blood lead level from 24 micrograms of lead per deciliter of whole blood (24 µg/dL) to 9 µg/dL. The Merck Index defines lead poisoning as a blood Pb content of > 0.05 mg %. Compare this value to the 9 µg/dL guideline issued by pediatricians. The density of blood is approximately 1.06 mg/mL.

 solution:

We need to convert 0.05 mg/100mg (mg %) to µg/dL :

$$\frac{0.05 \text{ mg}}{100 \text{ mg}} \times \frac{1.06 \text{ mg}}{\text{mL}} \times \frac{100 \text{ mL}}{\text{dL}} \times \frac{1000 \text{ µg}}{1 \text{ mg}} = 53 \text{ µg/dL}$$

Recent guidelines consider lower levels of lead to be more dangerous than was once thought. The level in The Merck Index refers to lead levels which would give noticeable symptoms such as weight loss, weakness, and anemia. The pediatrician guidelines are thought to indicate the level where slowed development may occur.

Radon limits:

Radon is a colorless, odorless, inert gas which enters houses from the underlying soil. The United States Environmental Protection Agency recommends keeping indoor air levels below 4.0 pCi/L (picocuries per liter). The Merck Index lists the maximum permissible concentration of radon in air as 10^{-8} µ-Curie/cc. How does this value compare to the value recommended by the EPA?

solution:

We will convert the units used by the EPA to the units listed in The Merck Index:

$$\frac{4.0 \text{ pCi}}{L} \times \frac{1 \text{ µCi}}{10^6 \text{ pCi}} \times \frac{1L}{1000 \text{ mL}} \times \frac{1 \text{ mL}}{1 \text{ cc}} = 4 \times 10^{-9} \text{ µCi/cc}$$

The acceptable level guidelined by the EPA is less than half the limit listed in The Merck Index. Since no one can no for sure what constitutes "safe" levels for people, the EPA tries to set levels which they consider to be acceptable exposures over a long period of time.

Methylene chloride exposure:

Many industrial processes use the solvent dichloromethane (or methylene chloride, CH_2Cl_2). It is a solvent for cellulose acetate (used for photographic films, waterproofing fabrics, phonograph records, insulating electric wires), degreasing and cleaning fluids, and as a solvent in food processing.

Many chlorinated organics are carcinogens, or suspected carcinogens. The EPA sets acceptable exposure to methylene chloride as 50 ppm (8 hour exposure). If the methylene chloride of your work place is measured to be 150 mg/m^3, is this an acceptable level according to EPA guidelines? (The density of air is 1.18 g/L).

solution:

We will start with 150 mg/m^3, and convert to ppm and compare the value to the EPA limit of 50 ppm.

$$\frac{150 \text{ mg CH}_2\text{Cl}_2}{\text{m}^3 \text{ air}} \times \left(\frac{\text{m}}{10\,\text{dm}}\right)^3 \times \frac{1\text{dm}^3}{\text{L air}} \times \frac{1 \text{ L air}}{1.18 \text{ g air}} \times \frac{1\,\text{g CH}_2\text{Cl}_2}{1000\,\text{mg CH}_2\text{Cl}_2} = 1.27 \times 10^{-4} \text{ g CH}_2\text{Cl}_2/\text{g air}$$

To convert to parts per million, we multiply the answer above by $10^6/10^6$:

$$\frac{1.27 \times 10^{-4} \text{ g CH}_2\text{Cl}_2}{1 \text{ g air}} \times \frac{10^6}{10^6} = \frac{127 \text{ g CH}_2\text{Cl}_2}{10^6 \text{ g air}} = 127\,\text{ppm}$$

The measured level for the workplace (150 mg/m^3) is an unacceptable level for 8 hours of exposure according to EPA limits.

Hydrochloric Acid:

Concentrated hydrochloric acid, HCl(aq), contains approximately 36.0% HCl by mass.

1) What is the <u>molarity</u> of concentrated hydrochloric acid? The density of concentrated hydrochloric acid is 1.18 g/mL.

<u>solution:</u>

In 100 g of concentrated hydrochloric acid, there is 36 g of HCl:

$$\frac{36.0 \text{ g HCl}}{100 \text{ g soln}} \times \frac{\text{mol HCl}}{36.5 \text{ g HCl}} \times \frac{1.18 \text{ g HCl}}{\text{mL soln}} \times \frac{1000 \text{ mL}}{\text{L}} = 11.6 \text{ mol/L}$$

2) What is the <u>molality</u> of concentrated hydrochloric acid?

<u>solution:</u>

We need to find the moles of solute per kg of solvent. In 100 g of solution, there is 36.0 g of HCl, and 64 g (.064 kg) of water.

$$\frac{36.0 \text{ g HCl}}{0.064 \text{ kg water}} \times \frac{\text{mol HCl}}{36.5 \text{ g HCl}} = 15.4 \text{ mol/kg}$$

3) What is the mole fraction of HCl?

<u>solution:</u>

In 100 g of concentrated hydrochloric acid, there is 36.0 g of HCl and 64 g of water.

$$\text{moles HCl} = 36.0 \text{ g} \times \frac{\text{mol}}{36.5 \text{ g}} = 0.986 \text{ mol}$$

$$\text{moles } H_2O = 64.0 \text{ g} \times \frac{\text{mol}}{18 \text{ g}} = 3.55 \text{ mol}$$

$$\text{mole fraction HCl} = \frac{\text{mol HCl}}{\text{total moles of solution}}$$

$$= \frac{0.986 \text{ mol HCl}}{3.55 \text{ mol } H_2O + 0.986 \text{ mol HCl}} = 0.217$$

SECTION 12.2: Energy changes during solution

The tendency for all systems (including dorm rooms) is for their disorder to increase. Entropy (S) is the measure of a system's disorder; entropy increases as systems become more disordered.

$$S_{disorder} > S_{order}$$

The entropy of the universe is always increasing. A spontaneous process may occur in which the system becomes more ordered, but the change in the entropy of the system + the surroundings (the universe) must be greater than zero:

$$\Delta S_{universe} = \Delta S_{sys} + \Delta S_{surr} > 0$$

Entropy favors solution formation, since entropy increases as pure substances mix together and become more disordered. However, the heat of solution (ΔH_{soln}) may prevent two substances from forming a solution because the entropy of the surroundings decreases if it requires heat to form a solution, and this decrease in the entropy of the surroundings may outweigh the increase in the entropy of the system.

When table salt, NaCl dissolves in water, polar water molecules attract the sodium and chloride ions breaking the crystal structure apart. The water molecules make **hydration shells** around the ions. For a sodium ion, Na^+, six water molecules octahedrally surround it. These water molecules hydrogen bond to water molecules farther away, forming a second, less structured hydration shell, and these in turn hydrogen bond to other molecules in the bulk solvent.

We can consider a solute and solvent forming a solution in steps:

1) The solid breaks apart into ions:

$$M^+X^-(s) \;\rightarrow\; M^+(g) + X^-(g) \qquad \Delta H > 0$$

2) Solvent molecules surround the ions (**solvation**, or **hydration** when the solvent is water):

$$M^+(g) \xrightarrow{\;\;H_2O\;\;} M^+(aq) \qquad \Delta H < 0$$

$$X^-(g) \xrightarrow{\;\;H_2O\;\;} X^-(aq) \qquad \Delta H < 0$$

The first step is the negative of the lattice energy of an ionic solid and always requires energy ($\Delta H > 0$). The second step is the heat of hydration of the cation and anion, and this step releases energy ($\Delta H < 0$).

SECTION 12.3: The effect of temperature and pressure on solubility

The solubility of one substance in another may depend on the external conditions of temperature and pressure. We can determine the effects of these two variables on solubility if we think of the solution process as an equilibrium:

$$NaCl(s) \underset{}{\overset{H_2O}{\rightleftharpoons}} Na^+(aq) + Cl^-(aq)$$

Temperature:

If a solute does not dissolve, one of the first things we might try to get it to go into solution is to heat it. <u>Most solids require heat for them to dissolve</u> because heat must be absorbed to break down the crystal lattice. Consequently, <u>the solubility of a solid in a liquid usually increases with temperature</u>. When ammonium nitrate dissolves in water, the solution noticeably cools down. The solubility of ammonium nitrate increases with increasing temperature. In some cases, heat is given off when solution occurs, and solubility decreases with temperature.

Bubbles of air form when we heat a beaker of water because oxygen is less soluble at high temperature. <u>Most gases give off heat when they dissolve</u>.

The effect of temperature on solubility:

Which of the following equilibrium equations near saturation represents sodium hydroxide (NaOH) dissolving in water and which represents ammonium nitrate (NH_4NO_3) dissolving in water? Do the solubilities of NaOH and NH_4NO_3 increase or decrease with temperature?

solute + solvent $\rightleftharpoons$ solution + **heat**

solute + solvent + **heat** $\rightleftharpoons$ solution

<u>solution</u>:

In the first reaction, heat is a product, and so it is an exothermic reaction ($\Delta H < 0$). NaOH releases heat when it dissolves in water. If we add heat, the equilibrium shifts in the direction of the solute and solvent, and so the solubility of NaOH decreases with temperature.

In the second reaction, heat is a reactant, and so it is an endothermic reaction ($\Delta H > 0$). NH_4NO_3 absorbs heat from the environment when it dissolves in water. If we add heat, the equilibrium shifts in the direction of the solution, and the solubility of NH_4NO_3 increases.

Gas solubility decreases as temperature increases because gases are only weakly attracted to solvents, and can overcome these weak forces as their kinetic energy increases with temperature.

Pressure:

Pressure has little effect on the solubility of liquids and solids since they are almost incompressible, but it has a major effect on the solubility of gases. At moderate pressures, the solubility of a gas is proportional to the partial pressure of the gas above the solution (Henry's Law):

$$S_{gas} = k \times P_{gas}$$

P_{gas} is the partial pressure of the gas over the solution, S_{gas} is its concentration in solution, and k is a constant characteristic of the particular gas-solvent combination.

Solubility of carbon dioxide:

The solubility of carbon dioxide in water at 25°C is 1.2×10^{-5} mol/L. If an unopened soft drink contains a CO_2 pressure of 4.0 atm over the liquid, calculate the concentration of CO_2 in the soda before the bottle is opened. The partial pressure of CO_2 in the atmosphere is 4.0×10^{-4} atm.

<u>solution:</u>

- First we use the solubility data to calculate Henry's law constant k:

$$S_{CO_2} = k \times P_{CO_2}$$

$$S_{CO_2} = k \times P_{CO_2}$$

- Now we can use the value for k to solve for the concentration of CO_2 at 5 atm pressure:

$$S_{CO_2} = k \times P_{CO_2}$$

$$S_{CO_2} = 0.030 \text{ mol/L atm} \times 4.0 \text{ atm}$$

$$= 0.12 \text{ mol/L}$$

Notice the large difference in the concentration of CO_2 in the open container (1.2×10^{-5} mol/L) and the pressurized container (0.12 mol/L). If a soft drink is left uncapped, the carbon dioxide slowly diffuses out of solution, and the beverage becomes "flat".

SECTION 12.4:   Colligative Properties of Solutions

Colligative properties depend only on the number of solute particles in an ideal solution, not on their identity.

Vapor pressure lowering:

Vapor pressures:

Arrange the following equilibrium solutions (25°C) in order of decreasing vapor pressure:

    a) water   b) water   c) acetone   d) sugar water

answer:

   c > a = b > d

Acetone (c) is a volatile organic compound that has the highest vapor pressure.
Water (a&b)is less volatile than acetone and has a lower vapor pressure.
Water with a solute (sugar water, d) as you will see, has a lower vapor pressure than pure water.

Evaporation experiment:

What would happen if you left a cup of water and a cup of water with dissolved sugar in a covered container for several days?

answer:

The vapor pressure of the pure water is higher than that of the sugar water.
As the pure water vaporizes to attempt to reach equilibrium, the sugar solution absorbs the vapor to try to lower the vapor pressure to reach its equilibrium value. Water

transfers from the pure water to the sugar water (via the vapor phase) until all the water transfers to the sugar solution, and the system reaches equilibrium:

Picture a covered glass of water at equilibrium. Water molecules escape from the surface at the same rate water molecules in the vapor state return to the liquid.

Now picture a covered glass of water with a dissolved solid (such as sugar) at equilibrium. Sugar is a nonvolatile solvent, and does not escape into the vapor state. Sugar molecules at or near the surface, block water molecules from escaping from the liquid, and so lower the vapor pressure of the solvent.

If half of the molecules are solute molecules, the vapor pressure is approximately half of that of the pure solvent. If one quarter of the molecules are solute molecules, they lower the pure solvent vapor pressure by approximately one fourth. Raoult's Law expresses this relationship mathematically:

$$P_{solution} = X_{solvent} \times P^o_{solvent}$$

$P_{solution}$ is the vapor pressure of the solution, $X_{solvent}$ is the mole fraction of solvent, and $P^o_{solvent}$ is the vapor pressure of the pure solvent.

Vapor pressure of a solution:

1) If one out of every ten molecules in a sugar water solution is a sugar (solute) molecule, what is the vapor pressure of the solution at 25°C? The vapor pressure of water at 25°C is 23.76 torr.

   <u>answer:</u>

   $$P_{solution} = X_{solvent} \times P^o_{solvent}$$

   $$P_{solution} = 9/10 \times 23.76 \text{ torr}$$

   $$= 21.38 \text{ torr}$$

2) What does a plot of the mole fraction of solvent vs. vapor pressure of solution look like?

   <u>answer:</u>

   When mole fraction of the solvent = 0 (only solute present), there is no vapor pressure. When mole fraction of the solvent = 1 (only solvent present), the vapor pressure is that of the pure solvent. When there is half solute, and half solvent, the vapor pressure is half that of the pure solvent:

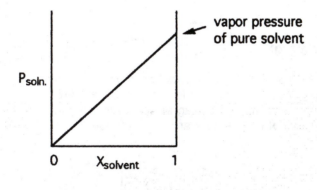

Raoult's law applies only to solutions that do not have strong interactions between the solvent and the solute ("ideal" solutions). We can imagine a case where a solute hydrogen bonds to the water molecules and prevents them from escaping as easily into the vapor state. In that case, the vapor pressure would be <u>lower</u> than Raoult's law predicts.

Now consider a mixture of two volatile components such as water and alcohol. Each liquid lowers the vapor pressure of the other liquid. The total vapor pressure is the sum of the two partial vapor pressures. In this case, Raoult's law becomes:

$$P_{solution} = P_{water} + P_{alcohol}$$

$$= (X_{water} \times P^o_{water}) + (X_{alcohol} \times P^o_{alcohol})$$

For an ideal solution of two volatile components, we can show the total vapor pressure as the sum of the partial pressures of each component:

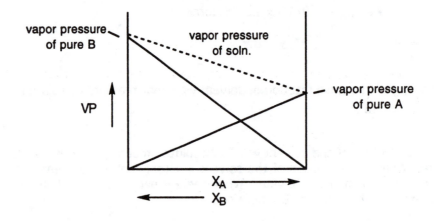

Vapor pressure lowering:

The diagram above shows the vapor pressure of an "ideal" solution where there is no interaction between components. What would you predict a similar diagram of a water-ethanol solution would look like? What might the diagram of a ethanol, hexane solution look like? The structures of ethanol and hexane are shown below.

ethanol

hexane

<u>answer</u>:

Ethanol and water molecules form hydrogen bonds with their -OH groups. The attraction between these solvents lowers the vapor pressure of each component from its ideal value, and therefore lowers the vapor pressure of the solution from that predicted by Raoult's law:

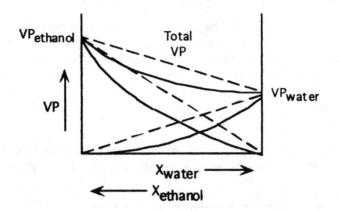

Polar ethanol molecules and nonpolar hexane molecules do not interact effectively. The solution shows a positive deviation from Raoult's law.

Vapor pressure of a benzene toluene mixture:

What is the vapor pressure of a 50:50 mixture of benzene and toluene at 100°C? $P^o_{benzene}$ = 1344 mm Hg, and $P^o_{toluene}$ = 557 mm Hg.

solution:

$$P_{solution} = (X_{benzene} \times P^o_{benzene}) + (X_{toluene} \times P^o_{toluene})$$

$$= (0.5 \times 1344 \text{ mm Hg}) + (0.5 \times 557 \text{ mm Hg})$$

$$= 951 \text{ mm Hg}$$

Note: Toluene is a benzene ring with a CH_3 group attached to one carbon atom. The similarity of the molecules causes the solution to be very close to ideal.

Boiling point elevation and freezing point depression:

We add salt to water to increase the boiling point of the water. It takes more heat to raise the vapor pressure of a salt water solution to atmospheric pressure (its boiling point) than to raise the vapor pressure of the pure solvent to atmospheric pressure. The more solute there is in the solution, the higher the boiling-point elevation:

$$\Delta T = K_b m_{solute}$$

ΔT is the difference between the boiling point of the solution and the pure solvent, K_b is the molal boiling-point elevation constant, and m_{solute} is the <u>molality</u> of the solute in the solution.

Boiling point elevation:

What is the boiling point of an aqueous solution containing 10 g of glucose ($C_6H_{12}O_6$) in 100 mL of water? $k_b = 0.52°C$ kg/mol.

<u>solution:</u>

- First we calculate the molality of the solution:

$$moles\ solute = \frac{10\ g}{180\ g/mol} = 0.056\ mol$$

mass of solvent = 0.1 kg (the density of water is 1 g/mL)

$$molality,\ m = \frac{0.056\ mol}{0.1\ kg} = 0.56\ mol / kg$$

- Substitute the molality into the equation for boiling point elevation and solve for ΔT:

$$\Delta T = k_b \times m_{solute}$$

$$\Delta T = 0.52°C\ kg/mol \times 0.56\ mol/kg = 0.29°C$$

$$\Delta T = T_{final} - T_{initial}$$

$$T_{final} = 0.29°C + 100.0°C = 100.29°C$$

We use salt on streets to lower the temperature at which water freezes. We can picture this phenomenon as follows:

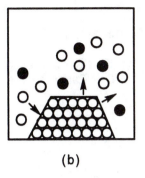

(a)   (b)

Diagram (a) shows ice in equilibrium with pure water (water molecules enter and leave the solid phase at the same rate). Diagram (b) shows ice in equilibrium with salt water. At equilibrium, the rate water molecules enter and leave the solid phase are equal. Salt molecules in the liquid water slow the rate that water molecules enter the solid phase. It requires a lower temperature to slow the rate that water molecules leave the solid. (Note that salt molecules do not "freeze" into the ice).

The equation for freezing point depression is analogous to that for boiling point elevation:

$$\Delta T = K_f m_{solute}$$

Freezing point depression:

Ethylene glycol ($C_2H_6O_2$) is a common antifreeze. Tetrahydrofuran (C_4H_8O) can also act as an antifreeze. How many grams of tetrahydrofuran would have to be added to water to produce the same freezing point lowering as one gram of ethylene glycol?

solution:

Freezing point lowering depends on the molality of the solution, so we need to determine how many grams of tetrahydrofuran contain the same number of moles as one gram of ethylene glycol:

$$\text{Moles of ethylene glycol} = \frac{1\ g}{62\ g/mol} = 0.016\ \text{moles}$$

Grams of tetrahydrofuran in 0.016 moles =

0.016 mol x 72 g/mol = 1.15 g

You would need to add a little more tetrahydrofuran to obtain the same freezing point lowering as ethylene glycol. Tetrahydrofuran is more expensive than ethylene glycol, so ethylene glycol is economically the preferable choice.

Molar mass determination:

7.8 kg of the antifreeze ethylene glycol is needed to lower the freezing point of 10.0 L of water to -10°F (-23.3°C). Calculate the molar mass of ethylene glycol. (K_f = 1.86°C kg/mol)

solution:

- We know:

    ΔT = freezing point lowering = 23.3°C

    10.0 L water = 10 kg, if we assume the density of water = 1 g/mL

- We can calculate the molality of the ethylene glycol solution:

$$m_{solute} = \frac{\Delta T}{K_f} = \frac{23.3 \ °C}{1.86 \ °C \ kg \ / \ mol} = 12.53 \ mol \ /kg$$

- In 10 kg of solvent, we would have 125.3 moles solute, so:

$$\text{molar mass of solute} = \frac{7800 \ g}{125.3 \ mol} = 62.2 \ g/mol$$

Check: The chemical formula for ethylene glycol is $C_2H_6O_2$, so molar mass =
(2 x 12) + (6 x 1) + (2 x 16) = 62 g/mol

Note: Freezing point lowering is more commonly used to measure molar masses than boiling point elevation because the temperature change is generally larger (compare K_f and K_b values for water: 1.86 and 0.52 respectively). Organic solvents are generally preferred to water because their K_f values tend to be larger.

Osmotic pressure:

If a pure solvent and a solution are separated by a **semipermeable membrane** (a membrane which allows only solvent molecules through), solute particles block solvent molecules in the solution from passing through the membrane. The rate of solvent moving from the solution into the pure solvent is slower than in the reverse direction, so there is a net transfer of solvent into the solution:

semipermeable
membrane

Osmotic pressure is the pressure which just stops osmosis. The dependence of the osmotic pressure on solution concentration is:

$$\pi = MRT$$

where π is the osmotic pressure (atm), M is the molarity of the solution, R is the gas law constant, and T is the temperature (Kelvin).

Osmotic pressure is useful for determining solute molar masses because a small concentration of solute causes a relatively large osmotic pressure.

Determining molar mass from osmotic pressure:

The osmotic pressure of a solution prepared by dissolving 10 g of insulin in 1000 g of water is 32.4 mm Hg at 25°C. What is the molar mass of insulin?

solution:

$$\pi = 32.4 \text{ mm Hg} \times \frac{1 \text{ atm}}{760 \text{ mm Hg}} = 0.0426 \text{ atm}$$

$$T = 25°C + 273 = 298 \text{ K}$$

$$M = \frac{\pi}{RT} = \frac{0.0426 \text{ atm}}{\left(0.0821 \text{ L atm/mol K}\right) \times \left(298 \text{ K}\right)} = 0.00174 \text{ mol/L}$$

We can now determine the moles of insulin in 1.00 L of solution (we neglect the minor change in the volume of 1000 g of water when 10 g of insulin is added):

1.00 L soln. x 0.00174 mol insulin/L soln. = .00174 mol of insulin.

The molar mass of insulin is therefore:

$$\frac{10 \text{ g}}{.00174 \text{ mol}} = 5750 \text{ g/mol}$$

Electrolyte solutions:

For electrolyte solutions, each ion counts as a particle. We incorporate the van't Hoff factor, i, into the equations for boiling point elevation, freezing point depression, and osmotic pressure:

$$\Delta T = iK_b m_{solute}$$

$$\Delta T = iK_f m_{solute}$$

$$\pi = iMRT$$

van't Hoff factor:

What is the approximate value for "i" for the following compounds?

a) $Al(NO_3)_3$ c) $Fe(NH_4)_2(SO_4)_2$
b) $MgCl_2$ d) $C_{12}H_{22}O_{11}$

answers:

a-4, b-3, c-5, d-1(nonelectrolyte)

   In the ideal case, "i" equals the number of ions per molecule. Most electrolyte solutions
are not ideal, and "i" is typically somewhat lower than expected from the number of ions. Ions
associate with one another and therefore do not behave as completely independent particles.
Generally, the higher the charge on the ion, the greater the deviation from the ideal value.

Osmotic pressure of an electrolyte solution:

1) What is the osmotic pressure for a 0.10 M solution of a nonelectrolyte at 25°C?

   solution:

   We can solve for the osmotic pressure, π, using the equation:

$$\pi = MRT$$

$$\pi = (0.1 \text{ moles/L}) \times (0.0821 \text{ L atm/mol K}) \times (298 \text{ K})$$

$$\pi = 2.45 \text{ atm.}$$

2) The osmotic pressure of a 0.10 M solution of an electrolyte is 10.8 atm. What is the
   value for i?

   solution:

$$i = \frac{\text{measured value for electrolyte}}{\text{expected value for nonelectrolyte}}$$

$$= \frac{10.8 \text{ atm}}{2.45 \text{ atm}} = 4.41$$

   The electrolyte in the above example is $Fe(NH_4)_2(SO_4)_2$, a compound which dissociates
in water to produce 5 ions. The actual value of "i" is less than five presumably because of ion
associations.

SECTION 12.5: Colloids

Colloids:

Choose words from the list below to complete the following paragraphs:

a) aggregate   d) scatter
b) coagulate   e) small
c) colloid   f) repel

A suspension of tiny particles in some medium is a _____(1). We can distinguish a colloid from a pure solution by shining light on it. The tiny particles (1 - 1000 nm) in a colloid _____(2) the light (the **Tyndall effect**) and make it visible if you look through the solution from the side. Ions and molecules are too _____(3) to scatter light, so the light going through a pure solution is invisible from the side.

Colloidal particles do not settle out because they attract from the medium ions of all like charge which cause the colloidal particles to _____(4) one another. An electrolyte can neutralize the adsorbed ion layers and cause the colloid to destruct, or _____(5). Heating a colloid can also destroy it by causing particles to _____(6) and settle out.

answers: 1-c; 2-d; 3-e; 4-f; 5-b; 6-a.

Chapter 12 exercises:

12.1) What is the relationship between the solubility of a substance and the concentration of a saturated solution of the substance at the same temperature?

12.2) a. In any group of ions, the most strongly solvated ion is the ion with
 a. the greatest charge and the largest radius
 b. the smallest charge and the largest radius
 c. the greatest charge and the smallest radius
 d. the smallest charge and the smallest radius

 b. Which one of the following ions is most strongly solvated by water in solution?

Na^+	SO_4^{2-}	K^+	Mg^{2+}
Sr^{2+}	Cs^+	Ba^{2+}	NO_3^-

12.3) a. As crystals of potassium nitrate are dissolved in water, the resulting solution feels colder. How can the solubility of KNO_3 be increased?

 b. When sodium hydroxide (NaOH) dissolves in water the overall process is strongly exothermic. However, when NaOH dissolves in a nearly saturated solution of NaOH, the process is endothermic. How can the solubility of NaOH be increased?

12.4) a. How does the solution process for a gas dissolving in water differ from the solution process for a salt dissolving in water?

 b. How is it possible to increase the concentration of a gas in water?

12.5) An ideal solution of 3 moles of carbon disulfide in 10 moles of ethyl acetate has a vapor pressure of 170 torr at 28°C. If the vapor pressure of pure carbon disulfide at 28°C is 390 torr, what is the vapor pressure of pure ethyl acetate at the same temperature?

12.6) An ideal solution is made up of a 1:4 molar mixture of ethanol and methanol. At room temperature, the vapor pressure of ethanol is 44 torr and the vapor pressure of methanol is 88 torr. What is the partial pressure of methanol in the vapor above the solution? What is the mole fraction of ethanol in the vapor above the solution?

12.7) a. What is a colligative property?

 b. Why does the freezing point of water decrease when salt is added to the water?

12.8) Two beakers under a bell jar, isolated from the surroundings, contain solutions of sugar in water. The first beaker initially contains 1.0 mol of sugar in 72 mL of water. The second beaker initially contains 0.50 mol of sugar in 72 mL of water. What is the mol fraction of sugar in the first beaker after the system has reached equilibrium?

12.9) a. How many grams of sodium hydroxide must be dissolved in 200 grams of water to make a 3.0 molal solution?

 b. What is the molality of a solution composed of 6.4 grams of naphthalene ($C_{10}H_8$) dissolved in 250 grams of benzene (C_6H_6)?

12.10) How would you prepare the following aqueous solutions?

a. 120 mL of a 0.200 M AgNO$_3$ solution.

b. 800 mL of a 0.010 M CuSO$_4$ solution from crystalline CuSO$_4$·5H$_2$O.

12.11) A solution is prepared by dissolving 50.0 g of moth balls, C$_6$H$_4$Cl$_2$ in 100 g of toluene, C$_6$H$_5$CH$_3$. What is the mole fraction of C$_6$H$_4$Cl$_2$ in the solution?

12.12) Jasmine perfume can be prepared by adding 5.00 g of methyl jasmonate, C$_{13}$H$_{18}$O$_3$, to 45.0 g of ethanol, C$_2$H$_5$OH. What is the molality of the resulting methyl jasmonate solution?

12.13) A dextrose intravenous solution is 5.50% glucose (C$_6$H$_{12}$O$_6$) by mass or 0.310 M. What is the density of this solution?

12.14) If a 3.0 m solution of naphthalene (C$_{10}$H$_8$) in cyclohexane boils at 89.4°C, what is the boiling point of pure cyclohexane? Assume that the vapor pressure of naphthalene is zero. k$_b$ for cyclohexane = 2.80.

12.15) A by-product of the production of bromobenzene, C$_6$H$_5$Br, is biphenyl, C$_{12}$H$_{10}$. A quality control technician analyzed the product and found it to contain 32.0 mg C$_{12}$H$_{10}$/gram C$_5$H$_5$Br. What would be the boiling point of the mixture? Bromobenzene: b.p. = 155.8°C; K$_b$ = 6.20°C/m.

12.16) 1.00 g of an unknown aromatic organic compound is dissolved in 10.0 g of biphenyl, C$_{12}$H$_{10}$. The freezing point of the resulting solution is 65.8°C. Elemental analysis of the unknown indicated the empirical formula to be C$_5$H$_4$. For biphenyl, the freezing point is 72.0°C and the freezing point constant, K$_f$, is 8.00°C/m .

a. What is the molar mass of the unknown compound?

b. What is the molecular formula?

12.17) Methanol, CH$_3$OH (b.p. 65°C) was used at one time as an antifreeze in radiators.

a. What is the freezing point of a 20% w/w methanol/water solution?

b. Explain why you would not want this solution in the radiator during the summer.

12.18) The osmotic pressure of blood is approximately 7.7 atm at 37°C (body temperature). What is the concentration of glucose (C$_6$H$_{12}$O$_6$) in an aqueous solution that is isotonic (same osmotic pressure) at this temperature?

Chapter 12 Answers:

12.1) The solubility of a substance is the amount of solute that can be dissolved in a particular solvent. In other words, the solubility of a solute is equal to the concentration of a saturated solution of that solute at the same temperature.

If the concentration is less than the solubility at the same temperature, then the solution is unsaturated. If the concentration exceeds the solubility at the same temperature then the solution is supersaturated.

12.2) a. A major type of intermolecular force in solution is the attraction between an ion and the dipole of the solvent molecule. The strength of this ion-dipole attraction depends upon the charge density on the ion. The charge density of an ion is the ratio of the charge on the ion to the size of the ion. The larger the charge and the smaller the size, the greater the attraction. The charge is usually more important than size. Hence, the correct answer is (c).

b. The doubly positive and doubly negative ions attract the solvent molecules more strongly than the corresponding singly charged ions. The smaller cations are more strongly solvated than the larger polyatomic anions. Of the cations, the smaller ions have a higher charge density, and are therefore more strongly solvated. Of the ions listed, Mg^{2+} will be the one most strongly solvated.

12.3) a. According to LeChatelier's Principle, endothermic processes are favored by an increase in temperature; addition of heat to the equilibrium system drives the process in the direction which absorbs heat (the endothermic process). An increase in temperature will increase the solubility of the potassium nitrate.

b. Even though sodium hydroxide dissolves in water with the liberation of energy initially, as the concentration of the solution increases, the heat released per mole of solute decreases. Eventually, the solution process becomes endothermic and an increase in temperature increases the solubility of sodium hydroxide.

12.4) a. Intermolecular bonding in the gaseous state is negligible. This means that heat is liberated when a gas dissolves in water because the bonds that are made are stronger than the bonds that must be broken. (The only bonds that have to be broken are the bonds between the solvent water molecules). The solution process for a gas is exothermic. Salts, on the other hand, often dissolve endothermically because considerable energy has to be supplied to break up the solid ionic lattice (the lattice energy) and insufficient energy is retrieved by solvation.

The other principal difference between the two cases is in the change in the disorder of the system (the entropy). For a gas dissolving in water, the entropy change is negative (unfavorable) because the gas is totally disordered and the solution is at least slightly ordered. For a salt dissolving in water, the entropy change is positive because the lattice structure of the solid salt is very ordered and the solution in comparison is disordered. A gas dissolving in water is an enthalpy-driven process. A salt dissolving in water is often an entropy-driven process.

b. Decrease the temperature, or increase the pressure of the gas above the solution (Henry's Law).

12.5) Raoult's Law relates the vapor pressure above a solution to the vapor pressures of the components in the solution and the composition of the solution. The total vapor pressure, $P = p_1 + p_2 = x_1 P_1^\circ + x_2 P_2^\circ$

$$170 = (3/13) \times 390 + (10/13) \times P_2^\circ$$

$$P_2^\circ = 104 \text{ torr}$$

12.6) In a distillation process, the vapor above a solution always becomes richer in the more volatile component. In other words, the mole fraction of the more volatile component is greater in the vapor than it is in the liquid because that component vaporizes more easily than the other component.

Raoult's Law: p_1(ethanol) $= x_1 P_1^\circ = (1/5) \times 44 = 8.8$ torr

Vapor pressure due to methanol above the solution $= (4/5) \times 88$ torr $= 70.4$ torr
Dalton's Law: mole fraction (ethanol) $= p_1/P = 8.8/79.2 = 0.11$

12.7) a. A colligative property is a property that depends upon the concentration of solute particles and not upon the nature of the solute particles. The particles may be ions, they may be molecules, they may be large, they may be small. A salt that dissociates into two ions per formula unit has approximately twice the effect as an equimolar concentration of a molecule that does not ionize because the concentration of solute particles is effectively twice as great. Examples of colligative properties are the depression of the freezing point, the elevation of the boiling point, the lowering of vapor pressure, and osmotic pressure.
In each of these examples there is an equilibrium across a phase boundary. The intrinsic cause for the colligative property is an increase or decrease in the disorder (entropy) of one of the two phases.

b. The freezing point of water decreases when salt is added because the addition of the salt increases the entropy (disorder) of the liquid phase. The change in the entropy of fusion (melting) is therefore greater and, in order to compensate, the melting point T_f must be lower since $T\Delta S = $ a constant for melting.

12.8) The system inside the bell jar will eventually reach equilibrium. Water will leave and re-enter the solutions in the beakers until both solutions have equal concentrations and the vapor pressures above the solutions therefore become equal.

Initial vapor pressure above the first beaker $= x_1 P_1^\circ = (4/5)P_1^\circ$

Initial vapor pressure above the second beaker $= x_1 P_1^\circ = (8/9)P_1^\circ$

Water will leave the second beaker at a higher rate than it will leave the first beaker. The solution in the first beaker will get more dilute and the solution in the second beaker will get more concentrated. The amount of sugar in each beaker remains the same-only water is transferred through the vapor phase.

Total amount of water $= 144$ mL $= 8$ mol
At equilibrium, 2/3 (16/3 mol) will be in beaker 1 and 1/3 (8/3 mol) will be in beaker 2.
Mole fraction of sugar in first beaker at equilibrium $= 1/(19/3) = 0.16$
The mole fraction of sugar in the second beaker will be the same.

12.9)   a. A 3.0 molal solution of sodium hydroxide contains 3.0 moles NaOH in 1 kg of water. This is the same as 3/5 mol in 200 grams of water.

3/5 mol = 3/5 mol x 40 g/mol = 24 grams

b. 0.20

12.10) a. Dissolve 4.08 g $AgNO_3$ in about 60 mL of distilled water, dilute to 120 mL and mix.
b. Dissolve 2.0 g of copper(II) sulfate pentahydrate to about 400 mL of distilled water, dilute to 800 mL, and mix.

12.11) X = 0.238 C_6H_4Cl

12.12) 1.67 M HNO_3; 1.76 m HNO_3

12.13) 0.99 g/mL

12.14) Naphthalene is a molecular solute; the molecule does not ionize or break up. The factor i = 1.
ΔT_b = 2.80 x 3.0 x 1 = 8.40°C
The boiling point of pure cyclohexane = 89.4 - 8.40 = 81°C.

12.15) new b.p. = 157.1°C

12.16) a. 129 g/mole
b. $C_{10}H_8$

12.17) a. -14.5°C
b. The boiling point of methanol is only 65°C, so the alcohol would boil off during the warm months.

12.18) molarity = 0.30 mol/L

CHAPTER 13

Periodic Patterns in the Main Group Elements

TOPICS:

- Hydrogen
- The Alkali metals (group 1A)
- The Alkaline earth metals (group 2A)
- The Boron family (group 3A)
- The Carbon family (group 4A)
- The Nitrogen family (group 5A)
- The Oxygen family (group 6A)
- The Halogens (group 7A)
- The Noble gases (group 8A)

Review of periodic trends:

Label the arrows on the periodic table diagram to indicate which of the following trends increase down and right to left across the table (arrows "A"), and which trends increase up, and left to right across the table (arrows "B").

a) atomic radius

c) ionization energy

b) electronegativity

d) metallic character

<u>answers</u>:

A: (a), (d). Cesium is the largest and most metallic element (excluding rare and radioactive Francium).

B: (c), (b). Fluorine is the smallest (except for helium and hydrogen), and most electronegative element with a high ionization energy. Small atoms hold their electrons more strongly than large atoms.

As we go across the periodic table, bonding character in the elements changes from:

metallic → **network covalent** → **individual molecules** → **separate atoms**
(metals)   (boron and carbon)   (N, O, F)   (noble gases)

The acid-base behavior of the oxides changes from <u>basic</u> to <u>amphoteric</u> (acts as an acid and a base) to <u>acidic</u> as the electronegativity of the atom becomes closer to that of oxygen, and the bond between the element and oxygen becomes more covalent.

SECTION 13.1: Hydrogen

What can we learn about elemental hydrogen with a tank of hydrogen gas, a balloon and a match?

- We can spray hydrogen into the air and observe that it is an <u>odorless and colorless gas</u> which is <u>unreactive in air and in the body</u> (i.e. nontoxic). In fact, the H-H bond energy is larger than almost any other single bond, and so is difficult to break.

- We can fill a balloon with hydrogen gas, let go of the balloon, and see that as the balloon rises, <u>hydrogen is less dense than air</u> as the balloon rises.

- If we ignite the balloon, we cause a spectacular fire ball and see that <u>hydrogen is a flammable gas</u>. It reacts with oxygen to produce water as its only combustion product:

$$H_2(g) + 1/2O_2(g) \rightarrow H_2O(g)$$

Hydrogen is a potentially useful, clean burning fuel.

Atomic Hydrogen has a ns^1 configuration and, like the alkali metals, it can obtain an empty shell by losing its electron. Because hydrogen is so small, its nucleus holds the electron more tightly than the ns^1 electron of the alkali metals, and it therefore has a higher ionization energy. Like the halogens, hydrogen can gain an electron to obtain a filled shell. However, hydrogen has a lower electronegativity than the halogens; H⁻ is reactive.

Hydrogen chemistry:

1) Classify each compound as 1) an ionic hydride, 2) a covalent hydride, or 3) an interstitial hydride.

        a) CaH_2 e) LiH
        b) CH_4 f) MgH_2
        c) HF   g) NH_3
        d) H_2O h) $TiH_{1.7}$

    <u>answers</u>: a-1, b-2, c-2, d-2, e-1, f-1, g-2, 3-h.

2) Fill in the blanks:

A) Ionic hydrides form between hydrogen and _____(1). Covalent hydrides form between hydrogen and _____(2). Interstitial hydrides form between hydrogen and _____(3). In interstitial hydrides, hydrogen atoms or molecules occupy _____(4) in the metal's crystal structure.

answers: 1-metals, 2-nonmetals, 3-transition metals, 4-holes

3) Fill in the blanks:

A) The reaction of lithium hydride with water:

$$LiH(s) + H_2O(l) \rightarrow Na^+(aq) + OH^-(aq) + H_2(g)$$

demonstrates that ionic hydrides are strong _____.

B) The reaction of LiH with $TiCl_4$:

$$TiCl_4(l) + 4LiH(s) \rightarrow Ti(s) + 4LiCl(s) + 2H_2(g)$$

demonstrates that ionic hydrides are strong _____.

answers: A-bases, B-reducing agents

SECTION 13.2: The Alkali Metals (Group 1A)

What can we learn about elemental sodium from a jar of sodium from the stockroom, some tweezers, and a beaker of water?

- We observe that sodium is a silvery-white metal that is likely reactive with oxygen since it is stored beneath a liquid. (Sodium is lustrous when freshly cut, but tarnishes on exposure to air becoming dull and gray).

- If we swirl the jar, we can see that it is likely reactive with water since it is stored beneath unreactive mineral oil.

- When we poke it with tweezers, we see that sodium is soft. Because of its softness, we might guess that it would have a low melting point (sodium melts below the boiling point of water, at 97.8°C).

- If we put sodium into water, we see that a low density (it floats on the water) and it violently decomposes water. Sodium reacts with water to form a basic solution:

$$2Na(s) + 2H_2O(l) \rightarrow 2NaOH(aq) + H_2(g)$$

It reduces the H in H_2O from the +1 to the zero oxidation state.

The characteristics of alkali metals can be explained by their ns^1 electron configuration. Since there is only one valence electron, the attraction between the delocalized electrons and the cationic cores is weak and the crystals deform easily (are "soft"). The low ionization energy of the outer s electron means that alkali metals are powerful reducing agents that are always found in nature as 1+ cations. They reduce halogens to form ionic solids in exothermic reactions:

$$2Na(s) + Cl_2(g) \rightarrow 2NaCl(s) + heat$$

The ionic salts (NaCl is this case) are water soluble as the high charge density cation attracts water molecules strongly to create a highly exothermic heat of hydration.

Reducing power of the alkali halides:

For each reaction below involving an alkali metal, determine the element which is reduced:

1) $Na(s) + H_2O(l) \rightarrow NaOH(aq) + H_2(g)$

2) $2Na(s) + Cl_2(g) \rightarrow 2NaCl(s) + heat$

3) $4Li(s) + O_2(g) \rightarrow 2Li_2O(s)$

4) $2Li(s) + H_2(g) \rightarrow 2LiH(s)$

answers:

1) H: from +1 in H_2O to 0 oxidation state in H_2

2) Cl: from 0 in Cl_2 to -1 oxidation state in NaCl

3) O: from 0 in O_2 to -2 oxidation state in Li_2O

4) H: from 0 in H_2 to -1 oxidation state in LiH

Alkali halides can reduce hydrogen in water, elemental hydrogen, oxygen, and halides.

The special case of lithium:

Lithium is the only alkali metal which forms salts that exhibit covalent bond character. LiCl and LiBr are more soluble in the polar solvent ethanol than the other halide salts because the high charge density of the small lithium cation (Li^+), deforms the polarizable electron cloud of the anion.

SECTION 13.3: The Alkaline Earth Metals (Group 2A)

Magnesium is one of the most common elements in the earth's crust. What can we learn about elemental magnesium from a ribbon of magnesium metal, a flame, a beaker of water, and some solid CO_2 (dry ice)?

- We see that magnesium is a silvery-white metal that is apparently not reactive in air (it only slowly oxidizes in moist air). It is in the form of a ribbon which is bendable (ductile).

- If we put it in water, we see no apparent reaction (magnesium reacts slowly with water at ordinary temperatures).

- If we hold magnesium in the flame, it burns with an intensely bright light in air and leaves a fine white odorless powder which is practically insoluble in water. (Magnesium oxide, the combustion product, has a melting point of 2800°C!). If we put magnesium oxide powder in water, it makes the solution alkaline (it forms magnesium hydroxide).

- If we ignite magnesium between two blocks of dry ice, we see that magnesium burns in carbon dioxide! Its affinity for oxygen is so great that it pulls the oxygen away from carbon to form MgO and carbon (soot).

$$2Mg(s) + CO_2(g, \text{ from sublimation of dry ice}) \rightarrow 2MgO(s) + C(s)$$

The alkaline earth metals are similar in behavior to the alkali metals. They are smaller, and therefore have higher ionization energies than the alkali metals. The attraction between the delocalized ns^2 electrons and the cationic cores is stronger than it is in the alkali metals, and the crystals are therefore harder with higher melting and boiling points. Compared with the transition metals, however, alkaline earth metals are still soft and lightweight.

The alkaline earth metals are strong reducing agents:

- They reduce oxygen in air to form strongly basic oxides (see discussion of magnesium oxide above).
- The larger metals reduce water at room temperature to form H_2.
- They reduce the halogens to form ionic compounds.
- Most reduce hydrogen to form ionic hydrides, and most reduce nitrogen to form ionic nitrides.

The special case of beryllium:

All beryllium compounds exhibit a degree of covalent bonding. The small size and high charge density of the Be^{2+} ion polarizes nearby electron clouds. The most ionic beryllium compound, BeF_2 has a low electrical conductivity when melted.

Recall from Chapter 9 that beryllium is often electron deficient in its compounds. Gaseous BeF_2, is a linear molecule with 4 valence electrons around beryllium:

$$F - Be - F$$

When the gas condenses, the molecules bond together into chains and each beryllium atom obtains an octet:

When beryllium bonds to hydrogen, hydrogen does not have enough electrons to donate two to each beryllium atom. Solid BeH_2 consists of three atom groups (Be–H–Be) bonded by only two electrons in a three center, two electron bond.

SECTION 13.4:   The Boron Family (Group 3A)

What can we infer about the element aluminum from our aluminum cook ware?

• Aluminum is a shiny metal that is machinable, and relatively soft.

• It is fairly inexpensive, and so must be readily available. (Aluminum is one of the most abundant metals in the earth's crust).

• It is not toxic to the body (recent studies that link aluminum to Alzheimer's disease are not definitive).

• It is a good conductor of heat (we assume, since it is used for cookware).

• It is lightweight, and unreactive with air and water at ordinary temperatures. (In moist air, an oxide film forms which protects the metal from corrosion).

Compare aluminum to gallium. Aluminum, with an atomic number of 13, contains 13 protons in its nucleus and has an electron configuration of $[Ne]3s^23p$. Gallium, with an atomic number of 31, contains 31 protons in its nucleus and has an electron configuration of $[Ar]3d^{10}4s^24p$. The 10 3d electrons do not effectively shield gallium's three valence electrons from the nuclear charge of +31, and gallium is smaller than aluminum. Many of the trends in group 3A reflect the existence of the d and f inner shell electrons of gallium, indium, and thallium.

Boron, like other first elements in a group, is unique. It is a network covalent solid: hard (its crystals are almost as hard as diamond), black, and high melting ($\approx 2200°C$). In contrast, aluminum melts at 660°, and gallium melts at 30°C (note: gallium boils at 2400°C and has the largest liquid range of any element).

Reactivity of group 3A elements:

• React sluggishly, if at all, with water

• Form oxides when heated in pure O_2. B_2O_3 is acidic; Al_2O_3 is amphoteric; Tl_2O is strongly basic

• Reduce halogens

Oxides of group 3A:

Match the oxide with its acid/base behavior:

1) B_2O_3	a) acidic
2) Al_2O_3	b) amphoteric
3) Tl_2O	c) basic

<u>answers</u>: 1-a, 2-b, 3-c. Basicity of the oxides increases down the group.

The special case of boron:

All boron compounds are covalent, often in the form of network covalent compounds or large molecules. Many boron compounds are electron deficient. Boron obtains an octet of electrons by either accepting a bonding pair from electron-rich atoms, or forming a three center, two electron bonds.

Boron compounds:

1) What is the Lewis structure for gaseous boron trifluoride (BF_3)?

<u>answer</u>:

The number of valence electrons = 3 + (3 x 7) = 24
Each fluorine atom is bonded to boron and has 3 lone pairs. Boron has 6 electrons around it in one of its resonance structures and so is electron deficient.

$$:\ddot{F}: \qquad \cdot\ddot{F}\cdot \qquad :\ddot{F}: \qquad :\ddot{F}:$$

2) Draw the Lewis structure for ammonia, and predict how it would react with BF_3 to give boron an octet of electrons.

   answer:

   Lewis structure of ammonia:

   Ammonia has a lone pair which it can "donate" to boron to form the compound $F_3B–NH_3$. Sometimes you may see this type of bond, where both electrons come from the same atom, drawn with an arrow from nitrogen to boron.

3) Compare the structure of ethane (C_2H_6) with amine-borane (BNH_6).

   answer:

   Boron has one less valence electron than carbon, and nitrogen has one more valence electron than carbon, so C–C and B–N bonds have the same number of valence electrons. Ethane and amine-borane have the same Lewis structures:

   ethane   amine-borane

   Each carbon atom in ethane donates one valence electron to the C–C bond. Nitrogen donates two valence electrons to form the B–N bond.

SECTION 13.5:   The Carbon Family (Group 4A)

Melting points of group 4A elements:

Match the melting points (°C) with its element:

1) carbon (diamond)		a) 232
2) silicon		b) 327
3) germanium		c) 945
4) tin		d) 1410
5) lead		e) 4100

<u>answers</u>: 1-e, 2-d, 3-c, 4-a, 5-b.

Carbon and silicon are network covalent solids and have high melting points. (Silicon has longer, weaker bonds than carbon which is reflected by silicon's lower melting point). The large decrease in the melting point of tin reflects the change to metallic bonding. Notice that the melting points of tin and lead are similar, and that tin has the lowest melting point in the group.

An unknown:

Name the element which is described by each list below:

<u>A</u>	<u>B</u>
hardest substance known	soft and "greasy"
colorless	black
poor conductor	conducts electricity
network covalent solid	sheets of extended 6-membered rings

<u>answer</u>:

List "A"-carbon; List "B"-carbon

Carbon can exist in different molecular forms (**allotropes**). Its most stable form at ordinary temperature and pressure is graphite (list B). At high pressure, diamond is the most stable form of carbon (list A).

Silicon is larger than carbon, and Si—Si bonds are weaker than C–C bonds. Silicon's chemistry centers around the silicon-oxygen bond, an unusually strong bond (368 kJ/mol). Silicon is the second most abundant element on earth (behind oxygen), and much of the silicon is found in silicates, minerals containing –Si–O– repeating units.

SECTION 13.6: The Nitrogen Family (Group 5A)

Group 5A contains the nonmetals nitrogen and phosphorus, the metalloids, arsenic and antimony, and the metal, bismuth. Nitrogen makes up 70% of the air we breathe. Liquid nitrogen can exist at -196°C at ordinary pressures, and is often used in laboratories to condense gases. Its triple bond is so strong that all of its six oxides have a positive heat of formation. Elemental phosphorus exists as P_4 molecules and exists in several allotropes.

Reactivity of group 5A members:

Consider the following reactions involving group 5A members:

a) $2P_4(s) + 6Cl_2(g) \rightarrow 4PCl_3(l)$

b) $P_4(s) + 3O_2(g) \rightarrow P_4O_6(s)$ (limited oxygen)

c) $P_4(s) + 5O_2(g) \rightarrow P_4O_{10}(s)$ (excess oxygen)

d) $P_4O_{10}(s) + 6H_2O(l) \rightarrow 4H_3PO_4(l)$

e) $PCl_3(l) + Cl_2(g) \rightarrow PCl_5(s)$

f) $PCl_5(s) + 4H_2O(l) \rightarrow H_3PO_4(l) + 5HCl(g)$

g) $N_2(g) + 3H_2(g) \rightarrow NH_3(g)$ (high P and T)

h) $3NO(g) + heat \rightarrow N_2O(g) + NO_2(g)$

i) $3NO_2(g) + H_2O(l) \rightarrow 2HNO_3(l) + NO(g)$

1) Which equation shows the **disproportionation** of an oxide to form two new oxides.

2) Which equation shows how to produce ammonia from the elements?

3) Which equation shows that the group 5A-halides form by direct combination of the elements?

4) Which equation requires phosphorus to expand its octet to form the product?

5) Which equations show methods for producing a weak acid used in fertilizer production, as a polishing agent for aluminum, and as an additive to soft drinks.

6) Which equations show the formation of the two important oxides of phosphorus?

7) Which equation shows the formation of an acid which is a strong oxidizing acid?

   answers: 1-h; 2-g; 3-a; 4-c,e; 5-d,f; 6-b,c; 7-i.

   Note: NO disproportionates to form NO_2 and N_2O, compounds with oxidation states higher and lower than in the original compound. The oxidation state of nitrogen in NO is +2. In NO_2 it is +4, and in N_2O, it is +1 (on the average).

SECTION 13.7: The Oxygen Family (Group 6A)

6A elements:

Identify the element from group 6A that each statement below describes:

1) The most abundant element on the earth's surface.

2) Supports combustion.

3) Is formed by plants during photosynthesis.

4) Exists as a low boiling diatomic gas, or as a pungent gas which decomposes in heat.

5) Exists in more than 10 forms; at room temperature it is an 8 membered crown-shaped ring.

6) Nonmetals which bond covalently to almost every other nonmetal.

7) Behaves like a metal in its salt-like compounds.

8) A strong oxidizer, second only in strength to fluorine.

9) Forms a poisonous gaseous hydride (H_2E) which smells like rotten eggs and is also known as swamp gas.

10) Its hydride is the most abundant compound on earth.

11) Its oxide in water produces sulfuric acid, a major contributor to acid rain.

Sulfur fluorides:

1) Draw a Lewis structures for sulfur tetrafluoride (SF_4) and sulfur hexafluoride (SF_6).

3) Which of the above compounds would you predict would be more reactive?

answer:

The lone pair on SF_4 makes it more reactive than SF_6. Sulfur hexafluoride is almost as inert as a noble gas. It has no odor or taste, is nontoxic, insoluble, and nonflammable.

Hydrides of oxygen and sulfur:

Consider the Lewis structures for the hydrides of sulfur (H_2S) and oxygen (H_2O):

Do these two molecules have similar properties?

answer:

Hardly!

H_2O is a high boiling liquid essential for life. H_2S is a vile-smelling, toxic gas formed during anaerobic decomposition of plant and animal matter. The bond angles in water = 104.5° reflecting sp^3 hybridization. Bond angles in hydrogen disulfide = ≈90° implying sulfur's use of unhybridized p orbitals.

The first element in a group always has different behavior than the rest of the group. Water, due to its relatively small and electronegative oxygen atom, has the ability to hydrogen bond and is therefore a unique substance.

SECTION 13.8: The Halogens (Group 7A)

The halogens are reactive because they require only one electron to obtain a noble gas configuration. They can either gain an electron from a metal, or share an electron pair with a nonmetal. Since halogens are readily reduced, they act as <u>oxidizing agents</u> in most of their reactions. Fluorine is such a powerful oxidizing agent that it oxidizes oxygen in water to O_2 and some O_3. The other halogens disproportionate in water:

$$Br_2(l) + H_2O(l) \rightleftharpoons HBr(aq) + HBrO(aq)$$

The oxidation state of Br in HBr is -1; that of Br in HBrO is +1.

The Halogens:

Match each halogen on the left with the properties that describe it:

<u>halogen</u>	<u>description</u>	<u>bp (°C)</u>	<u>reactivity</u>
1) F_2	a) faint yellow gas	e) -188.1	i) forms a weak acid with hydrogen
2) Cl_2	b) yellow-green gas	f) -34.6	j) the most reactive element
3) Br_2	c) dark red liquid	g) 58.8	k) the most electronegative element
4) I_2	d) purple-black solid	h) 184.3	l) the smallest halogen

<u>answers</u>: 1-a,e,i,j,k,l; 2-b,f; 3-c,g; 4-d,h.

The halogens are all diatomic molecules whose intermolecular dispersion forces increase in strength as the atoms become larger. As we move down the group, we change from gases (F_2, Cl_2), to a liquid (Br_2), to a solid (I_2). Fluorine is the most electronegative, and the most reactive element. Fluorine, with its short strong H-F bond, is the only compound to form a weak acid with hydrogen. HCl, HBr, and HI are all strong acids which dissociate completely in water.

Halogen oxoacids:

Reaction of halogens and their oxides with water produce halogen oxoacids. Arrange each group of oxoacids in order of decreasing acid strength:

<u>A</u>	<u>B</u>
a) HOCl	a) $HOClO_2$
b) HOClO	b) $HOBrO_2$
c) $HOClO_2$	c) $HOIO_2$
d) $HOClO_3$	

The strength of an acid depends on how easily the proton dissociates. The less electron density associated with the O-H bond, the easier it is to remove the proton. Another way to look at this is to consider the stability of the resulting anion once the proton is removed. For group "A" oxoacids, $HOClO_3$ contains three oxygen atoms which stabilize the anion by giving it three electronegative sites to absorb the negative charge. $HOClO_3$ loses its proton most readily (is the strongest acid) because its anion is the most stable.

$HOClO_2$ is the strongest acid in group "B" because the electronegative chlorine atom removes electron density from the O-H bond to a greater degree than the less electronegative bromine and iodine atoms, and so the proton more easily dissociates from $HOClO_2$.

SECTION 13.9:   The Noble Gases (Group 8A)

The filled valence shells in noble gases (ns^2np^6) explains their unreactivity. The noble gases are individual atoms with extremely weak intermolecular forces.

The noble gases:

Match each noble gas on the left with the properties that describe it:

halogen	bp (°C)	reactivity
1) He	a) -268.9	g) smallest atom except for hydrogen
2) Ne	b) -246.1	h) cannot be frozen at ordinary pressure.
3) Ar	c) -185.9	i) used in balloons to make them lighter than air
4) Kr	d) -153.4	j) used in light tubes to produce colored light
5) Xe	e) -108.1	k) makes up ≈1% of the atmosphere
6) Rn	f) -62	l) colorless, odorless, inert gas
		m) radioactive element enters houses from rock.

answers: 1-a,g,h,i,l;   2-b,j,l;   3-c,k,l;   4-d,l;   5-e,l;   6-f,l,m.

> **The chemistry of xenon:**
>
> **Xenon difluoride, XeF_2:** prepared from the elements; colorless crystals; solubility in water at 0°: 25 g/L
>
> **Xenon tetrafluoride, XeF_4:** prepared from the elements at 6 atm. and 400°; colorless crystals.
>
> **Xenon hexafluoride, XeF_6:** colorless solid with greenish-yellow vapor; more powerful oxidizing and fluorinating agent than XeF_2 and XeF_4.
>
> **Xenon trioxide, XeO_3:** colorless, hygroscopic solid; powerful explosive.

Diagonal Relationships:

We have seen in this chapter how the same valence electron configuration of atoms in a group imparts similar valency to the group members, and that trends based on size and atomic charge exist within a group. We have also seen unique behavior exhibited by some elements, particularly those in the 2nd period: Li, Be, B, C, N, O, F, and He. Lithium, beryllium, and boron, the first three members of the 2nd period, exhibit behavior very similar to elements in the 3rd period immediately below and to the right. These **diagonal relationships** are due to their similar atomic and ionic sizes.

Chapter 13 Exercises:

13.1) a. According to a dictionary, the word "halogen" means what?

b. Sodium is a soft, shiny, low-melting m _ _ _ _.

c. Chlorine is a toxic, pale green g _ _.

e. But NaCl is a white, high-melting s _ _ _.

13.2) a. What does the word "noble" mean in chemistry?

b. Name several metals that are sometimes found "free", uncombined in nature.

c. Cite several metals that are used in making jewelry and coins.

b. In ancient times, what do you suppose were known as the metals of kings and queens - and nobles?

13.3) a. From atoms of what element is it most difficult to remove electrons? I.e., what element has the highest "ionization energy"?

b. What is the most "noble" of the gases?

c. What family of elements is found in nature never combined with other elements, or with itself, but as monatomic "molecules"?

13.4) a. What are the members of the next to the last family in the 10-membered "d-block"?

b. What letters of the alphabet are used to identify "blocks" of elements in periodic tables?

c. If a new block of super-heavy elements were created, what would it be called?

13.5) a. How many families are there in the s,p,d, and f blocks?

b. What series of integers is obtained when the "lengths" of the s,p,d,and f blocks are divided by 2?

c. How many families of elements would you expect there to be in a complete g block?

d. What series of numbers is generated by the rule:
$2l + 1$ with $l = 0,1,2,3,4...$?

e. What does the "quantum number" l represent in the modern electronic model of atoms?

f. Why is the number of families in a block <u>twice</u> the block's value of $2l + 1$?

13.6) a. Excepting hydrogen, in which block(s) of periodic tables are all the elements metals?

b. Metallicity increases in what direction in a family?

c. For which families are the first members nonmetals and the last member, at least, a metal? (Name the families by their first members).

d. For which families are the first members so nonmetallic that, in fact, <u>all</u> family members are nonmetallic?

13.7) a. What are the simplest chemical formulas for the highest oxides of the elements in sodium's row?

b. Since oxygen has a combining capacity of 2 (consider, for example, the formula

H_2O), an element's maximum combining capacity is given by the number of oxygen atoms per 2 atoms of the element in the element's highest oxide. What are the formulas of the highest oxides of the elements M in sodium's row, written in the form M_2O_x?

c. What are the maximum combining capacities (or "valencies") of the elements in sodium's row?

d. How many electrons are there in neutral atoms of the elements in sodium's row?

e. What are the charges of the "cores" of atoms of elements in sodium's row? (An atomic core = atom - valence-shell electrons).

f. What are the number of the groups that the following elements are in: Na, Mg, Al, Si. P, S, Cl?

13.8) In what way, electronically, is a hydrogen atom like:

a. an atom of sodium?   b. an atom of carbon?   c. an atom of chlorine?

13.9) True of False? At ordinary conditions:

a. Most compounds of metals are solids.
b. Compounds of nonmetals with nonmetals are often gaseous, or low boiling liquids.
c. Few compounds of metals are gaseous.

13.10) Identify the following substances as Ionic, Covalent, or Metallic.

a. Conducts in the solid state without occurrence of chemical changes at the electrodes.
b. Is an insulator in the solid state, but conducts fairly well in the molten state with occurrence of chemical changes at the electrodes.
c. Is an insulator (poor conductor) in both the solid and liquid states.

13.11) Indicate with the symbols M, I, and C (exercise 13.10) the type of substances formed by atoms of Na, C, Cl, and H reacting with Na, C, Cl, and H.

	Na	C	Cl	H
Na				
C				
Cl				
H				

13.12) From a chemical point of view (Exercise 13.11), hydrogen is most like which element(s)?

13.13) What is sulfur's oxidation number in the following compounds?

a. SF_2, SF_4, SF_6; b. SO, SO_2, SO_3; c. SOF_2, SOF_4; d. SO_2F_2

13.14) a. Li^+ is about the same size as what ion in Group 2?
b. Before atomic weights were well established, Be was often placed in the same family as the element ___.
c. Exercises a and b cite examples of the D _ _ _ _ _ _ _ Relationship.

Chapter 13 Answers:

13.1)   a. "Salt-generator" (the best known example is NaCl);   b. metal;   c. gas;   d. salt

13.2)   a. Relatively <u>unreactive</u>;   b. Gold and silver; also copper
             c. Copper, silver, and gold; also platinum;   d. The "noble metals"

13.3)   a. He;   b. He;   c. the noble gases

13.4)   a. Cu, Ag, Au;   b. s, p, d, f;
             c. The "g block" (<u>S</u>ober <u>p</u>hysicists <u>d</u>on't <u>f</u>ind <u>g</u>iraffes <u>h</u>iding <u>i</u>n <u>k</u>itchens)

13.5   a. 2, 6, 10, 14;   b. 1, 3, 5, 7;   c. $2 \times 9 = 18$;   d. 1, 3, 5, 7, 9,...
            e. An electron's angular momentum.
            f. Because of electrons' classically nondescribable two-valuedness called "spin".
              Each orbital with distinct values of n, l, and m_1 can have up to 2 electrons.

13.6)   a. s, d, f;   b. downward;   c. The boron, carbon, nitrogen, and oxygen families
             d. The halogen and noble gas families.

13.7)   a. Na_2O, MgO, Al_2O_3, SiO_2, P_2O_5, SO_3, Cl_2O_7
             b. Na_2O, Mg_2O_2, Al_2O_3, Si_2O_4, P_2O_5, S_2O_6, Cl_2O_7
             c. Na 1, Mg 2, Al 3, Si 4, P 5, S 6, Cl 7
             d. Na 1, Mg 2, Al 3, Si 4, P 5, S 6, Cl 7, Ar 8
             e. Na^{+1}, Mg^{+2}, Al^{+3}, Si^{+4}, P^{+5}, S^{+6}, Cl^{+7};   f. 1, 2, 3, 4, 5, 6, 7

13.8)   a. Each atom has one outer electron (which hydrogen, unlike sodium, never
              loses in chemical compounds).
             b. Each atom has a half-filled shell (often filled by sharing, as in H_2, diamond,
              and CH_4).
             c. Each atom is one electron short of a filled shell (often filled by electron capture
              from a good electron donor such as sodium, as in Na^+Cl^- (sodium <u>chloride</u>) and
              Na^+H^- (sodium hydride) - both high melting salts).

13.9)   a. T; b. T; c. T.

13.10)   a. M; b. I; c. C

13.11)

	Na	C	Cl	H
Na	M	I	I	I
C	I	C	C	C
Cl	I	C	C	C
H	I	C	C	C

13.12)   C, Cl (Compare four rows in the above exercise; cf. exercise 13.8)

13.13)   a. 2, 4, 6; b. 2, 4, 6; c. 4, 6; d. 6

13.14)   a. Mg^{+2}; b. Al; c. Diagonal

CHAPTER 14

Organic Chemistry

TOPICS:

- Structures and classes of hydrocarbons
  - alkanes
  - alkenes
  - alkynes
  - isomers
  - aromatic hydrocarbons
- Classes of organic reactions
  - addition
  - elimination
  - substitution
  - redox
- Reactivity of functional groups
- Polymers

INTRODUCTION:

Carbon chemistry is so diverse that there is a whole branch of chemistry named for it. Carbon is a small atom with an intermediate electronegativity (2.5) which forms millions of covalent compounds. It almost always has four bonds in its compounds and makes chains, rings, branched compounds, and compounds with single, double, and triple bonds. Most reactions of organic compounds take place at a <u>functional group</u>, an oxygen, nitrogen, or halogen atom, for example. C-C and C-H bonds are nonpolar, and so are not reactive groups. C-C single bonds are shielded by C-H bonds, and hence are not accessible to reagents. C=C double bonds, in contrast, are exposed and readily attacked, e.g. by proton donors and halogens.

SECTION 14.1: The structures and classes of hydrocarbons

Hydrocarbons contain hydrogen and carbon. **Alkanes** are hydrocarbons with only single bonds (saturated hydrocarbons), **alkenes** contain one or more double bonds, and **alkynes** contain triple bonds. Alkenes and alkynes are referred to as unsaturated hydrocarbons.

Paraffins

Paraffin wax is a mixture of solid hydrocarbons obtained from petroleum. Par-affin means without-affinity. The hydrocarbons have little affinity for other reagents because their C-C and C-H bonds are unreactive.

Naming alkanes:

Carbon - carbon single bonded skeletons often survive reaction conditions, hence alkanes are named by the number of carbon atoms.

How Chemists Count:

methane,	ethane,	propane,	butane,	pentane,
(1)	(2)	(3)	(4)	(5)

hexane,	heptane,	octane,	nonane,	decane.
(6)	(7)	(8)	(9)	(10)

We name the position of substituents on the hydrocarbon chain by numbering the carbon atoms along the main chain starting from the end that will put the substituents on the lowest number carbon atom.

Naming alkanes:

1) Name the following hydrocarbons:

1) CH_4

2) CH_3CH_3

3) $CH_3CH_2CH_3$

4) $CH_3CH_2CH_2CH_3$

5) $CH_3CH_2CH_2CH_2CH_3$

10) C_5H_{12}

answers:

1) methane

2) ethane

3) propane

4) n-butane

5) n-pentane

6) n-pentane   7) 2-methylbutane

8) cyclopentane   9) 2,2-dimethylpropane   10) ?

C_5H_{12} does not tell us the specific compound. C_5H_{12} could be n-pentane, 2-methylbutane, or 2,2-dimethylpropane, all of which have five carbon atoms and 12 hydrogen atoms. These compounds, which have the same formula but different structures, are called **isomers**.

Note: 2-methylbutane is also called isopentane (common name); 2,2-dimethylpropane is also called neopentane (common name). The common names are based on the total number of carbon atoms: 5 carbon atoms is a pentane. The IUPAC name uses the longest continuous carbon chain in the molecule as the root.

Linear alkanes:

How does the number of hydrogen atoms relate to the number of carbon atoms for linear alkanes?

answer:

• Let's consider the first 4 members of the series:

methane   ethane   propane   n-butane
CH_4 C_2H_6 C_3H_8 C_4H_{10}

• Each successive member adds a CH_2 group to the previous member.

propane   n-butane

• What is the formula for pentane?

n-butane   n-pentane

(answer: $C_4H_{10} + CH_2 = C_5H_{12}$)

- We can see from these examples that the general formula for linear alkanes is:

$$H_3C-(CH_2)_n-CH_3$$

Linear alkanes therefore have "n" carbon atoms from the internal CH_2 groups, and 2 additional carbon atoms from the terminal CH_3 groups.

The number of hydrogen atoms is "2n" from the internal CH_2 groups, and 6 additional hydrogen atoms from the terminal CH_3 groups:

$$C_{n+2}H_{2n+6}$$

We can simplify the formula by letting n+2 = m.

Then, n = m-2, and we have:

$$C_{(m-2)+2}H_{2(m-2)+6} =$$

$$C_mH_{2m+2}$$

Boiling points and molar mass of alkanes:

1) Arrange the following hydrocarbons in order of increasing molar mass.

1) n-butane	5) methane
2) cyclopentane	6) 2-methylbutane
3) 2,2-dimethylpropane	7) n-pentane
4) ethane	8) propane

answer: see part "b"

2) Match the hydrocarbons (listed below in order of increasing molar mass) to their boiling points (°C).

1) methane (16.04)	a) -164
2) ethane (30.07)	b) -88.6
3) propane (44.11)	c) -42.1
4) n-butane (58.12)	d) -0.5
5) cyclopentane (70.14)	e) 9.5
6) n-pentane (72.15)	f) 27.8
7) 2-methylbutane (72.15)	g) 36.1
8) 2,2-dimethylpropane (72.15)	h) 49.2

answers: 1-a; 2-b; 3-c; 4-d; 5-h; 6-g; 7-f; 8-e.

The boiling points of linear hydrocarbons increase with their molar mass. Cylopentane boils at a higher temperature than the other 5-carbon hydrocarbons because its disc-shape allows its molecules to have more intermolecular contact. The branched 5-carbon molecules have lower boiling points than n-pentane because they make less intermolecular contact.

3) Which of the alkanes above are gases at room temperature (25°C)?

answer: methane, ethane, propane, butane, and 2,2-dimethylpropane

Linear hydrocarbons with five or more carbon atoms are liquids at room temperature. The primary component of gasoline, octane, is a liquid at ambient temperature. Substances such as margarine, grease and wax contain long hydrocarbon chains and are solids at room temperature.

Naming alkenes:

We name alkenes the same way we name alkanes, except the suffix is -ene instead of -ane. We often refer to alkenes by their common names. Ethylene is the common name for ethene, C_2H_4. Propylene is the common name for propene, C_3H_6. Polyethylene and polypropylene are polymers composed of ethylene and propylene chains respectively.

Exercise:

What is wrong with the names for the following alkenes?

      1) 3-butene   3) 4-ethyl-1-pentene

      2) 3-methyl-5-hexene   4) 4-ethene-heptane

answers:

1) We indicate the position of the double bond by the number of the first C atom in it. The correct name is 2-butene

2) We number the chain from the end closest to the double bond. The correct name is 4-methyl-1-hexene.

3) The longest root chain which contains both carbon atoms of the double bond is the 6-carbon chain. The correct name is 4-methyl-1-hexene.

4) The root chain must contain the carbons of the double bond, even though it is not the longest chain. The correct name is 3-propyl-1-hexene.

Naming alkynes:

We name alkynes the same way we name alkenes, except the suffix is *-yne* instead of *-ene*. Alkynes are often called by their common names.

Exercise:

Name the following alkynes:

1) $HC \equiv CH$

2) $H_3C - C \equiv CH$

3) $BrH_2C - C \equiv CBr$

4) $H_3C - C \equiv C - \hexagon$

answers:

1) ethyne, or acetylene

2) propyne, or methyl acetylene

3) 1,3-dibromopropyne

4) 1-phenylpropyne

Ethyne is almost always called by its common name, acetylene. A benzene ring attached to a hydrocarbon chain is called a **phenyl** group.

Isomerism:

Structural isomers are compounds which have the same molecular formula, but a different sequence of bonded atoms.

Structural isomers:

Draw 3 structural isomers for the formula C_5H_{12}.

answer:

n-pentane, 2-methylbutane, and 2,2-dimethylpropane each has 5 carbon atoms and 12 hydrogen atoms, and so are isomers.

$H_3C - \underset{H_2}{C} - \underset{H_2}{C} - \underset{H_2}{C} - CH_3$

n-pentane

$H_3C - \underset{H}{C} - \underset{H_2}{\overset{CH_3}{C}} - CH_3$

2-methylbutane

$H_3C - \underset{CH_3}{\overset{CH_3}{C}} - CH_3$

2,2-dimethylpropane

We saw in the exercise on naming alkanes that the structural isomers n-pentane, 2-methylbutane, and 2,2-dimethylpropane all with the molecular formula C_5H_{12} have different physical properties:

	bp
n-pentane	27.8°C
2-methylbutane	36.1°C
2,2-dimethylpropane	49.2°C

Stereoisomers have the same molecular formula and the same sequence of bonded atoms, but different orientations of groups in space. There are two types of steroisomers.

- **Geometric isomers** (or *cis-trans* isomers) have atoms which take different positions around a rigid ring or bond.

- **Optical isomers** are mirror images which cannot be superimposed on each other (a good example of mirror images which cannot be superimposed is your hands). Optical isomers differ only in the direction that each isomer rotates the plane of polarized light.

Geometric Isomers:

1) Why can 2-butene have *cis-trans* isomers, while butane cannot?

    <u>answer</u>:

    2-butene contains a double bond which does not allow rotation of the groups at room temperature. Carbon-carbon single bonds allow free rotation of the groups about them at room temperature.

2) Draw the *cis* and *trans* isomers for 2-butene.

    <u>solution</u>:

    The *trans* isomer has the methyl groups on <u>opposite</u> sides of the carbon-carbon double bond; the *cis* isomer has the methyl groups on the same side of the C=C double bond:

       cis-2-butene   *trans*-2-butene

Optical Isomers:

An organic molecule is asymmetric (not superimposable on its mirror image) if it contains a carbon atom that is bonded to four different groups.

Which of the following compounds contain an asymmetric carbon atom?

1)
$$Cl - \overset{\overset{\displaystyle Br}{|}}{\underset{\underset{\displaystyle H}{|}}{C}} - H$$

bromo-chloromethane

2)
$$Cl - \overset{\overset{\displaystyle Br}{|}}{\underset{\underset{\displaystyle F}{|}}{C}} - H$$

bromo-chloro-fluoromethane

3)
$$H_3C - \overset{\overset{\displaystyle CH_3}{|}}{\underset{\underset{\displaystyle H}{|}}{C}} - \underset{\underset{\displaystyle H_2}{}}{C} - CH_3$$

2-methylbutane

4)
$$H_3C - \overset{\overset{\displaystyle CH_3}{|}}{\underset{\underset{\displaystyle CH_3}{|}}{C}} - \underset{\underset{\displaystyle H_2}{}}{C} - CH_3$$

2,2-dimethylbutane

5)
$$H_3C - \overset{\overset{\displaystyle Br}{|}}{\underset{\underset{\displaystyle H}{|}}{C}} - \underset{\underset{\displaystyle H_2}{}}{C} - CH_3$$

2-bromobutane

answers:

Compounds #2 and #5 are asymmetric:

• There are four different groups attached to the carbon atom (a bromine, a chlorine, a fluorine, and a hydrogen atom), so bromo-chloro-fluoro-methane is asymmetric.

• The second carbon atom contains four different groups (a bromine, a methyl group, an ethyl group, and a hydrogen atom), so 2-bromobutane is asymmetric.

Drugs are often asymmetric molecules, which, in order to bind to sites in our body, must have the correct "handedness". The wrong isomer may not only be ineffective, but it could be detrimental. Drug companies therefore must be able to synthesize these drugs so that they are optically pure.

Aromatic hydrocarbons contain carbon atoms with delocalized electrons (often benzene rings). Usually benzene is the parent compound, and attached groups are named as prefixes.

Naming aromatic compounds:

Name the following compounds:

(1)   (2)   (3)   (4)

answers:

1) methylbenzene (toluene)

2) 1,2-dimethylbenzene (*o*-xylene)

3) 1,3-dimethylbenzene (*m*-xylene)

4) 1,4-dimethylbenzene (*p*-xylene)

Xylene is the common name for dimethyl benzene. We indicate the positions of the methyl groups by:

o- (ortho) for adjacent groups

m- (meta) for groups separated by one ring carbon atom

p- (para) for groups on opposite ring carbon atoms

SECTION 14.2:   Classes of Organic Reactions

There are three general classes of organic reactions into which many reactions fall. Two of the reaction types (addition reactions and elimination reactions) involve carbon-carbon double (or triple) bonds which are exposed and readily attacked.

- **Addition reactions** occur when the π bond of an alkene or alkyne breaks, and new sigma bonds form to the atoms which "add" to the molecule.

- **Elimination reactions** are the opposite of addition reactions. A molecule is eliminated, and a double or triple bond forms in the hydrocarbon product. The driving force for this type of reaction is usually the loss of a small, stable molecule such as HBr or water.

- **Substitution reactions** occur in alkanes when an atom from an added reagent substitutes for an atom in a hydrocarbon molecule. Usually in these reactions, halogen atoms replace hydrogen atoms.

Classes of reactions:

Classify the following as addition, elimination, or substitution reactions:

1) Br_2 + $H_2C=CH_2$ $\longrightarrow$ BrH_2C-CH_2Br

2) $CH_3CH_2CH_2Br$ $\longrightarrow$ $H_3C-\underset{H}{C}=CH_2$ + HBr

3) CH_3CH_2Cl + CH_3ONa $\longrightarrow$ $H_3CCH_2OCH_3$ + $NaCl$

answers:

1) Addition reaction: bromine atoms add across the double bond in ethylene.

2) Elimination reaction: HBr is eliminated and a double bond forms in the hydrocarbon product.

3) Substitution reaction: a methoxy group ($-OCH_3$) substitutes for the chlorine atom in chloroethane.

Oxidation and reduction of organic groups:

When we assign an oxidation number to a carbon atom in an organic compound, we must decide if the attached groups are more or less electronegative than carbon. For example, the oxidation number of the carbon atom in carbon tetrachloride, CCl_4, is +4 because chlorine is more electronegative than carbon, and is assigned an oxidation number of -1. The oxidation number of carbon in methane, CH_4, is -4 because hydrogen is less electronegative than carbon and is assigned an oxidation number of +1.

When hydrogen atoms add to a carbon atom, the carbon atom is reduced. Conversely, the loss of hydrogen atoms from a carbon atom results in its oxidation. Compare the oxidation number of the carbon atoms in methane, CH_4, and ethylene, $H_2C=CH_2$. The oxidation number of carbon in ethylene is -2, the oxidation number of carbon in methane is -4. Carbon is reduced on addition of hydrogen atoms.

The addition of oxygen to carbon results in oxidation of the carbon atom since oxygen is more electronegative than carbon.

Redox reactions:

For each reaction, determine if the carbon atom is oxidized or reduced:

$$1)\ CH_4 + 2O_2 \rightarrow CO_2 + 2H_2O$$

$$2)\ CH_3CH(OH)CH_3 \rightarrow CH_3C(O)CH_3$$

$$3)\ CH_2CH_2 + Br_2 \rightarrow BrCH_2CH_2Br$$

$$4)\ CH_2CH_2 + HCl \rightarrow ClCH_2CH_3$$

answers:

1) Addition of oxygen and loss of hydrogen to carbon results in its oxidation. Carbon is oxidized from an oxidation state of -4 in CH_4 to an oxidation state of +4 in CO_2. (Hydrogen atoms are assigned an oxidation number of +1 in hydrocarbons because they are less electronegative than carbon).

2) Loss of a hydrogen atom from the central carbon results in its oxidation.

3) Addition of bromine atoms to carbon result in its oxidation; the bromine atoms are reduced.

4) The carbon which gains a hydrogen atom is reduced, and the carbon which gains the chlorine atom is oxidized.

SECTION 14.3: Reactivity of Common Functional Groups

A **functional group** is some group other than a C-H single bond on a hydrocarbon. Functional groups have names that become familiar with use. We have already discussed alkenes and alkynes with double and triple bonds respectively. Other functional groups involve other atoms such as oxygen, nitrogen, and halogen atoms.

Functional groups:

Name each type of compound drawn below:

1) $H_2C = CH_2$

2) $HC \equiv CH$

3) $H_3C - OH$ (wood alcohol)

4) $H_3C - Br$

7) $H_3C - \overset{\overset{O}{\|}}{C} - CH_3$ (acetone)

8) $H_3C - \overset{\overset{O}{\|}}{C} - OH$ (vinegar)

9) $H_3C - \overset{\overset{O}{\|}}{C} - OCH_3$

5) H_3C-NH_2

6) $H_3C-\overset{\overset{\displaystyle O}{\|}}{\underset{\displaystyle H}{C}}$

10) $H_3C-\overset{\overset{\displaystyle O}{\|}}{C}-NH_2$

11) $CH_3C\equiv N$

answers:

1) alkene $(R=R)$

2) alkyne $(R\equiv R)$

3) alcohol (R-OH)

4) haloalkane (R-X)

5) amine (R-N)

6) aldehyde (CH=O)

7) ketone (C=O bonded to 2 C's)

8) carboxylic acid (C=O bonded to OH)

9) ester (C=O bonded to -OR)

10) amide (C=O bonded to N)

11) nitrile $(C\equiv N)$

The word aldehyde comes from alcohol dehydrogenated.

Functional groups:

Determine what class of compound is described in each case below:

1) They are named by adding the suffix -ol to the parent hydrocarbon name (example: methanol). Their -OH group allows them to hydrogen bond and gives them similar properties to water. They lose H and -OH groups (dehydration) to form alkenes. With hydrohalic acids, they react to give haloalkanes:.

$$R-OH \ + \ HBr \ \longrightarrow \ R-Br \ + \ H_2O$$

2) They form when two alcohol molecules combine eliminating water. They have no -OH group, and so melt and boil at lower temperatures than alcohols.

3) They consist of a carbon atom connected to a halogen. In the presence of base, alcohols form with the elimination of the halogen:

$$R-Br \ + \ OH^- \ \longrightarrow \ R-OH$$

4) They are derivatives of ammonia which causes fish to smell "fishy". They are weakly basic and can be neutralized by the acid from a lemon (citric acid). They form hydrogen bonds, and so have fairly high melting and boiling points. They can undergo substitution reactions to form larger ammonia derivatives:

$$2\ R-NH_2 \ + \ R''-Cl \ \longrightarrow \ R-\underset{\displaystyle H}{N}-R'' \ + \ RNH_3Cl$$

5) Hydrocarbons which typically undergo addition reactions due to their electron rich multiple bonds.

6) They have delocalized electrons, and therefore are less reactive than compounds with localized multiple bonds.

7) They contain an electron rich **carbonyl group** at the end of a hydrocarbon chain. They are named by adding -*al* to the parent hydrocarbon name (example: propanal). They are reduced to alcohols.

8) They contain an electron rich carbonyl group within a hydrocarbon chain. They are named by adding -*one* to the parent hydrocarbon name (example: propanone). They can be reduced to alcohols:

9) They are named by adding -*oic* acid to the parent alkane (example: propanoic acid). They are weak acids in water which are neutralized completely by a base to give a salt (on evaporation) and water:

10) Long chains of these are **fatty acids**.

11) They form from an alcohol and a carboxylic acid with loss of water (dehydration):

12) They form from a substitution reaction of an ester and an amine:

13) The most important example of this group for life is the **peptide bond** which links amino acids in a protein.

<u>answers</u>:

1-alcohols; 2-ethers; 3-haloalkanes; 4-amines; 5-alkenes, alkynes; 6-aromatic compounds; 7-aldehydes; 8-ketones; 9,10-carboxylic acids; 11-ester; 12,13-amides.

Small **monomer** units combine to form **Polymers** (poly-mer = many molecules). Synthetic polymers form from addition reactions, or condensation reactions. Polyethylene (used for plastic bags, bottles, and toys) forms from the addition of ethylene molecules to each other. A **free radical** (species with an unpaired electron, usually from a peroxide) initiates the reaction. The polymerization reaction stops when two free radicals form a covalent bond. Polymerization can also be initiated by formation of an ion instead of a free radical. **Catalysts** are often used to allow polymerization reactions to occur under mild conditions.

Condensation polymers form when two functional groups undergo a dehydration-condensation reaction. Condensation of carboxylic acid and amines form polyamides (nylon); condensation of carboxylic acid and alcohols form polyesters.

$$H_2N-R-NH_2 \;+\; \underset{HO}{}\overset{O}{\underset{}{\overset{\|}{C}}}{-}R{-}\overset{O}{\underset{}{\overset{\|}{C}}}{-}OH \longrightarrow \left(\!\!\begin{array}{c} \underset{H}{N}{-}R{-}\underset{H}{N}{-}\overset{O}{\overset{\|}{C}}{-}R{-}\overset{O}{\overset{\|}{C}} \end{array}\!\!\right)_n$$

polyamide

$$HO-R-OH \;+\; \underset{HO}{}\overset{O}{\underset{}{\overset{\|}{C}}}{-}R{-}\overset{O}{\underset{}{\overset{\|}{C}}}{-}OH \longrightarrow \left(\!\!\begin{array}{c} O{-}R{-}O{-}\overset{O}{\overset{\|}{C}}{-}R{-}\overset{O}{\overset{\|}{C}} \end{array}\!\!\right)_n$$

polyester

Don't throw that gunk out!

The discovery of nylon and Teflon:

A dicarboxylic acid and a diamine react to form a sticky material with little apparent use. However, one day a chemist in the Carothers research group in the DuPont Company, put a small ball of the sticky stuff on the end of a stirring rod and formed a string as he pulled it away from the sticky mass. He noticed the strength and silky appearance of the thread and realized that it could be a useful fiber.

In 1938, a DuPont chemist named Roy Plunkett synthesized about 100 pounds of gaseous perfluoroethylene and stored it in steel cylinders.

$$\overset{F}{\underset{F}{>}}C=C\overset{F}{\underset{F}{<}}$$

Unexpectedly, a cylinder produced no perfluoroethylene gas when the valve was opened. Apparently, a technician said it was empty. Strange, thought Plunkett, we filled it recently and haven't used much. How about taking the tank over to the machine shop and having them cut it open with their large hack saw? The cylinder was cut open to reveal a white powder. The powder was a polymer of perfluoroethylene which was developed into Teflon.

Synthetic polymers are ever present in our lives: Teflon is used for cookware and insulation; polypropylene is used for carpeting and bottles; polyvinyl chloride is used for plastic wrap and indoor (lead-free) plumbing; polystyrene is used for insulation and furniture; polyacrylonitrile is used for yarns, and the fabrics Orlon, and Acrilon; polyvinyl acetate is used for adhesives, paints, textile coatings, and computer disks; polyvinylidine chloride is used in Saran wrap; polymethyl methacrylate is used as a glass substitute (e.g., Plexiglas), and bowling balls. Nylon, lycra, and all synthetic yarns are polymers.

Biopolymers make up the most abundant organic chemicals on earth. Polysaccharides form from individual sugar molecules. Cellulose (found in plants), starch (an energy store for plants), and glycogen (the energy storage molecule for animals) all form from repeated glucose units. Proteins are polyamides made up of amino acids, monomers which contain a carboxyl and an amine group on the same molecule. Two amino acids combine to form a peptide (amide) bond with the elimination of a water molecule (dehydration):

peptide bond

Nucleic acids are unbranched polymers which contain the chemical information that guides the designing of proteins. The monomers (mononucleotides) which form nucleic acids have three parts: an N-containing base, a sugar, and a phosphate group. Your text shows these groups on p.631. The two types of nucleic acids, **ribonucleic acid** (RNA) and **deoxyribonucleic acid** (DNA) differ in the sugar portions of their mononucleotides. Linear segments of the DNA molecule act as **genes**, chemical blueprints for constructing an organism's proteins.

Chapter 14 Exercises:

14.1) Circle the functional group(s) on the following compounds.

Vanillin
(vanilla beans)

Citric Acid
(citrus juice)

Dichlorodiphenyltrichloroethane
(DDT insecticide)

14.2) Circle the functional group(s) on the following compounds.

Northeindrone
(the pill)

Caffeine

14.3) Fill in the needed reactant(s) to complete the reactions.

a.

b. $H_2C= C- C= CH_2$ ⟶ $H_3C- C- C- CH_3$
         $\overset{|}{H}$ $\overset{|}{H}$ $\overset{|}{H_2}$ $\overset{|}{H_2}$

c. $H_3C- \overset{CH_3}{\underset{H}{C}}- \overset{}{\underset{H_2}{C}}- Br$ ⟶ $H_3C- \overset{CH_3}{C}= CH_2$

d. ⟶

14.4) Complete the following reactions.

a. $H_3C-\underset{H}{C}=\overset{CH_3}{C}-CH_3$ $\xrightarrow{HBr}$

b. $H_3C-\overset{\displaystyle O}{\overset{\|}{C}}-OH$ $\xrightarrow[H^+]{(CH_3)_2CHOH}$

14.5) How many monomers make up this segment of polyvinyl chloride (PVC)?

$$-\underset{H_2}{C}-\underset{H}{\overset{Cl}{C}}-\underset{H_2}{C}-\underset{H}{\overset{Cl}{C}}-\underset{H_2}{C}-\underset{H}{\overset{Cl}{C}}-\underset{H_2}{C}-\underset{H}{\overset{Cl}{C}}-$$

Give the structure of the monomer.

14.6) The structure below is a peptide.

$$H_3\overset{+}{N}-\underset{\underset{\underset{CH}{|}}{\underset{CH_2}{|}}}{CH}-\overset{\displaystyle O}{\overset{\|}{C}}-\underset{H}{N}-\underset{\underset{CH_3}{|}}{CH}-\overset{\displaystyle O}{\overset{\|}{C}}-\underset{H}{N}-\underset{\underset{CH_2}{|}}{CH}-\overset{\displaystyle O}{\overset{\|}{C}}-O^-$$

a. How many amino acids make up the peptide?

b. Circle the peptide bonds.

Chapter 14 Answers:

14.1)

14.2)

14.3) a. CrO₃/H₂SO₄; b. H₂/Pt; c. KOH; d. NaOH

$$a. CrO_3/H_2SO_4; \quad b. H_2/Pt; \quad c. KOH; \quad d. NaOH$$

14.4) a.

$$H_3C-\underset{H_2}{C}-\underset{\underset{Br}{|}}{\overset{\overset{CH_3}{|}}{C}H}-CH_3$$

b.

$$H_3C-\overset{O}{\overset{||}{C}}-O-CH\overset{CH_3}{\underset{CH_3}{<}}$$

14.5) a. four monomer units;   b.

$$H_2C=\underset{\underset{Cl}{|}}{\overset{\overset{H}{|}}{C}}$$

14.6) a.

b. three amino acids

CHAPTER 15

Kinetics: Rates and Mechanisms of Chemical Reactions
"It's About Time"

TOPICS:

- Reaction rate
- Rate laws
- Integrated rate laws
  (relating concentration and time)
- Dependence of Reaction Rate on Temperature
    - Collision Theory
- Reaction mechanisms
- Catalysts

SECTION 15.1: Reaction Rate

Chemical kinetics is the study of reaction rates. Diamond spontaneously changes to graphite, but the reaction is extremely slow. "Diamonds may be a girl's best friend".....but not to eternity. We saw in Chapter 6 that the combustion of sugar in our bodies occurs at a much slower rate (it is metabolized) than the combustion of gasoline in our cars. We study kinetics to determine a possible "reaction mechanism" - what might be happening at the molecular level.

Reaction rates:

Molecules must physically collide with each other with enough energy in order for them to react. Which of the following would you expect might change the rate of a chemical reaction?

1) concentration of reactants

2) concentration of products

3) temperature

4) mixing speed

5) surface area of solid particles reacting with a liquid or gas

answer:

All of the above.

The rate of a chemical reaction is a positive quantity that tells us how the concentration of a reactant or product changes with time. Lets consider the reaction between nitrogen dioxide and ozone which destroys ozone in our atmosphere:

$$2NO_2(g) + O_3(g) \rightarrow N_2O_5(g) + O_2(g)$$

The rate of this reaction can be taken to be the disappearance of one of the reactants with time, or the appearance of one of the products with time:

$$\text{Rate} = \frac{\text{decrease in conc. of } O_3}{\text{time interval}} = \frac{-\Delta \text{conc. } O_3}{\Delta t} = \frac{-\Delta[O_3]}{\Delta t}$$

$$\text{Rate} = \frac{\text{increase in conc. of } O_2}{\text{time interval}} = \frac{\Delta \text{conc. } O_2}{\Delta t} = \frac{\Delta[O_2]}{\Delta t}$$

Since reactants disappear over time (conc. of O_3 at time 2 < conc. of O_3 at time 1), the rate equation contains a negative sign to make the reaction rate a positive number. We use square brackets to express <u>concentration</u> in moles per liter. The units of reaction rate depend on the unit of time:

mol/L sec, mol/L min, mol/L day

Rate of reaction:

Refer to the rates of reaction for the disappearance of O_3 and the appearance of O_2 (above) to write the rate of reaction in terms of the disappearance of NO_2 and the appearance of N_2O_5.

<u>solution</u>:

For every molecule of ozone that disappears, two molecules of nitrogen dioxide disappear (nitrogen dioxide disappears at twice the rate of ozone). The rate of the reaction written in terms of NO_2 is therefore 1/2 the rate of the reaction written in terms of O_3:

$$\text{Rate} = \frac{-\Delta \text{conc. } O_3}{\Delta t} = -\frac{1}{2} \frac{\Delta \text{conc. } NO_2}{\Delta t}$$

For every molecule of oxygen that appears, one molecule of N_2O_5 appears. Therefore, the rate of the reaction written in terms of N_2O_5 = the rate of the reaction written in terms of O_2:

$$\text{Rate} = \frac{\Delta \text{conc. } O_2}{\Delta t} = \frac{\Delta \text{conc. } N_2O_5}{\Delta t}$$

The value for the rate of reaction depends on which reaction component we monitor. The convention is to use a reactant or product with a coefficient of 1 as the reference.

Example:

1) Express the reaction rate in terms of a change in concentration of each substance for the reaction of NO(g) with oxygen to produce N_2O_3(g).

   solution:

   • The balanced equation is:

   $$4NO(g) + O_2(g) \rightarrow 2N_2O_3(g)$$

   • Let's work through this problem slowly. First, determine the change in moles of NO, O_2, and N_2O_3 between reactants and products:

   $$\Delta n_{NO} = -4 \text{ mol}, \quad \Delta n_{O_2} = -1 \text{ mol}, \quad \Delta n_{N_2O_3} = +2 \text{ mol}$$
   $$\text{(reference)}$$

   • Based on the above values, complete the equations:

   $$-\Delta n_{O_2} = -\left(?\right)\Delta n_{NO} \quad \text{and,} \quad -\Delta n_{O_2} = \left(?\right)\Delta n_{N_2O_3}$$

   (answers: 1/4, 1/2)

   • Thus, $\quad \dfrac{-\Delta n_{O_2}}{\Delta t} = \dfrac{-\left(?\right)\Delta n_{NO}}{\Delta t} \quad$ and, $\quad \dfrac{-\Delta n_{O_2}}{\Delta t} = \dfrac{\left(?\right)\Delta n_{N_2O_3}}{\Delta t}$

   (answers: 1/4, 1/2)

   • We can now express the rate of the reaction in terms of each substance. O_2 is the reference since its coefficient is 1:

   $$\text{Rate} = -\frac{\left[O_2\right]}{\Delta t} = -\frac{1}{4}\frac{\left[NO\right]}{\Delta t} = \frac{1}{2}\frac{\left[N_2O_3\right]}{\Delta t}$$

2) How fast does the concentration of oxygen decrease when N_2O_3 forms at 8.0×10^{-5} mol/L s?

   solution:

   Oxygen disappears at 1/2 the rate that N_2O_3 forms:

   $$\text{Rate} = 1/2 \times (8.0 \times 10^{-5} \text{ mol/L s})$$
   $$= 4.0 \times 10^{-5} \text{ mol/L s}$$

Reaction rates generally slow down during the course of a reaction as reactants are used up. (The higher the concentration of reactants, the more frequently they will collide and be converted to products). The **instantaneous rate** is the rate of a reaction at a particular instant during the reaction. Often we measure the rate of a reaction when the reactants are mixed before any product accumulates so that the reverse reaction is negligible. This is the **initial rate** of a reaction, the rate of the reaction when t=0.

Instantaneous reaction rate:

If you measure the average rate of a reaction over the first 5 second period to be 0.006 mol/L s, and in the period between 5 and 10 seconds to be 0.002 mol/L s, what would you estimate would be the rate of the reaction at t=5 seconds?

<u>answer</u>:

We might guess that the rate of the reaction at t=5 seconds would be halfway between the two rates, or

$$\frac{0.006 + 0.002}{2} = 0.004 \text{ mol/L s}$$

The more accurate way to estimate the reaction rate at this point would be to draw a curve of the change in reaction rate with the change in time, and find the slope of the line tangent to the curve at t=5 s:

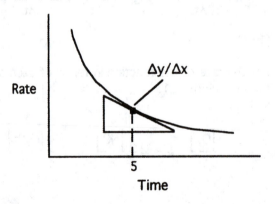

SECTION 15.2: Rate Laws

The **rate law** for a reaction tells us how reaction rate changes with the concentrations of the reactants and products. Consider the reaction between carbon monoxide and nitrogen dioxide to form carbon dioxide and nitrogen monoxide:

$$CO(g) + NO_2(g) \rightarrow CO_2(g) + NO(g)$$

We could determine how the rate of the reaction changes with, say, the concentration of carbon monoxide by measuring the initial rate of the reaction at several different concentrations of CO, holding the concentration of nitrogen dioxide constant. Likewise, we could study the dependence of the reaction rate on the concentration of NO_2 by changing its concentration and keeping the concentration of CO constant.

Rate Law:

1) Study the data from two sets of experiments to determine how the rate depends on the concentration of CO and NO_2. The data was taken at 400°C, and the rates have units of mol/L s.

	Experiment #1			Experiment #2	
[CO]	[NO$_2$]	rate	[CO]	[NO$_2$]	rate
0.10	0.10	0.005	0.10	0.20	0.010
0.20	0.10	0.010	0.20	0.20	0.020
0.30	0.10	0.015	0.30	0.20	0.030
0.40	0.10	0.020	0.40	0.20	0.040

solution:

- We see from the first experiment that when the concentration of CO doubles, the rate doubles: when [CO] increases from 0.10 to 0.20, the rate doubles from 0.005 to 0.010 mol/L s; when [CO] increases from 0.20 to 0.40, the rate doubles from 0.010 to 0.020 mol/L s.

- We see from comparing experiments 1 and 2 that, in each case, when the concentration of NO_2 doubles, the rate also doubles.

- We can conclude that the rate of the reaction is directly proportional to the concentrations of both CO and NO_2:

$$rate = k[CO][NO_2]$$

We say the reaction is <u>first order</u> with respect to CO and NO_2. The overall reaction order is the sum of the individual orders, so the reaction is <u>second order</u> overall. k is the rate constant for the reaction and is a function of the temperature.

2) What is the value for k at 400°C for this reaction?

solution:

We calculate k by substituting experimental values into the rate equation. We could choose any set of conditions from experiments #1 and #2 above, and it is a good check on your calculated value of k to use a more than one set of conditions. We'll use the values from the first set of conditions of experiment #2 (rate = 0.010 mol/L s; [CO] = 0.10 mol/L; [NO$_2$] = 0.20 mol/L):

$$\text{rate} = k[CO][NO_2]$$

$$0.010 \text{ mol/L s} = k(0.10 \text{ mol/L}) \times (0.20 \text{ mol/L})$$

$$k = \frac{0.010 \text{ mol/L s}}{(0.20 \text{ mol/L}) \times (0.10 \text{ mol/L})} = 0.5 \text{ L/mol s (or } 0.5 \text{ } M^{-1}s^{-1})$$

For the general reaction,

$$aA(g) + bB(g) \rightarrow \text{products}$$

the rate law takes the form

$$\text{Rate} = k[A]^m[B]^n$$

Rate laws are determined by experiment; we cannot predict the rate law from the balanced equation. The exponents m and n, called reaction orders, are not necessarily related to the coefficients of the reaction a and b.

How can we measure the concentration of the reactants at various times while the reaction is in progress? We could remove a sample of the reaction mixture and titrate it, or, if one of the substances has a color, we could monitor the change in color over time. If either the reactants or the products is ionic, we could monitor the change in conductivity of the reaction mixture with electrodes. If a reaction involves a change in the number of moles of a gas, the rate could be determined from the change in pressure with time.

Following reactions:

What method would you choose to study the change in concentration of the reactants or products for each of the following reactions?

1) $CH_3CHO(g) \rightarrow CH_4(g) + CO(g)$

2) $NO(g) + O_3(g) \rightarrow O_2(g) + NO_2(g)$

3) $CH_3COCH_3(l) + Br_2(l) \rightarrow CH_3COCH_3Br(l) + H^+(aq) + Br^-(aq)$

answers:

1) The rate of this reaction could be followed by measuring the change in pressure with time since 1 mole of gas reacts to form 2 moles of gas.

2) The rate of this reaction could be followed spectrometrically be measuring the formation of brown $NO_2(g)$.

3) This reaction rate could be monitored by measuring the increase in conductivity as H^+ and Br^- ions form. More likely, this reaction would be monitored by measuring the disappearance of the dark red color from liquid bromine.

Determining rate laws:

The Williamson synthesis is a major means to producing organic ethers from alkyl halides. The following Williamson synthesis at 25°C produced the rate data below.

$$CH_3I + NaOC_2H_5 \rightarrow CH_3OC_2H_5 + NaI$$

Use the data to determine the rate law for the reaction and the rate constant (k).

Expt.	Initial concentrations		Initial rate
	$[CH_3I]$ (M)	$[NaOC_2H_5]$ (M)	(M/s)
1	0.20	0.10	2.0×10^{-5}
2	0.40	0.10	4.0×10^{-5}
3	0.40	0.20	8.0×10^{-5}
4	0.10	0.20	2.0×10^{-5}

solution:

- The general form of the rate equation is:

$$\text{Rate} = k[CH_3I]^m[NaOC_2H_5]^n$$

- We can determine the value for m by comparing the rates from experiments 1 and 2, in which only the concentration of CH_3I changes:

$$\frac{\text{Rate}_2}{\text{Rate}_1} = \frac{4.0 \times 10^{-5} \ M \ s^{-1}}{2.0 \times 10^{-5} \ M \ s^{-1}} = \frac{k\left(0.40 \ M\right)^m \left(0.10 \ M\right)^n}{k\left(0.20 \ M\right)^m \left(0.10 \ M\right)^n}$$

$$2 = \frac{\left(0.40 \ M\right)^m}{\left(0.20 \ M\right)^m} = 2^m$$

$$m = 1$$

- We can determine the value for n by comparing the rates from experiments 2 and 3, in which only the concentration of $NaOC_2H_5$ changes:

$$\frac{\text{Rate}_3}{\text{Rate}_2} = \frac{8.0 \times 10^{-5} \ M \ s^{-1}}{4.0 \times 10^{-5} \ M \ s^{-1}} = \frac{k\left(0.40 \ M\right)^m \left(0.20 \ M\right)^n}{k\left(0.40 M\right)^m \left(0.10 M\right)^n}$$

$$2 = \frac{\left(0.20 \ M\right)^n}{\left(0.10 \ M\right)^n} = 2^n$$

$$n = 1$$

- The rate of this reaction is first order in both CH_3I and $NaOC_2H_5$. The overall reaction order is $n + m = 2$. The rate law is:

$$\text{Rate} = k[CH_3I][NaOC_2H_5]$$

- We can calculate the rate constant k from the results of any of the four experiments. Using the results from experiment 1 we have:

$$2 \times 10^{-5} \ M \ /s = k(0.20 \ M)(0.10 \ M)$$

$$2 \times 10^{-5} \ M \ /s = k(0.020 \ M^2)$$

$$k = \frac{2 \times 10^{-5} \ M \ s^{-1}}{0.020 \ M^2}$$

$$k = 1.0 \times 10^{-3} \ M^{-1} \ s^{-1}$$

It is a good idea to check the value for k with data from the other experiments. Note that the units for k depend on the order of the reaction and the units of time.

Often the exponents for rate equations are integers (0,1,2); in many reactions, however, they are fractions. When the product influences the reaction rate, the exponents for the rate equation may be negative meaning that an increase in that component decreases the reaction rate.

SECTION 15.3: Integrated Rate Laws

The rate equation in the example above tells us how the rate of the reaction changes with concentration, but it does not tell us how the rate and concentrations change with time. Usually this question is the one of most practical interest: how much carbon monoxide is left after several minutes? several days? The form of the equations relating concentration to time depends on the reaction order. We can integrate the rate laws over time to obtain the integrated rate laws.

Integrated Rate Laws:

Zero order (rate = k):

$$[A]_0 - [A]_t = kt$$

First order (rate = k[A]):

$$\ln [A]_0 - \ln [A]_t = kt$$

Second order (rate = k[A]^2):

$$\frac{1}{[A]_t} - \frac{1}{[A]_o} = kt$$

Integrated rate laws:

Rearrange the integrated rate laws to put them in the form of an equation for a straight line, y = mx + b. For each equation give the slope and the y-axis intercept.

solutions:

- Zero order: $[A]_t = -kt + [A]_0$

    A plot of $[A]_t$ vs. t will have a slope of -k and a y-axis intercept of $[A]_0$.

- First order: $\ln [A]_t = -kt + \ln [A]_0$

    A plot of $\ln [A]_t$ vs. t will have a slope of -k and a y-axis intercept of $\ln [A]_0$.

- Second order: $\frac{1}{[A]_t} = kt + \frac{1}{[A]_o}$

    A plot of $\frac{1}{[A]_t}$ vs. t will have a slope of -k and a y-axis intercept of $\frac{1}{[A]_o}$.

If you wish to determine the reaction order from concentration-time data, it may take several plots. If you plot reactant vs. time and obtain a straight line, the reaction is zero order with respect to that reactant. If you plot ln [reactant] vs. time and obtain a straight line, the reaction is first order with respect to that reactant. If you plot 1/[reactant] vs. time, the reaction is second order with respect to that reactant.

The decomposition of hydrogen peroxide:

The decomposition of hydrogen peroxide is a useful way to obtain oxygen:.

$$2 H_2O_2(l) \rightarrow 2 H_2O(l) + O_2(g)$$

It is a first order reaction with a rate constant of 0.0410 min^{-1}.

1) If we start with a 3% hydrogen peroxide solution (a typical solution sold in grocery stores), what will be its concentration after 30 minutes?

2) How long will it take for its concentration to drop to 2%?

3) How long will it take for one-half of the sample to decompose?

solutions:

1) • Since it is a first order reaction, the integrated rate law is:

$$\ln [A]_0 - \ln [A]_t = 0.0410 \text{ min}^{-1} \times t, \text{ or:}$$

$$\ln[A]_t - \ln[A]_0 - \left(0.0410 \text{ min}^{-1} \times t\right)$$

• The concentration in mol/L of a 3% H_2O_2 solution is:

$$\frac{3 \text{ g } H_2O_2}{100 \text{ g water}} \times \frac{\text{mol}}{34 \text{ g } H_2O_2} \times \frac{1g}{1 \text{ mL}} \times \frac{1000 \text{ mL}}{L} - 0.882 \text{ mol/L}$$

• Its concentration after 30 min. would be:

$$\ln[A]_{30 \text{ min}} - \ln\left(0.882 \text{ mol/L}\right) - \left(0.0410 \text{ min}^{-1} \times 30 \text{ min}\right)$$

$$[A]_{30 \text{ min}} - e^{\left[\ln(0.882)-(0.0410)(30)\right]} - 0.258 \text{ mol/L}$$

2) • The concentration of a 2% H_2O_2 solution is two-thirds of the concentration of a 3% solution:

$$0.882 \text{ mol/L} \times \frac{2}{3} - 0.588 \text{ mol/L}$$

• The time it would take for its concentration to drop from 3% to 2% is:

$$t - \frac{\ln\left(0.882 \text{ mol/L}\right) - \ln\left(0.588 \text{ mol/L}\right)}{0.0410 \text{ min}^{-1}}$$

$$t = 9.89 \text{ min.}$$

3) The time it takes for half the sample (0.441 mol/L) to decompose is:

$$t = \frac{\ln (0.882 \text{ mol/L}) - \ln (0.441 \text{ mol/L})}{0.0410 \text{ min}^{-1}}$$

t = 16.9 min.

How useful is hydrogen peroxide as a way to obtain oxygen? It doesn't appear to be useful since the reaction is slow. However, the rate of decomposition of hydrogen peroxide can be increased greatly with yeast which catalyzes the reaction. Try it.

The time it takes for half of the hydrogen peroxide to decompose is its half-life. The half-life of a first-order reaction is independent of reactant concentration. We can see this from the integrated first-order rate law:

$$\ln [A]_0 - \ln [A]_t = kt$$

$$\ln [A]_0 - \ln \frac{1}{2}[A]_0 = kt_{\frac{1}{2}}$$

$$\ln \frac{[A]_0}{\frac{1}{2}[A]_0} = kt_{\frac{1}{2}}$$

$$\ln 2 = kt_{\frac{1}{2}}$$

$$t_{\frac{1}{2}} = \ln \frac{2}{k} = \frac{0.693}{k} \quad \text{(first-order process)}$$

Half-life of hydrogen peroxide solution:

1) How long does it take for one-half of a 0.882 M hydrogen peroxide solution to decompose? (The decomposition of hydrogen peroxide is a first order reaction with k = 0.0410 min^{-1}).

2) How long does it take for one-half of a 0.441 M hydrogen peroxide solution to decompose?

solution:

1) Since the decomposition of hydrogen peroxide is a first order reaction, we use the equation:

$$t_{\frac{1}{2}} = \frac{0.693}{k}$$

$$t_{\frac{1}{2}} = \frac{0.693}{0.0410 \text{ min}^{-1}}$$

$$t_{\frac{1}{2}} = 16.9 \text{ min}$$

2) The half-life is independent of starting concentration, so the amount of time it takes for half of <u>any</u> concentration hydrogen peroxide to decompose is 16.9 min.

Half-life equations:

Zero order (rate = k):

$$t_{\frac{1}{2}} = \frac{1}{2} \frac{[A]_o}{k}$$

First order (rate = k[A]):

$$t_{\frac{1}{2}} = \frac{0.693}{k}$$

Second order (rate = k[A]^2):

$$t_{\frac{1}{2}} = \frac{1}{k[A]_o}$$

SECTION 15.4: Dependence of Reaction Rates on Temperature

Most chemical reactions proceed faster as temperature increases. We store food in the refrigerator to slow down chemical reactions which spoil food. We use pressure cookers to increase the temperature at which food cooks. An <u>approximate</u> rule is that an increase of 10°C doubles the reaction rate.

Recall from Chapter 5 that an increase in temperature increases the average kinetic energy of molecules. At higher temperatures, a greater fraction of collisions between molecules have enough energy to overcome the <u>energy barrier</u> to reaction. We can think of this energy barrier as the amount of energy required to weaken the bonds holding reactant molecules together allowing them to rearrange to form products. This energy barrier is called **activation energy, E_a.**

Not all collisions which have enough energy to overcome the energy barrier result in reaction. The molecules must have the correct <u>orientation</u> to react. The Arrhenius equation relates temperature, activation energy, and molecular orientation to the rate constant, k:

$$\ln\ k = \frac{-E_a}{R} \times \frac{1}{T} + \ln\ A$$

where: k is the rate constant

T is the absolute temperature

R is the universal gas constant

E_a is the activation energy

A is a term which takes into account the frequency of collisions, and the fraction of collisions with an effective orientation.

If we measure the rate constant of a reaction at different temperatures, we can determine the energy of activation by plotting ln k versus 1/T. The slope of this line is $-E_a/R$.

Exercise:

The reaction:

$$2NOCl(g)\ \rightarrow\ 2NO(g)\ +\ Cl_2(g)$$

has an activation energy, E_a, of 1.00×10^2 kJ/mol, and a rate constant of 0.286 s^{-1} at 500 K. What is the rate constant at 400 K?

<u>solution</u>:

• We know: T_1 = 500 K
  T_2 = 400 K
  k_1 = 0.286 s^{-1}
  E_a = 100 kJ/mol = 1.00×10^5 J/mol

• Use the Arrhenius equation to write equations for k_1 and k_2.

$$\ln\ k_2 = \frac{-E_a}{R} \times \frac{1}{T_2} + \ln\ A , \text{ and } \ln\ k_1 = \frac{-E_a}{R} \times \frac{1}{T_1} + \ln\ A$$

• If we subtract k_1 from k_2, we obtain the expression:

$$\ln \frac{k_2}{k_1} = \frac{-E_a}{R} \left(\frac{1}{T_2} - \frac{1}{T_1} \right)$$

• Now we can substitute values in to the equation and solve for k_2:
  (Make sure you feel comfortable with the math in these problems)

$$\ln\frac{k_2}{0.286\ s^{-1}} = -\frac{1.00 \times 10^5\ J/mol}{8.3148\ J/K\,mol}\left(\frac{1}{400\ K} - \frac{1}{500\ K}\right)$$

$$\ln\frac{k_2}{0.286\ s^{-1}} = -6.01$$

$$\frac{k_2}{0.286\ s^{-1}} = 2.44$$

$$k_2 = 6.99 \times 10^{-4}\ s^{-1}$$

Let's see how well the approximation holds that a 10°C increase in temperature doubles the rate of the reaction . The temperature change between 500 K and 400 K is 100 K, or ten 10 degree temperature changes. If we double the rate constant k_2 (6.99×10^{-4}) 10 times, we obtain 0.715. This number does not equal k_1 (0.286), but it is the same order of magnitude.

Collision theory and transition state theory:

Choose words from the list below to complete the following paragraphs:

a) activated complex   f) orientation
b) activation energy   g) products
c) collide   h) reversible
d) energy   i) speed
e) one

Molecules must _____(1) in order to react. Most collisions do not result in the formation of _____(2). The number of collisions is usually much, much larger than the rate of the reaction. Molecules must collide with enough _____(3) and with the proper _____(4) to react. An increase in temperature increases the average _____(5) of particles, and therefore the frequency which they _____(6). The effect of temperature on reaction rate, however, is primarily due to the increase in the fraction of collisions with enough energy to exceed the _____(7) threshold. In the Arrhenius equation, the factor "A" is the product of the collision frequency (Z) and an orientation probability factor (p) which is specific for each reaction. Collisions between individual atoms have p values near _____(8) since almost no matter how they collide, they react.

When molecules collide in an effective orientation and have enough energy to overcome the activation energy, they form a high energy species with partial bonds called an _____(9), or transition state. Every reaction goes through a transition state from which point it can either continue on to form product, or fall back apart into reactants. This theory implies that all reactions are _____(10).

<u>answers</u>: 1-c; 2-g; 3-d; 4-f; 5-i; 6-c; 7-b; 8-e; 9-a; 10-h.

SECTION 15.5: Reaction Mechanisms

A reaction mechanism describes a sequence of elementary steps by which a reaction occurs. Usually the elementary steps are **unimolecular**, (one molecule decomposing), or **bimolecular** (two molecules combining). A third less likely event is a **termolecular** reaction (three particles colliding). <u>For elementary reactions, the reaction order equals the number of reactants</u>. The slowest elementary step limits how fast the overall reaction proceeds and is called the **rate-limiting step**.

Atmospheric chemistry:

Nitrogen oxide, a pollutant from automobile gas, reacts with oxygen to produce the colorless gas, N_2O_4, which is in equilibrium with the brownish gas, NO_2. The reaction mechanism is:

$$NO(g) + O_2(g) \rightleftharpoons NO_3(g) \qquad \text{pre-equilibrium}$$
$$NO(g) + NO_3(g) \rightarrow N_2O_4(g) \qquad \text{rate determining step}$$
$$N_2O_4(g) \rightleftharpoons 2NO_2(g) \qquad \text{equilibrium}$$

What is the rate law for this reaction mechanism?

<u>solution</u>:

• We use the rate limiting step to write an expression for the rate:

$$\text{Rate} = k_2\,[NO][NO_3]$$

• The rate law should not contain intermediate species (such as NO_3), so we use the pre-equilibrium to write the concentration of NO_3 in terms of reactants. In an equilibrium, the forward and reverse reaction rates are equal:

$$k_1[NO][O_2] = k_{-1}[NO_3]$$

$$\left[NO_3\right] = \frac{k_1}{k_{-1}}\left[NO\right]\left[O_2\right]$$

- Now substitute this expression for $[NO_3]$ into the rate law for the slow step:

$$Rate = \frac{k_2 k_1}{k_{-1}} [NO]^2 [O_2]$$

- Often all the constants are grouped together into one constant: $K = \frac{k_2 k_1}{k_{-1}}$, and

$$Rate = K[NO]^2[O_2]$$

The rate law is second order in NO. In the lab, if you squirt NO into the air, the brown color of NO_2 appears almost immediately. The reaction is _much_ slower in the atmosphere owing to the low concentration of NO, and the second order dependence of reaction.

Reaction pathway:

The reaction of hydrogen with iodine produces hydrogen iodide or hydroiodic acid:

$$H_2(g) + I_2(g) \rightarrow 2HI(g)$$

The experimentally determined rate law is: rate = $k[H_2][I_2]$. Which of the following mechanisms is consistent with this rate law?

1) proposed pathway #1:

$$H_2 + I_2 \rightarrow 2HI$$

2) proposed pathway #2:

$$I_2 \rightarrow 2I \text{ (slow)}$$
$$H_2 + 2I \rightarrow 2HI \text{ (fast)}$$

3) proposed pathway #3:

$$I_2 \rightleftharpoons 2I \text{ (fast)}$$
$$H_2 + 2I \rightarrow 2HI \text{ (slow)}$$

4) proposed pathway #4:

$$I_2 \rightleftharpoons 2I \text{ (fast)}$$
$$H_2 + I \rightarrow H_2I \text{ (fast)}$$
$$H_2I + I \rightarrow 2HI \text{ (slow)}$$

<u>answer</u>:

The rate laws for the three proposed pathways are:

1) rate = $k[H_2][I_2]$

2) rate = $k[I_2]$

3) The rate law for the slow step is: rate = $k_2[H_2][I]^2$
   For the first step, the forward and reverse reaction rates are equal at equilibrium:

$$k_1[I_2] = k_{-1}[I]^2$$

$$[I]^2 = \frac{k_1}{k_{-1}}[I_2]$$

Substituting the expression for $[I]^2$ into the rate law for the slow step gives:

$$\text{rate} = k_2[H_2]\frac{k_1}{k_{-1}}[I_2]$$

$$\text{rate} = k[H_2][I_2] \text{ where } k = \frac{k_2 k_1}{k_{-1}}$$

4) The rate law for the slow step is: rate = $k_3[H_2I][I]$

The first step is reversible, so we have:

$$k_1[I_2] = k_{-1}[I]^2$$

$$[I]^2 = \frac{k_1}{k_{-1}}[I_2] \quad \text{or} \quad [I] = \left(\frac{k_1}{k_{-1}}[I_2]\right)^{\frac{1}{2}}$$

Using the same approach with the second step we have:

$$[H_2I] = \frac{k_2}{k_{-2}}[H_2][I]$$

and substituting the relationship for [I]:

$$[H_2I] = \frac{k_2}{k_{-2}}[H_2]\left(\frac{k_1}{k_{-1}}[I_2]\right)^{\frac{1}{2}}$$

Substituting the expressions for the intermediates into the rate law for the slow step gives:

$$\text{rate} = \frac{k_3 k_2}{k_{-2}}[H_2]\left(\frac{k_1}{k_{-1}}[I_2]\right)^{\frac{1}{2}} \times \left(\frac{k_1}{k_{-1}}[I_2]\right)^{\frac{1}{2}}$$

$$\text{rate} = k[H_2][I_2] \text{ where } k = \frac{k_3 \, k_2 \, k_1}{k_{-1} \, k_{-2}}$$

Proposed pathways #1, #3, and #4 are consistent with the experimentally determined rate law. Which pathways are consistent with the spectroscopic detection of free iodine atoms?

<u>answer</u>:

Mechanisms #3 and #4 both propose the existence of free iodine atoms. Mechanism #4 is a more likely reaction pathway than #3 because pathway #3 includes a less likely termolecular reaction between a hydrogen molecule and two free iodine atoms. We cannot <u>prove</u> that a particular mechanism is the one actually taken during the reaction, we simply use mechanisms which have no data to refute them.

- The overall rate law includes only the species up to and including those in the rate-determining step. For reaction pathway #2, in the example above, the species in the fast step <u>after</u> the slow step are not included in the rate law.

- Reaction intermediates do not appear in the overall rate law.

SECTION 15.6: Catalysts

A catalyst is a substance that increases the rate of reaction without being consumed in the reaction. A catalyst speeds up a reaction by providing a different reaction mechanism with lower activation energy:

Homogeneous catalysts exist in solution with the reaction mixture. . Heterogeneous catalysts exist as a separate phase from the reaction. Often they are solids interacting with gases or liquids. Enzymes are incredibly efficient biological catalysts which have features of both homogeneous and heterogeneous catalysts.

Chapter 15 Exercises:

15.1) The portable breathalyzers used to check suspects for drunk driving uses the oxidation of ethyl alcohol by dichromate ion in the test.

$$3C_2H_5OH + Cr_2O_7^{2-} + 8H^+ \rightarrow 3CH_3CHO + 2Cr^{3+} + 7H_2O$$

The results of the rate law study, at 25°C are given below.

Exptl. Trial	Initial concentration (M)			Initial Rate ($M min^{-1}$)
	$[C_2H_5OH]$	$[Cr_2O_7^{2-}]$	$[H^+]$	
1	0.15	0.05	0.010	5.0×10^{-4}
2	0.30	0.05	0.010	1.0×10^{-3}
3	0.30	0.10	0.010	2.0×10^{-3}
4	0.30	0.05	0.100	1.0×10^{-3}

a. Determine the rate law, including the rate constant (k) for the reaction.
b. What is the overall order of the reaction?

15.2) The hydrolysis reaction:

produces the experimental rate law data:

Exptl. Trial	Initial concentration (M)		Initial Rate ($M min^{-1}$)
	$[C_9H_{11}Cl]$	$[H_2O]$	
1	0.12	0.20	1.4×10^{-3}
2	0.36	0.20	4.3×10^{-3}
3	0.36	0.30	4.3×10^{-3}
4	0.24	0.10	2.9×10^{-3}

a. Determine the rate law, including the rate constant, k for the reaction.
b. What is the overall order of the reaction?

15.3) Under high temperature conditions, the following rate law data was obtained for the reaction below:

Exptl. Trial	Initial concentration (M)		Initial Rate ($M min^{-1}$)
	$[NO_2]$	$[CO]$	
1	0.050	0.050	0.031
2	0.075	0.050	0.070
3	0.075	0.10	0.070

$$NO_2(g) + CO(g) \rightarrow NO(g) + CO_2(g)$$

a. Determine the rate law, including the rate constant, k, for the reaction.
b. What is the overall order of the reaction?

15.4) The rate of gas phase dehydration of t-butyl alcohol is temperature dependent.

$$C_4H_{10}O(g) \rightarrow C_4H_8(g) + H_2O(g)$$

Temperature (K)	k (1/sec)
600	4.5×10^{-9}
800	4.1×10^{-4}
1000	0.39
1200	38

Calculate the activation energy, E_a, and the Arrhenius constant, A, for the reaction.

15.5) The rate of decarboxylation of ethyl chloroformate is temperature dependent.

$$ClCOOC_2H_5 \rightarrow CO_2 + C_2H_5Cl$$

Temperature (K)	k (1/sec)
473	0.0013
523	0.026
573	0.31
623	2.43

Calculate the activation energy, E_a, and the Arrhenius constant, A, for the reaction.

15.6) Radioactive isotopes undergo a first-order rate of decay. Iodine-131 has a half life of 8.0 days.

a. What is the rate constant for the decay?
b. Calculate the time required for a 200 mg sample of ^{131}I to decay to 10 mg.

15.7) The dimerization of tetrafluoroethylene at 403 K proceeds according to the rate law:
Rate = 1.6×10^{-3} $M^{-1}s^{-1}[C_2F_4]^2$

a. 0.80 mole of C_2F_4 is injected into a 1.0 L reaction chamber at 403 K. How much C_2F_4 is left after 1 hour?
b. What is the half-life of the reaction?

15.8) Oxygen difluoride undergoes decomposition at elevated temperatures. The concentration vs. time data at 523 K is given in the table below.

Time (min)	0	2	4	6	8	10	12
$[OF_2]$ M	0.200	0.150	0.120	0.100	0.085	0.075	0.066

Graph the data to determine the order of reaction and the rate constant, k.

15.9) Sulfuryl chloride undergoes decomposition at elevated temperatures. The concentration vs. time data at 593 K is given in the table below.

Time (hr)	0	2	4	6	8	10	12
$[SO_2Cl_2]$ M	0.50	0.43	0.36	0.31	0.27	0.23	0.19

Graph the data to determine the order of reaction and the rate constant, k.

15.10) For many exothermic reactions, a 10°C increase in temperature will double the rate of the reaction. Explain using the collision theory and transition rate theory.

15.11) The addition of a catalyst will increase the rate of a chemical reaction. Use the collision theory and transition state theory to explain why this occurs.

Chapter 15 Answers:

15.1) a. Rate = $6.7 \times 10^{-2} \ M^{-1}min^{-1}[C_2H_5OH][Cr_2O_7{}^{2-}]$
       b. second order

15.2) a. Rate = $1.2 \times 10^{-2} \ min^{-1}[C_9H_{11}Cl]$
       b. first order

15.3) a. Rate = $12 \ M^{-1}min^{-1}[NO_2]^2$
       b. second order

15.4) $E_a = 2.27 \times 10^5$ J or 227 kJ; A = 2.8×10^{11}

15.5) $E_a = 120$ kJ; A = 5.0×10^{10}

15.6) a. k = $8.7 \times 10^{-2} \ day^{-1}$
       b. 34 days

15.7) a. 0.14 mol
       b. 780 s

15.8) Plot of $1/[OF_2]$ vs. Time (min) gives slope = k = $0.846 \ M^{-1}min^{-1}$

15.9) Plot of $\ln[SO_2Cl_2]$ vs. Time (hr) gives slope = -k = $-0.078 \ hr^{-1}$

15.10) a. An increase in temperature will increase the kinetic energy thus an increase in molecular speed. The increase in molecular speed will increase the frequency of collision which is one of the factors in the Collision Theory of reaction rates.

       b. An increase in temperature will increase the fraction of molecules with sufficient energy to break and/or make bonds (E_a).

15.11) A catalyst will increase the rate of a reaction by enhancing the proper orientation of the reactants and/or lowering the activation energy by weakening existing bonds.

CHAPTER 16
Equilibrium: The Extent of Chemical Reactions

TOPICS:

- The Equilibrium Constant
  - Equilibrium calculations
  - Approximations
  - Relationship between K_c and K_p
  - K_c for overall reactions

- Reaction progress
  (predicting the direction of reaction)

- Le Chatelier's Principle

INTRODUCTION:

We saw in Chapters 11 and 12 that a closed container of water is at equilibrium when the rate that water molecules escape from the liquid equals the rate that water molecules in the vapor state return to the liquid water. Although it is a dynamic system (water molecules continually move between the liquid and gaseous states), we see the system as static since the water level does not change. The chemical equation for the equilibrium is:

$$H_2O(l) \rightleftharpoons H_2O(g)$$

The position of this equilibrium (the relative amounts of the liquid and gas) depends on the temperature of the system. At a particular temperature, the number of gaseous water molecules is fixed.

Vapor pressure of water:

1) What is the equilibrium vapor pressure in each 1 L beaker shown below at 50°C? at 100°C? What is the equilibrium concentration of gaseous water in each beaker at 50°C? at 100°C?

900 mL liquid water
(a)

10 mL liquid water
(b)

1 mL liquid water
(c)

<u>solutions</u>:

- The vapor pressure in each beaker at 50°C is 92.51 mm Hg (0.122 atm.)

- The vapor pressure in each beaker at 100°C is 760 mm Hg (1 atm.)
  (You can look these values up in the Handbook of Chemistry and Physics)

- We can use the ideal gas law to calculate the equilibrium concentration of $H_2O(g)$ at 50°C and 100°C:

  $$PV = nRT \text{ so } n = PV/RT$$

  at 50°C, $\quad n = \dfrac{(0.122\,atm)\,(1\,L)}{(0.0821L\ atm/mol\ K)\,(323\ K)}$

  $$n = 0.00460 \text{ mol}$$

  The concentration of $H_2O(g)$ is therefore 0.0046 mol/L.

  at 100°C, $\quad n = \dfrac{(1atm)\,(1\,L)}{(0.0821L\ atm/mol\ K)\,(373\ K)}$

  $$n = 0.0327 \text{ mol}$$

  Concentration = 0.0327 mol/L

2) Does 1 mL of water in a 1 liter container at 100°C maintain an equilibrium between the liquid and gaseous states, or does all the liquid water evaporate?

  solution:

  - Moles of water molecules in 1 mL of water are:

    $$1\ mL \times \frac{1\,g}{mL} \times \frac{1\,mol}{18\,g} = 0.055 \text{ mol}$$

  - Since the equilibrium concentration of gaseous water at 100°C is 0.0327 mol/L (problem #1 above), a mL of liquid water contains more than enough water molecules to provide this concentration, and not all of the water evaporates.

3) What is the vapor pressure of 1/2 mL of water at 100°C in a 1 liter flask?

  solution:

  - Moles of water molecules in 0.5 mL water are:

    $$0.5\ mL \times \frac{1\,g}{mL} \times \frac{1\,mol}{18\,g} = 0.0278 \text{ mol}$$

  - We use the ideal gas law to solve for pressure, P

$$P = \frac{nRT}{V}$$

$$P = \frac{(0.0278 \text{ mol})(0.0821 \text{ L atm/mol K})(373 \text{ K})}{1 \text{ L}}$$

$$P = 0.851 \text{ atm.}$$

The vapor pressure falls below 1 atm. because all the water in the flask evaporates, and there is not enough to reach a pressure of 1 atm. (There must be about 0.6 mL of water in a 1 L flask to establish an equilibrium between the liquid and gaseous phases).

For a given temperature, the equilibrium position does not depend on the amount of liquid water present, and we need to know only the concentration of gaseous water molecules to give us the equilibrium position.

SECTION 16.1:   The Equilibrium Constant

Many chemical systems have more than one gaseous species present. In these systems, we cannot express the equilibrium concentration of one species independent from the others. The equilibrium between N_2O_4, (a colorless gas), and NO_2 (a brown gas) is described in your text on page 696.

$$N_2O_4(g) \rightleftharpoons 2NO_2(g)$$

At a given temperature, we could obtain an infinite number of equilibrium concentrations of N_2O_4 and NO_2 depending on how we prepare the system. It turns out that there is a mathematical relationship between all sets of concentrations we could obtain. The quotient,

$$\frac{\left[NO_2\right]^2}{\left[N_2O_4\right]}$$

equals 0.36 at 100°C regardless of the amounts of reaction species we start with, the volume of the container, or the total pressure. We call this constant K_c, the equilibrium constant. At 150°C, the equilibrium constant for this reaction is 3.2. The brackets indicate concentrations are in moles per liter (M), and the subscript "c" on the equilibrium constant tells us the equilibrium is expressed in terms of concentrations.

The numerator of the quotient contains the concentration of the product raised to the power of its coefficient in the balanced equation (2), and the denominator of the quotient contains the concentration of the reactant raised to the power of its coefficient in the balanced equation (1). For every gaseous system, we can write an analogous expression for the equilibrium constant.

Equilibrium constant expressions:

1) Write the equilibrium constant expression for each of the following reactions:

   1) $N_2(g) + 3H_2(g) \rightleftharpoons 2NH_3(g)$

   2) $C_2H_6(g) + 7/2O_2(g) \rightleftharpoons 2CO_2(g) + 3H_2O(g)$

   3) $2C_2H_6(g) + 7O_2(g) \rightleftharpoons 4CO_2(g) + 6H_2O(g)$

   4) $2HI(g) \rightleftharpoons H_2(g) + I_2(g)$

   5) $H_2(g) + I_2(g) \rightleftharpoons 2HI(g)$

   6) $CaCO_3(s) \rightleftharpoons CaO(s) + CO_2(g)$

   answers:

   1) $K_c = \dfrac{[NH_3]^2}{[N_2][H_2]^3}$

   2) $K_c = \dfrac{[CO_2]^2[H_2O]^3}{[C_2H_6][O_2]^{\frac{7}{2}}}$

   3) $K_c = \dfrac{[CO_2]^4[H_2O]^6}{[C_2H_6]^2[O_2]^7}$

   4) $K_c = \dfrac{[H_2][I_2]}{[HI]^2}$

   5) $K_c = \dfrac{[HI]^2}{[H_2][I_2]}$

   6) $K_c = [CO_2]$

2) What is the relationship between the equilibrium expressions for equations 2 and 3?

   solution:

   K_c (eqn. 3) = K_c^2 (eqn. 2)

3) What is the relationship between the equilibrium expressions for equations 4 and 5?

   solution:

   K_c (eqn. 4) = $1/K_c$ (eqn. 5)

The above example illustrates two important points about writing expressions for K_c:

1) The concentration of a pure liquid or solid does not appear in the expression of K_c.

2) The numerical value of K_c is meaningful only when it is associated with a particular chemical equation.

Equilibrium calculations:

1) For the reaction of nitrogen with hydrogen to produce ammonia,

$$N_2(g) + 3H_2(g) \rightleftharpoons 2NH_3(g)$$

the equilibrium concentrations of N_2, H_2, and NH_3 are 0.15, 0.80, and 0.20 mol/L respectively at 400°C. What is the value of K_c?

solution:

$$K_c = \frac{[NH_3]^2}{[N_2][H_2]^3} = \frac{(0.20)^2}{(0.15)(0.80)^3} = 0.52$$

2) For the same reaction at a different temperature, if one starts with 0.18 mol/L of both hydrogen and nitrogen, the equilibrium concentration of ammonia is 0.040 mol/L. Calculate K_c.

solution:

- We know the equilibrium concentration of ammonia, but we know only the starting concentrations of hydrogen and nitrogen. In cases such as these, it can be helpful to make a table.

	N_2	$+$	$3H_2$	$=$	$2NH_3$
Initial conc.	0.18		0.18		0
Change	$-x$		$-3x$		$+2x$
Equilibrium conc.	0.18-x		0.18-3x		0+2x

- We can solve for x since we know the equilibrium concentration of NH_3:

$$2x = 0.040, \text{ so } x = 0.020,$$

and solve for moles of N_2 and H_2 at equilibrium:

$$\text{conc. } N_2 = 0.18 - x = 0.18 - 0.020 = 0.16$$

$$\text{conc. } H_2 = 0.18 - 3x = 0.18 - 0.060 = 0.12$$

- Now calculate the value of K_c at the new temperature:

$$K_c = \frac{[NH_3]^2}{[N_2][H_2]^3} = \frac{(0.040)^2}{(0.16)(0.12)^3} = 5.79$$

Determining equilibrium concentrations:

When gasoline burns in an automobile, the temperature is high enough that nitric oxide forms from the elements nitrogen and oxygen.

$$N_2(g) + O_2(g) \rightleftharpoons 2NO(g)$$

At 2000°C, K_c for this reaction is 0.10. What are the equilibrium concentrations of NO, N_2 and O_2, when the starting concentration of each reactant is 1 mol/L?

solution:

• We can use the same table as above to express initial and equilibrium concentrations for all species.

	N_2	$+$	O_2	$=$	$2NO$
Initial conc.	1		1		0
Change	- x		- x		+2x
Equilibrium conc.	1 - x		1 - x		2x

• In this case, we do not know any of the equilibrium concentrations, but we know the value for the equilibrium constant, K_c:

$$K_c = 0.10 = \frac{[NO]^2}{[N_2][O_2]} = \frac{(2x)^2}{(1 - x)(1 - x)}$$

We could solve this equation using the "quadratic formula", but often there is a way out of that. In this case, the right-hand side of the equation is a perfect square, so we can simply take the square root of both sides to solve for x:

$$(0.10)^{\frac{1}{2}} = 0.32 = \frac{2x}{1 - x}$$

$$0.32 - 0.32x = 2x$$

$$x = 0.14$$

• Now we can use values from our table to solve for the concentrations of all species:

$$[N_2] = [O_2] = 1-x = 0.86 \text{ mol/L}$$

$$[NO] = 2x = 0.28 \text{ mol/L}$$

Approximations:

At a certain temperature, K_c for the reaction between N_2 and O_2 to form NO is 3.2×10^{-4}. What is the equilibrium concentration of NO when 0.150 moles of N_2 and 0.30 moles of O_2 are added to a 1.50 L container?

solution:

$$N_2(g) + O_2(g) \rightleftharpoons 2NO(g)$$

	N_2	+	O_2	=	2NO
Initial conc.	0.10		0.20		0
Change	- x		- x		+2x
Equilibrium conc.	0.10-x		0.20-x		2x

$$K_c = 3.2 \times 10^{-4} = \frac{[NO]^2}{[N_2][O_2]} = \frac{(2x)^2}{(0.10-x)(0.20-x)}$$

• Instead of using the quadratic formula to solve this equation, lets assume that since K_c is small, very little NO forms and x is very small compared to 0.10 or 0.20. We will assume that 0.10-x ≈ 0.10, and 0.20-x ≈ 0.20. Now the equation to solve becomes:

$$K_c = 3.2 \times 10^{-4} = \frac{(2x)^2}{(0.10)(0.20)}$$

Solving for x gives us:

$$4x^2 = 6.4 \times 10^{-6}$$

$$x = 1.3 \times 10^{-3}$$

• Our assumption seems to be OK because x is small compared to 0.1 and 0.2 (1.3×10^{-3} is less than 5% of 0.1).

$$[NO] = 2x = 2 \times (1.3 \times 10^{-3}) = 2.6 \times 10^{-3}$$

Relationship between K_c and K_p:

Often it is easier to measure the pressure of a gas than its concentration. K_p is the equilibrium constant obtained when the concentrations of all species are expressed as their partial pressures in atmospheres. The ideal gas law can be used to related K_c and K_p:

$$K_p = K_c(RT)^{\Delta n(gas)}$$

where $\Delta n(gas)$ is moles of gaseous products - moles gaseous reactants.

Calculating K_p from K_c:

Calculate K_p for the following reaction:

$$PCl_3(g) + Cl_2(g) \rightleftharpoons PCl_5(g) \qquad K_c = 1.67 \text{ (at 500 K)}$$

solution:

There is one mole of gaseous products and 2 moles of gaseous reactants, so $\Delta n_{gas} = 1-2 = -1$.

$$K_p = K_c(RT)^{\Delta n(gas)} = 1.67[(0.0821)(500)]^{-1} = 0.0407$$

K_c for an overall reaction:

Consider the following reactions:

$$SO_2(g) + 1/2O_2(g) \rightleftharpoons SO_3(g)$$

$$NC_2(g) \rightleftharpoons NO(g) + 1/2O_2(g)$$

1) What is the expression for K_c for each reaction?

solution:

$$K_c = \frac{[SO_3]}{[SO_2][O_2]^{\frac{1}{2}}} \qquad\qquad K_c = \frac{[NO][O_2]^{\frac{1}{2}}}{[NC_2]}$$

2) Show the equation which is the sum of the above two equations and the K_c expression for it.

solution:

$$SO_2(g) + NO_2(g) \rightleftharpoons SO_3(g) + NO(g)$$

$$K_c = \frac{[SO_3][NO]}{[SC_2][NC_2]}$$

3) What is the relationship between the K_c for the overall reaction and the K_c's for the two reactions in step #1 which add to give the overall reaction?

solution:

If we multiply the K_c's for the two reactions in step #1, we obtain K_c for the overall reaction in step #2:

$$\frac{[SO_3]}{[SO_2][O_2]^{\frac{1}{2}}} \times \frac{[NO][O_2]^{\frac{1}{2}}}{[NO_2]} = \frac{[SO_3][NO]}{[SO_2][NO_2]}$$

If an overall reaction is the sum of two or more reactions, K for the overall reaction is the product of the equilibrium constants of the individual reactions.

SECTION 16.2:   Reaction quotient

How can we tell if a reaction mixture is at equilibrium, and if it is not at equilibrium, how can we tell which direction the reaction proceeds in order to reach equilibrium? We use the mass-action expression (Q), often called the reaction quotient. Q has the same form as the expression for K, but it contains actual concentrations in place of equilibrium concentrations. For the reaction,

$$aA + bB \rightleftharpoons cC + dD$$

the reaction quotient (Q) is:

$$Q_c = \frac{[C]^c[D]^d}{[A]^a[B]^b}$$

When the system is at equilibrium, $Q_c = K_c$. If Q_c is less than K_c, the reaction proceeds to the right to consume reactants (decreasing the denominator in the above quotient), and produce products (increasing the numerator). If Q_c is greater than K_c, the reverse reaction occurs to produce reactants and consume products.

Predicting the direction of reaction:

We saw earlier that a reaction which takes place in automobiles is:

$$N_2(g) + O_2(g) \rightleftharpoons 2NO(g); \quad K_c = 0.10 \ (2000°C)$$

Predict the direction the system will move to attain equilibrium at 2000°C if we add:

a) 1 mole of nitrogen and 1 mole of oxygen to a 1 L container.

b) 1 mole of nitrogen, 1 mole of oxygen, and 1 mole of NO to a 10 L container.

solutions:

• The Q_c expression for the reaction is:

$$Q_c = \frac{[NO]^2}{[N_2][O_2]}$$

• For situation a), we have:

initial concentration of N_2 = 1 mol/L

initial concentration of O_2 = 1 mol/L

initial concentration of NO = 0

$Q_c = 0$, since [NO] = 0. The reaction will proceed to the right to produce NO.

• For situation b), we have:

initial concentration of N_2 = 1 mol/10 L = 0.1 mol/L

initial concentration of O_2 = 1 mol/10 L = 0.1 mol/L

initial concentration of NO = 1 mol/10 L = 0.1 mol/L

$$Q_c = \frac{(0.1)^2}{(0.1)(0.1)} = 1$$

$Q_c > K_c$ so the reverse reaction will take place. The concentration of NO decreases and the concentration of N_2 and O_2 increase until the quotient equals 0.10, and the system attains equilibrium.

Le Chatelier's Principle states that if we impose a change on a system at equilibrium, the system shifts to minimize the effect of the change. We will discuss how a system at equilibrium shifts to minimize the effects of:

- changing the concentration of reactants or products
- changing the volume of the system
- changing the temperature

Shifts in equilibrium:

Use Le Chatelier's Principle to work through the following exercise.

Consider nitrogen and hydrogen in equilibrium with ammonia:

$$N_2(g) + 3H_2(g) \rightleftharpoons 2NH_3(g) + heat$$

What would be the effect of:

1) adding more nitrogen to the system?

2) removing ammonia from the system?

3) decreasing the volume of the reaction vessel?

4) increasing the temperature?

answers:

1&2) The reaction shifts to restore the concentration ratio to the value required by K_c. Reaction would occur in the forward direction to remove reactants and produce products. Equilibrium would be established with new concentrations of nitrogen, oxygen, and nitric oxide.

3) Decreasing volume increases pressure. The equilibrium shifts to decrease pressure. There are 4 moles of gaseous reactants and 2 moles of gaseous products, so reaction occurs in the forward direction, the direction which produces fewer moles of gaseous species. For reactions that have no net changes in the number of moles of gas, a volume change has no effect on the position of the equilibrium.

4) Heat is produced as ammonia forms (the reaction is exothermic). If heat is added to increase the temperature, the reverse reaction occurs to consume some of the added energy. The equilibrium constant does change with a change in temperature.

The van't Hoff equation describes how changes in temperature affect the equilibrium constant:

$$\ln \frac{K_2}{K_1} = \frac{-\Delta H^o_{rxn}}{R}\left(\frac{1}{T_2} - \frac{1}{T_1}\right)$$

Calculating the change in K with a change in temperature:

For the Haber process in which nitrogen is converted to ammonia,

$$N_2(g) + 3H_2(g) \rightleftharpoons 2NH_3(g)$$

ΔH = -92.0 kJ. If the equilibrium constant at room temperature is 5×10^8, calculate its value at 200°C.

solution:

- Will K_2 be smaller or larger than K_1? The reaction is exothermic (gives off heat), so heat is a product. If we add heat to increase the temperature, the reaction will shift to the left to consume some of the added energy, and K_2 will be smaller than K_1 as the concentration of products decrease and reactants increase.

- We know:

    T_2 = (200+ 273) = 473 K

    T_1 = (25 + 273) = 298 K

    ΔH = -92 kJ/(4.184 kJ/kcal) = -22.0 kcal = -22,000 cal

- If we substitute these values into the van't Hoff equation, we obtain:

$$\ln\frac{K_2}{K_1} = \frac{-92,000 \text{ J}}{8.3148 \text{ J/K mol}}\left(\frac{1}{473 \text{ K}} - \frac{1}{298 \text{ K}}\right)$$

$$\ln\frac{K_2}{K_1} = -13.737$$

$$\frac{K_2}{K_1} = 1.0815 \times 10^{-6}$$

$$K_2 = \left(1.0815 \times 10^{-6}\right)\left(5 \times 10^8\right) = 540$$

Although one could obtain higher yields of ammonia by decreasing temperature, the reaction rate decreases with temperature to a point where it is no longer practical. In practice, the temperature is raised to a moderate level, and a catalyst is used to increase the reaction rate. Catalysts affect the time to reach equilibrium, but not the equilibrium position.

Chapter 16 Exercises:

16.1) Methyl alcohol can be prepared commercially by the reaction below.

$$CO(g) + 2H_2(g) \rightleftharpoons CH_3OH(g)$$

For the reaction at 200°C, the equilibrium concentrations are [CO] = 0.60 M, [H_2] = 0.13 M, and [CH_3OH] = 0.43 M. What is K_c for the reaction?

16.2) For the reaction below at 25°C, the equilibrium concentrations for P_4, H_2, and PH_3 are 0.933 M, 0.600 M, and 0.267 M respectively.

$$P_4(g) + 6H_2(g) \rightleftharpoons 4PH_3(g)$$

What is K_c for the reaction? What is K_p?

16.3) The following reaction occurs within an erupting volcano.

$$2H_2S(g) + SO_2(g) \rightleftharpoons 3S(s) + 2H_2O(g)$$

In an effort to understand the equilibrium of the reaction, a researcher analyzed the equilibrium mixture in a reaction vessel. At 650 K, the partial pressures of H_2S, SO_2, and H_2O are 1.17 atm, 1.07 atm, and 0.83 atm respectively. Calculate K_p for the reaction.

16.4) For the reaction below at 50°C, the equilibrium partial pressures for N_2O, O_2, and NO_2 are 0.904 atm, 1.36 atm, and 2.19 atm respectively.

$$2N_2O(g) + 3O_2(g) \rightleftharpoons 4NO_2(g)$$

Calculate K_p for the reaction. What is K_c?

16.5) 34.3 g of phosphorous pentachloride is placed in a 500 mL reaction vessel at 300°C.

$$PCl_5(g) \rightleftharpoons PCl_3(g) + Cl_2(g)$$

Chemical analysis of the equilibrium mixture indicates 52.0% of the reactant decomposes. Calculate K_c for the reaction.

16.6) 1.00 mole of N_2O_4 is placed in a 1.00 L reaction vessel at 298 K.

$$N_2O_4(g) \rightleftharpoons 2NO_2(g)$$

Chemical analysis of the equilibrium mixture indicates only 3.5% of the reactant decomposes. Calculate K_c for the reaction.

16.7) K_c for the reaction below at 100°C is 1.41.

$$CH_3CH{=}CH_2(g) + H_2O(g) \rightleftharpoons C_3H_8O(g)$$

The equilibrium concentrations are [H_2O] = 0.35 M and [C_3H_8O] = 0.65 M. What is the equilibrium concentration of $CH_3CH=CH_2$?

16.8) Toluene, $C_6H_5CH_3$, can be prepared at an oil refinery from heptane, C_7H_{16} and a catalyst at high temperature.

$$C_7H_{16}(g) \rightleftharpoons C_6H_5CH_3(g) + 4H_2(g)$$

K_p for the reaction at 600 K is 11.9. At equilibrium the partial pressures of C_7H_{16} and $C_6H_5CH_3$ are 4.28 atm and 0.723 atm respectively. What is the equilibrium partial pressure of H_2?

16.9) Urea, $CO(NH_2)_2$ is a major nitrogen fertilizer produced in the U.S.

$$2NH_3(g) + CO_2(g) \rightleftharpoons CO(NH_2)_2(s) + H_2O(g) \qquad \Delta H_{rxn} = -21.4 \text{ kJ}$$

What direction, if any, will the following changes have on the equilibrium?
a. Addition of CO_2 d. increase in the temperature
b. removal of $CO(NH_2)_2$ e. increase in the pressure
c. addition of H_2O f. addition of a catalyst

16.10) 0.50 mole of CH_3Br and 0.75 mole of H_2S are injected into a 1.00 L reaction vessel at 500 K.

$$CH_3Br(g) + H_2S(g) \rightleftharpoons CH_3SH(g) + HBr(g)$$

K_c for the reaction at 500 K is 0.33. What are the equilibrium concentrations for each of the species in the reaction?

16.11) Ethyl chloride, C_2H_5Cl, undergoes thermal decomposition at 250°C.

$$C_2H_5Cl(g) \rightleftharpoons CH_2=CH_2(g) + HCl(g)$$

K_p at 250°C is 2.09. If the initial partial pressure of C_2H_5Cl is 1.00 atm, what is the partial pressure of C_2H_5Cl, $CH_2=CH_2$, and HCl at equilibrium?

16.12) Dimethyl ether can be produced by the dehydration of methyl alcohol.

$$2CH_3OH(g) \rightleftharpoons CH_3OCH_3(g) + H_2O(g)$$

K_c for the reaction at 600 K is 12.9. 1.00 mole each of CH_3OH, CH_3OCH_3, and H_2O are placed in a 2.00 L reaction vessel. What are the equilibrium concentrations of each species?

16.13) K_c for the following reaction is 0.068 at 273 K.

$$CO(g) + H_2(g) \rightleftharpoons CH_2O(g)$$

The initial concentrations of CO, H_2, and CH_2O are 1.25 M, 2.00 M, and 1.00 M respectively. What are the equilibrium concentrations of each species?

Chapter 16 Answers:

16.1) $K_c = 42$

16.2) $K_c = 0.117$; $K_p = 7.99 \times 10^{-6}$

16.3) $K_p = 0.47$

16.4) $K_p = 11.2$; $K_c = 297$

16.5) $K_c = 0.185$

16.6) $K_c = 5.1 \times 10^{-3}$

16.7) $[C_3H_8O] = 1.3$ M

16.8) $[H_2] = 2.90$ M

16.9) a. shifts towards product
b. no effect; $CO(NH_2)_2$ is a solid
c. shifts towards reactant
d. shifts towards reactant; reaction is exothermic
e. shifts towards product
f. no effect; catalysts change the reaction rate

16.10) $[CH_3Br] = 0.28$ M; $[H_2S] = 0.53$ M; $[CH_3SH] = 0.22$ M; $[HBr] = 0.22$ M

16.11) $C_2H_5Cl = 0.26$ atm; $C_2H_4 = 0.74$ atm; $HCl = 0.74$ atm

16.12) $[CH_3OH] = 0.18$ M; $[CH_3OCH_3] = 0.66$ M; $[H_2O] = 0.66$ M

16.13) $[CH_2O] = 0.34$ M; $[CO] = 1.9$ M; $[H_2] = 2.7$ M

CHAPTER 17
Acid-Base Equilibria

TOPICS:

- Water and pH
- Strong acids and bases
- Weak acids and bases
- Salt solutions
- Polyprotic acids
- Theories of acids and bases

SECTION 17.1: Water and pH

We saw in the last chapter how to write the equilibrium expression, K_c, for a chemical equilibrium. Water molecules are in equilibrium with hydronium ion (H_3O^+) and hydroxide ion (OH^-):

$$H_2O(l) + H_2O(l) \rightleftharpoons H_3O^+(aq) + OH^-(aq)$$

Water is **amphoteric** because it may act as both an acid (donates a proton), and a base (accepts a proton). Chemists sometimes abbreviate this reaction as:

$$H_2O(l) \rightleftharpoons H^+(aq) + OH^-(aq)$$

which focuses on the **dissociation** of water. Water dissociates to a very small extent: only about one in every <u>555 million</u> water molecules exists in its dissociated form.

The **equilibrium expression for water (K_w):**

1) Write the equilibrium expression for the dissociation of water shown above.

   <u>solution:</u>

   When the species dissociating is an acid, we call the equilibrium constant K_a:

   $$K_a = \frac{\left[H_3O^+\right]\left[OH^-\right]}{\left[H_2O\right]^2}$$

2) What is the concentration of water ($[H_2O]$) in mol/L assuming the concentration of dissociated ions is negligible?

---

solution:

One liter of water contains 1000 g of water (density of water = 1 g/mL); we also know the molar mass of water is 18 g/mole:

$$1000g \times \frac{1\ mol}{18\ g} = 55.5\ mol$$

The molarity of water is therefore 55.5 mol/L

Since the molarity of water is large, and remains essentially constant for dilute solutions, we include its concentration with the constant, K_a:

$$K_a[H_2O]^2 = K_w = [H_3O^+] \times [OH^-] = 1.0 \times 10^{-14} \quad (25°C)$$

The dissociation constant of water has the symbol K_w. Any water solution contains H_3O^+ and OH^- at concentrations such that their product is 1.0×10^{-14}. Although we sometimes refer to protons in solution as H^+, this notation is an abbreviation. The small proton with its high charge density bonds covalently to one of the lone electron pairs on an oxygen atom of water to form the hydronium ion, H_3O^+.

The dissociation constant of water:

1) Given that $K_w = [H^+] \times [OH^-] = 1.0 \times 10^{-14}$, what is the concentration of H_3O^+ and OH^- ions in pure water?

   solution:

   In pure water, H_3O^+ ions and OH^- ions form in equal amounts:

   $$\left[H_3O^+\right] = \left[OH^-\right] = \sqrt{1.0 \times 10^{-14}} = 1.0 \times 10^{-7}\ M$$

2) What percent of water molecules exist in the dissociated form?

   solution:

   We calculated in the problem above that the concentration of water is 55.5 mol/L. The concentration of H_3O^+ and OH^- ions in pure water is 1×10^{-7} mol/L. Therefore, the percent of water molecules in the dissociated form is:

   $$\frac{1.0 \times 10^{-7}\ mol/L}{55.5\ mol/L} \times 100\% = 1.8 \times 10^{-7}\%$$

   or one out of every approximately 555 million water molecules exists in the dissociated form.

When the concentration of H_3O^+ ions equals the concentration of OH^- ions, the solution is **neutral**. In most cases, water solutions are not neutral. Drinking water is not neutral because carbon dioxide from the air dissolves in water to form carbonic acid which dissociates in water:

$$H_2O(l) + CO_2(g) \rightarrow H_2CO_3(aq) \overset{H_2O}{\rightleftharpoons} H_3O^+(aq) + HCO_3^-(aq)$$

Solutions which have higher H_3O^+ concentrations than OH^- concentrations are **acidic**. Solutions which have higher OH^- concentrations than H_3O^+ concentrations are **basic**. Are drinking water solutions typically acidic or basic? (answer: acidic)

pH:

We saw in the problem above that the concentration of H_3O^+ and OH^- ions in pure (neutral) water is 1×10^{-7}. The concentration of H_3O^+ ions is small and is expressed using negative exponents. We can use a logarithmic scale to express hydrogen ion concentration without negative exponents because a logarithm is an exponent:

$$\log 10^5 = 5$$
$$\log 10 = 1$$
$$\log 1 = \log 10^0 = 0$$
$$\log 10^{-7} = -7$$
$$\log 10^{-8.3} = -8.3$$

For pure water, the log of the hydrogen ion concentration (1×10^{-7}) is -7. In order to work with positive numbers, we take <u>minus</u> the log of the hydrogen ion concentration to obtain **pH**. The "p" in pH means "minus the log of". pH = "minus the log of the hydrogen ion concentration":

$$pH = -\log[H_3O^+]$$

The pH of neutral water = $-\log(1 \times 10^{-7}) = 7$.

---

pH and pOH:

We use pH to express the acidity of a solution, but we could just as easily express the acidity as pOH, or "minus the log of the hydroxide ion concentration". Use the expression for K_w to derive the relationship between pH and pOH.

<u>solution</u>:

• The expression for K_w is:

$$K_w = [H_3O^+] \times [OH^-] = 1.0 \times 10^{-14} \quad (25°C)$$

• We can take the negative log of each side to obtain:

$$-\log K_w = -\log[H_3O^+] \times -\log[OH^-] = -\log(1.0 \times 10^{-14})$$

$$pK_w = pH + pOH = 14.00 \quad (25°C)$$

pH problems:

1) What is the pH and the pOH of neutral water?

solution:

In neutral water, $[H_3O^+] = [OH^-] = \sqrt{1.0 \times 10^{-14}} = 1.0 \times 10^{-7} \, M$

$$pH = pOH = -\log(1 \times 10^{-7}) = 7.$$

The pH (and the pOH) of a neutral solution is 7.0.

2) If the pH of a sample of drinking water is 6.5, what is the hydrogen ion concentration? What is the concentration of OH⁻ ions? What is the pOH?

solution:

- We need to take the antilog of the negative of the pH to solve for $[H^+]$:

$$[H_3O^+] = 10^{-6.5} = 3.2 \times 10^{-7}$$

- We can solve for the concentration of OH⁻ using either relationship:

$$[H_3O^+] \times [OH^-] = 1.0 \times 10^{-14}$$

or, $pH + pOH = 14$

Using the first relationship: $[H_3O^+] = 3.2 \times 10^{-7}$, and

$$[OH^-] = \frac{1.0 \times 10^{-14}}{3.16 \times 10^{-7}} = 3.16 \times 10^{-8} \, M$$

Using the second equation, $pOH = 14 - 6.5 = 7.5$, and $[OH^-] = 10^{-7.5} = 3.2 \times 10^{-8}$

(The answers are slightly different due to rounding off the number for $[H_3O^+]$).

3) Human blood has a pH of about 7.4. What is the concentration of H_3O^+ ions in blood?

solution:

We take the antilog of the negative of the pH to solve for $[H_3O^+]$:

$$[H_3O^+] = 10^{-7.4} = 4.0 \times 10^{-8}$$

As hydrogen ion concentration <u>increases</u>, pH <u>decreases</u>. When pH < 7, the solution is <u>acidic</u>. When pH > 7, the solution is <u>basic</u>.

Basicity and acidity:

Which of the following solutions is acidic, and which is basic?

1) beer (pH = 4-5)

2) blood (pH = 7.4)

3) household ammonia (pH = 11.9)

4) milk (pH = 6.4)

5) milk of magnesia (pH = 10.5)

6) 1 M HCl (pH = 0.0)

7) 1 M NaOH (pH = 14.0)

8) sea water (pH = 8.3)

9) stomach acid (pH = 2.2-2.4)

10) vinegar ( )

<u>answers</u>:   acidic (pH < 7): 1,4,6,9,10
                  basic (pH > 7): 2,3,5,7,8

Acids are generally sour, react with active metals to produce hydrogen gas, and cause "burns". Bases are usually bitter and feel slippery (because they dissolve a layer of your skin).

The pH scale:

Many aqueous solutions have pH values which fall between 1 and 14 (hydrogen ion concentrations between 0.1 and 1×10^{-14}). As you saw in the above problem, the pH of 1 M HCl is 0.0. Can you have a negative pH? Yes. Concentrated hydrochloric acid is about 12M. Hydrochloric acid dissociates completely in water. What is the pH of concentrated HCl? (Answer: The concentration of hydronium ions in concentrated hydrochloric acid is 12M. Therefore its pH = -log(12) = -1.08).

SECTION 17.2: Strong acids and bases

Hydrochloric acid dissociates in water to produce hydrogen ions and chloride ions:

$$HCl(aq) + H_2O(l) \rightarrow H_3O^+(aq) + Cl^-(aq)$$

The reaction goes essentially 100% to completion. No HCl molecules exist because they all dissociate into ions. We do not write an equilibrium expression for this reaction because we assume K_a is infinite. Acids such as hydrochloric acid are called **strong acids**. Sometimes you may see the dissociation of hydrochloric acid written as:

$$HCl(aq) \rightarrow H^+(aq) + Cl^-(aq)$$

This equation does not indicate the association of the proton to an oxygen atom of water to form the hydronium ion, H_3O^+.

---

pH of solutions containing a strong acid:

1) What is the pH of an aqueous solution containing 0.050 M HCl?

  solution:

  • Hydrochloric acid is a strong acid which dissociates completely in water:

  $$HCl(aq) + H_2O(l) \rightarrow H_3O^+(aq) + Cl^-(aq)$$

  • Since all the HCl forms hydronium ions and chloride ions, $[H_3O^+] = 0.050$.

  $$pH = -\log(0.05) = 1.3$$

2) What is the pH of an aqueous solution containing 1.00×10^{-8} M HCl?

  solution:

  If we try to solve this problem analogous to the one above, we run into a problem:

  $$[H_3O^+] = 1.00 \times 10^{-8}, \text{ and } pH = -\log(1.00 \times 10^{-8}) = 8.00$$

  There is such a small amount of acid, that even though it is a strong acid, there is not enough of it for it to be the major source of hydrogen ions. Water itself is the major source of hydrogen ions: $[H_3O^+]$ from water $= 1.0 \times 10^{-7}$. We find the pH of this solution by adding together the hydrogen ion concentration from the dissociation of water and the hydrogen ion concentration from the strong acid, HCl:

  $$pH = -\log(1.00 \times 10^{-7} + 1.00 \times 10^{-8}) = 6.96$$

  In aqueous solutions, water always contributes a small number of hydronium ions. Often another source (such as an acid) will swamp out the contribution from water, but keep in mind that the contribution from water will be significant whenever $[H_3O^+]$ is close to, or less than, 1.0×10^{-7}.

Analogous to the strong acids are the **strong bases**. The most common strong base is sodium hydroxide:

$$NaOH(aq) + H_2O(l) \rightarrow Na^+(aq) + OH^-(aq)$$

Sodium hydroxide is 100% ionized in solution to sodium and hydroxide ions. The concentration of undissociated NaOH is virtually zero. Strong acids and bases behave in solution as strong electrolytes. Compare a solution of 0.1 M sodium hydroxide and 0.1 M sodium chloride. Both conduct electric current well and behave as completely ionized substances in their colligative properties. One has a pH of 13, and the other has a pH of close to 7.

We saw in Chapter 4 that there are not very many strong acids and bases. The most important ones are listed in the table below.

THE STRONG ACIDS AND BASES

ACIDS		BASES
Hydrochloric acid	HCl	Hydroxides of the 1A and 2A metals:
Hydrobromic acid	HBr	LiOH, NaOH*, KOH*, RbOH, CsOH
Hydroiodic acid	HI	$Mg(OH)_2$, $Ca(OH)_2$, $Sr(OH)_2$,
Nitric acid	HNO_3	$Ba(OH)_2$
Perchloric acid	$HClO_4$	Metal oxides
Sulfuric acid	H_2SO_4	* most commonly used in the lab
	(first H_3O^+)	

Concentrated hydrochloric acid (HCl) is about $12M$; concentrated nitric acid (HNO_3) is about $16\ M$; sulfuric acid is provided as essentially pure H_2SO_4.

Reaction of a strong acid (such as hydrochloric acid) with a strong base (such as sodium hydroxide) quickly produces water as its product in an exothermic neutralization reaction:

Molecular equation:
$$HCl(aq) + NaOH(aq) \rightarrow H_2O(l) + NaCl(aq)$$

Total ionic equation:
$$H_3O^+(aq) + Cl^-(aq) + Na^+(aq) + OH^-(aq) \rightarrow 2H_2O(l) + Na^+(aq) + Cl^-(aq)$$

Net ionic equation:
$$H_3O^+(aq) + OH^-(aq) \rightarrow 2H_2O(l)$$

Reaction of a strong acid with a strong base:

1) What is the pH of a 2.0 L solution containing 50 mL of 0.10 M HCl, and 10 mL of 1.0 M NaOH?

solution:

- First we calculate initial concentrations of HCl and NaOH:

$$[HCl] = [H_3O^+] = \frac{0.050 \text{ L} \times 0.10 \text{ mol/L}}{2.0 \text{ L}} = 0.0025 \text{ } M$$

$$[NaOH] = [OH^-] = \frac{0.010 \text{ L} \times 1.0 \text{ mol/L}}{2.0 \text{ L}} = 0.005 \text{ } M$$

- Next we determine the equilibrium concentrations of H_3O^+ and OH^-:

$$H_3O^+(aq) \quad + \quad OH^-(aq) \quad \rightarrow \quad H_2O(l)$$

start conc.	0.0025	0.005
change	-0.0025	-0.0025
equil. conc.	0	0.0025

- Calculate pH from the pOH:

$$pOH = -\log(.0025) = 2.6$$
$$pH = 14 - 2.6 = 11.4$$

2) How would you prepare 2 L of a solution with a pH of 3 from a 1 M HCl stock solution?

solution:

- A solution with a pH of 3 has $[H_3O^+] = 1\times10^{-3} \text{ } M$. 2 L of that solution contains:

$$2 \text{ L} \times (1\times10^{-3} \text{ mol/L}) = 2 \times 10^{-3} \text{ mol } H_3O^+.$$

- Since HCl completely dissociates, we need 2×10^{-3} mol HCl from our 1 M stock solution. The volume of stock solution that contains 2×10^{-3} mol HCl is:

$$\frac{2 \times 10^{-3} \text{ mol HCl}}{1 \text{ mol/L}} = 2 \times 10^{-3} \text{ L stock solution}$$

- We would dilute 2×10^{-3} L (2 mL) of the stock solution to 2 L to obtain a solution with pH = 3.

SECTION 17.3: Weak acids and bases

Weak acids:

Most acids do not dissociate completely in water. The acid which makes vinegar acidic, acetic acid (CH_3COOH), dissociates only a small extent in water:

$$CH_3COOH(aq) + H_2O(l) \rightleftharpoons CH_3COO^-(aq) + H_3O^+(aq)$$

$$K_a = \frac{\left[CH_3COO^-\right]\left[H_3O^+\right]}{\left[CH_3COOH\right]} = 1.8 \times 10^{-5}$$

Most of the acetic acid molecules are in the undissociated form CH_3COOH. In a 0.1 M solution of acetic acid for example, only about 1.3% of the acetic acid molecules exist in ionized form.

We can think of this reaction as a competition between acetate ions (CH_3COO^-) and water molecules for a proton. $K_a = 1.8 \times 10^{-5}$, so the equilibrium favors the reactants. Acetate wins the competition for the proton, and therefore acetate is a stronger base than water.

Acetic acid:

1) Calculate the pH of a 0.1 M acetic acid solution. K_a for acetic acid = 1.8×10^{-5}.

solution:

It may be helpful to make a table:

$$CH_3COOH(aq) + H_2O \rightleftharpoons CH_3COO^-(aq) + H_3O^+(aq)$$

start conc.:	0.1	0	0
equil. conc.:	0.1-X	X	X

$$K_a = \frac{\left[CH_3COO^-\right]\left[H_3O^+\right]}{\left[CH_3COOH\right]} = 1.8 \times 10^{-5} = \frac{x^2}{0.1 - x}$$

We could now solve for x by rearranging the equation to the form $x^2 + bx + c = 0$ and using the quadratic formula. However, since acetic acid dissociates to only a small extent, we know that the concentration of H_3O^+ will be small, and the concentration of CH_3COOH will be close to its original concentration. We make the approximation that x is much less than 0.1 and therefore 0.1-x $\approx$ 0.1. Now we can solve for x easily:

$$\frac{x^2}{0.1} = 1.8 \times 10^{-5}$$

$$x = 1.34 \times 10^{-3} = [H_3O^+]$$
(notice that x is only 1.3% of 0.1, so our approximation was OK)*

$$pH = -\log (1.34 \times 10^{-3}) = 2.9$$

2) Calculate the pH of a 1 M acetic acid solution.

solution:

As above:

$$CH_3COOH(aq) + H_2O \rightleftharpoons CH_3COO^-(aq) + H_3O^+(aq)$$

start conc.:	0.1	0	0
equil. conc.:	1 - X	x	x

$$K_a = \frac{\left[CH_3COO^-\right]\left[H_3O^+\right]}{\left[CH_3COOH\right]} = 1.8 \times 10^{-5} = \frac{x^2}{1.0 - x}$$

We make the approximation that x is << 1.0 and therefore 1.0-x $\approx$ 1.0. Now we can solve for x easily:

$$\frac{x^2}{1.0} = 1.8 \times 10^{-5}$$

$$x = 4.24 \times 10^{-3} = [H^+]$$
(notice that x is only 0.42% of 1.0, so our approximation was OK)

$$pH = -\log(4.24 \times 10^{-3}) = 2.4$$

*Note: You are justified in setting molarity-x = molarity as long as this approximation does not introduce an error of more than about 5%. It is a reasonable start to make this approximation for all weak acid and weak base problems, and then determine if your calculated x is less than 5% of the original molarity. If it is more than 5% of the original molarity, you can repeat the calculation subtracting the value just calculated for x from the original molarity. If the calculated value for x is still more than 5% of the original molarity, you could repeat the procedure. This method of successive approximations is usually simpler than solving the quadratic formula.

The above example illustrates that the percent of dissociated molecules of a weak acid increases as the acid becomes more dilute. Acetic acid is about 1.3% dissociated in a 0.1 M solution, but only 0.42% dissociated in a 1 M solution.

Relative strengths of acids:

Arrange the following acids in order of increasing strength:
1) NH_4^+, $K_a = 5.6 \times 10^{-10}$ 4) CH_3COOH, $K_a = 1.8 \times 10^{-5}$
2) $HClO_2$, $K_a = 1.12 \times 10^{-2}$ 5) HCN, $K_a = 6.2 \times 10^{-10}$
3) HF, $K_a = 6.8 \times 10^{-4}$ 6) HIO, $K_a = 2.3 \times 10^{-11}$

solution:

The strongest acid has the largest K_a, the weakest acid has the smallest K_a:
$$HClO_2 > HF > CH_3COOH > HCN > NH_4^+ > HIO$$

Strong vs. weak acids:

Assign of each of the following properties to:

      a) a 0.1 M HCl solution, or

      b) a 0.1 M acetic acid solution.

1) has a higher conductivity than a 0.1 M NaCl solution

2) is a relatively poor conductor

3) has a pH of 2.9

4) has a pH of 1.0

5) has a freezing point almost the same as a 0.10 M NaCl solution.

6) has a freezing point comparable to a 0.10 M sugar solution.

<u>answers</u>: 1-a, 2-b, 3-b, 4-a, 5-a, 6-b.

Strong acids and bases, because they completely dissociate into ions, behave in solution as strong electrolytes. They are good electrical conductors and lower the freezing point of water to about the same extent as a strong electrolyte. Weak acids are poor conductors because there are few ions present, and they lower the freezing point of water to about the same extent as a nonelectrolyte such as sugar.

Periodic trends in acid strength:

Consider the dissociation of an acid, HA:

$$HA + H_2O \rightleftharpoons H_3O^+ + A^-$$

As the electronegativity of A increases, acidity of the acid HA increases:

$$HF > H_2O > NH_3$$

The electronegative fluorine atoms in hydrofluoric acid displace electron density away from the hydrogen atoms allowing them to dissociate into H^+ ions.

As the atomic radius of A increases, for example down a column in the periodic table, the HA bond becomes weaker, and acidity increases:

$$H_2Se > H_2S > H_2O$$

Weak bases:

Weak bases produce hydroxide ions in water, but to a much smaller extent than do strong bases. One of the most common weak bases is ammonia, NH_3. Ammonia produces hydroxide ions in water:

$$NH_3(aq) + H_2O(l) \rightleftharpoons NH_4^+(aq) + OH^-(aq)$$

Most of the ammonia molecules are undissociated, but a small percent of them dissociate to produce ammonium ion and hydroxide ion. For weak bases, we can write a K_b expression that is analogous to the K_a expression for weak acids:

$$K_b = \frac{\left[NH_4^+\right]\left[OH^-\right]}{\left[NH_3\right]} = 1.8 \times 10^{-5}$$

It is only coincidence that K_a for the common weak acid, acetic acid, is the same as K_b for the common weak base, ammonia! Without working out the math, we know from our exercises with acetic acid, that a 0.1 M ammonia solution is 1.3% dissociated, and a 1 M NH_3 solution is 0.42% dissociated. It may be confusing to see ammonia solutions labeled as NH_4OH. There is no evidence that this species exists. Think of these solutions as mainly ammonia (NH_3) and water (H_2O) molecules, with a small percent of NH_4^+ and OH^- ions.

Any molecule that removes a proton from water to produce hydroxide ions is a base. A salt which contains the anion of a weak acid is a weak base. $NaCH_3COO$ is a weak base because the acetate anion removes protons from water to form acetic acid:

$$CH_3COO^-(aq) + H_2O(l) \rightleftharpoons CH_3COOH(aq) + OH^-(aq)$$

The acetate anion is called the conjugate base of acetic acid. Acetic acid is the conjugate acid of the acetate anion. All weak acid/weak base pairs can be related this way. What is the other conjugate acid/conjugate base pair in this equilibrium? (Answer: OH^- is the conjugate base of the conjugate acid, H_2O.)

Relation between K_a and K_b:

Consider the dissociation reactions of ammonia (NH_3) and its conjugate acid, the ammonium ion (NH_4^+):

$$NH_3(aq) + H_2O(l) \rightleftharpoons NH_4^+(aq) + OH^-(aq); \qquad K_b = 1.8 \times 10^{-5}$$

$$NH_4^+(aq) + H_2O(l) \rightleftharpoons NH_3(aq) + H_3O^+(aq) \qquad K_a = ?$$

1) Write K_a and K_b expressions for these dissociation reactions.

solution:

$$K_b = \frac{\left[NH_4^+\right]\left[OH^-\right]}{\left[NH_3\right]} = 1.8 \times 10^{-5}$$

$$K_a = \frac{[NH_3][H_3O^+]}{[NH_4^+]} = ?$$

2) What is $K_a \times K_b$?

solution:

$$K_a \times K_b = \frac{[NH_3][H_3O^+]}{[NH_4^+]} \times \frac{[NH_4^+][OH^-]}{[NH_3]} = [H_3O^+][OH^-] = K_w$$

$$K_a \times K_b = K_w$$

3) What is K_a for the ammonium ion?

$$K_a \times K_b = K_w, \text{ or}$$

$$K_a = \frac{K_w}{K_b} = \frac{1 \times 10^{-14}}{1.8 \times 10^{-5}} = 5.6 \times 10^{-10}$$

4) Consider the reaction between ammonium, NH_4^+, and base:

$$NH_4^+(aq) + OH^-(aq) \rightleftharpoons NH_3(aq) + H_2O(aq)$$

a) Identify the conjugate acid/base pairs for this reaction
b) Is water (H_2O) or NH_4^+ a stronger acid?
c) Would you expect this equilibrium to lie farther to the right or to the left?

solution:

a) NH_4^+ = conj. acid; NH_3 = conj. base
   H_2O = conj. acid; OH^- = conj. base

b) The dissociation constant for water is 1×10^{-14}. The dissociation constant for NH_4^+ is 5.6×10^{-10}. The ammonium ion dissociates more than water, and so the ammonium ion is a stronger acid than water.

c) Since NH_4^+ is a stronger acid than water, the equilibrium lies to the right.

The relationship $K_a \times K_b = K_w$ tells us that the stronger the conjugate acid, the weaker the conjugate base and vice versa.

Reaction of a weak acid and a strong base:

What is the pH of 100 mL of a 0.10 M acetic acid solution to which 50 mL of 0.10 M NaOH has been added?

solution:

- The initial concentrations of acetic acid and NaOH are:

$$[CH_3COOH] = \frac{0.100 \text{ L} \times 0.10 \text{ mol/L}}{0.150 \text{ L}} = 0.067 \text{ } M$$

$$[NaOH] = \frac{0.050 \text{ L} \times 0.10 \text{ mol/L}}{0.150 \text{ L}} = 0.033 \text{ } M$$

- The strong base (hydroxide ion) deprotonates the weak acid until the base is used up. The reaction goes essentially to completion:

$$CH_3COOH(aq) + OH^-(aq) \rightarrow CH_3COO^-(aq) + H_2O(aq)$$

initial conc.	0.067	0.033	0
change	-0.033	-0.033	+0.033
end conc.	0.033	0	0.033

- Now we can solve for the pH of the acetic acid/acetate ion solution:

$$CH_3COOH(aq) + H_2O(aq) \rightleftharpoons CH_3COO^-(aq) + H_3O^+(aq)$$

$$K_a = \frac{\left[CH_3COO^-\right]\left[H_3O^+\right]}{\left[CH_3COOH\right]} \quad \text{or} \quad \left[H_3O^+\right] = \frac{K_a\left[CH_3COOH\right]}{\left[CH_3COO^-\right]}$$

$$\left[H_3O^+\right] = \frac{\left(1.8 \times 10^{-5}\right)\left(0.033\right)}{\left(0.033\right)} = 1.8 \times 10^{-5}$$

$$pH = 4.7$$

We calculated in the acetic acid exercise that a 0.1 M acetic acid solution has a pH of 2.9. The pH of the acetic acid solution to which base has been added is higher (less acidic).

pH of a conjugate acid/base pair:

To find the pH of weak acid, HA , it is helpful to set up a table which shows the initial concentration of the acid HA (say 0.1 M), the change in the concentration when some HA dissociates into H_3O^+ and A^- , and the equilibrium concentrations of HA, H_3O^+, and A^- :

$$HA(aq) \rightleftharpoons H_3O^+(aq) + A^-(aq)$$

	HA	H_3O^+	A^-
start conc.:	0.1	0	0
change:	- x	x	x
equil. conc.:	0.1 - x	x	x

The K_a expression allows us to solve for the concentration of H_3O^+ ions, and hence to find the pH of the weak acid:

$$K_a = \frac{\left[H_3O^+\right]\left[A^-\right]}{\left[HA\right]} = \frac{(x)(x)}{(0.1 - x)}$$

We usually can make the approximation that 0.1-x≈0.1, and solve for x, which is the concentration of H_3O^+ ions.

In problems such as the one above where a strong base reacts with a weak acid (CH_3COOH), there is a source of the conjugate base (CH_3COO^-) which does not come from the dissociation of acetic acid in water, but from <u>deprotonation</u> of acetic acid by OH^-. The amount of CH_3COO^- formed in this manner (0.033 M in the above exercise), is much, much more than the amount which dissociates from acetic acid, and so we disregard the dissociated amount. Another example of an "independent source" of CH_3COO^- would be the addition of a salt such as $NaCH_3COO$ to an acetic acid solution.

Caution:

Don't confuse conjugate acid/base pairs with the K_a and K_b expressions for amphoteric substances, substances that behave as either acids or bases. The sodium bicarbonate anion may act as either an acid or a base:

$$HCO_3^-(aq) + H_2O(l) \rightleftharpoons CO_3^{2-}(aq) + H_3O^+(aq); \quad K_a = 4.8 \times 10^{-11}$$

$$HCO_3^-(aq) + H_2O(l) \rightleftharpoons H_2CO_3(aq) + OH^-(aq); \quad K_b = 2.4 \times 10^{-8}$$

$$K_a = \frac{\left[CO_3^{2-}\right]\left[H_3O^+\right]}{\left[HCO_3^-\right]} \qquad K_b = \frac{\left[H_2CO_3^-\right]\left[OH^-\right]}{\left[HCO_3^-\right]}$$

In these two equations, K_a and K_b are __not__ related by $K_a \times K_b = K_w$ since they are not expressions for a conjugate acid/conjugate base pair.

Will a solution containing HCO_3^- be acidic or basic? (Answer: The solution will be slightly basic since $K_b > K_a$).

SECTION 17.4: Acid-base behavior of salt solutions

When salts (strong electrolytes) dissolve in water, they dissociate into ions. The ions may be acidic, basic, or neutral. __Neutral anions are derived from strong acids; neutral cations are derived from strong bases__. To see why this is the case, consider the strong acid, HCl:

$$HCl(aq) + H_2O(l) \rightarrow H_3O^+(aq) + Cl^-(aq)$$

The anion, Cl^-, has no tendency to combine with H^+ to form HCl since HCl is a strong acid and completely dissociates in water. Cl^- does not affect the pH of a solution and is a neutral ion. Likewise, the sodium cation, Na^+, has no tendency to combine with OH^- since NaOH is a strong base, and so Na^+ is a neutral cation. Conjugate bases of weak acids are basic anions. The acetate ion, CH_3COO^- is the conjugate base of acetic acid, and is an example of a basic anion. Examples of acidic cations are the ammonium ion, NH_4^+, Al^{3+}, and the transition metal ions.

Acidity and basicity of salt solutions:

Would you expect each of the following solutions to lower the pH, raise the pH, or have no effect on the pH of a neutral solution?

1) NaCl
2) NH_4Cl
3) NH_4NO_3
4) $NaCH_3COO$

5) $KClO_2$
6) KCN
7) $AlCl_3$
8) Na_2CO_3

answers:

1) neutral cation and neutral anion; no pH effect
2) acidic cation and neutral anion; lowers pH
3) acidic cation and neutral anion; lowers pH
4) neutral cation and basic anion; raises pH

5) neutral cation and basic anion; raises pH

6) neutral cation and basic anion; raises pH

7) acidic cation and neutral anion; lowers pH

8) neutral cation and basic anion; raises pH

SECTION 17.5: Polyprotic acids

Polyprotic acids have more than one ionizable proton. Acetic acid (CH_3COOH) is not a polyprotic acid because it has only one acidic hydrogen atom. (The hydrogen atoms bonded to carbon are not acidic). Sulfuric acid (H_2SO_4), sulfurous acid (H_2SO_3), phosphoric acid (H_3PO_4), phosphorous acid (H_3PO_3), carbonic acid (H_2CO_3), and hydrosulfuric acid (H_2S) are all polyprotic acids. It is always more difficult to pull off each successive proton: successive dissociation constants usually differ by several orders of magnitude. In calculations, we can usually assume that the first ionization produces almost all the hydrogen ion. Consider phosphoric acid:

$$H_3PO_4(aq) + H_2O(l) \rightleftharpoons H_2PO_4^-(aq) + H_3O^+(aq); \qquad K_{a_1} = 7.2 \times 10^{-3}$$

$$H_2PO_4^-(aq) + H_2O(l) \rightleftharpoons HPO_4^{2-}(aq) + H_3O^+(aq); \qquad K_{a_2} = 6.3 \times 10^{-8}$$

$$HPO_4^{2-}(aq) + H_2O(l) \rightleftharpoons PO_4^{3-}(aq) + H_3O^+(aq); \qquad K_{a_3} = 4.2 \times 10^{-13}$$

There is 5 orders of magnitude difference between the first and second dissociation constants, and another 5 orders of magnitude difference between the second and third dissociation constants.

A polyprotic acid:

Calculate the hydrogen ion concentration in a 0.050 M phosphoric acid solution.

solution:

• We will begin by assuming that the only significant source of hydrogen ions comes from the first dissociation of phosphoric acid. We will check this assumption after our calculation.

$$H_3PO_4(aq) + H_2O(l) \rightleftharpoons H_2PO_4^-(aq) + H_3O^+(aq)$$

	H_3PO_4	$H_2PO_4^-$	H_3O^+
initial conc.	0.050	0	0
change	- x	x	x
equil. conc.	0.050-x	x	x

$$K_a = \frac{\left[H_2PO_4^-\right]\left[H_3O^+\right]}{\left[H_3PO_4\right]} = \frac{x^2}{0.050 - x} = 7.2 \times 10^{-3}$$

- We'll assume that x is small compared to 0.050 (0.050-x≈0.050) and solve for x:

$$\frac{x^2}{0.050} = 7.2 \times 10^{-3}$$

$$x = \left[H_3O^+\right] = 0.019 M$$

- Was our assumption that x is small compared to 0.050 valid?

$$\frac{0.019 M}{0.05 M} \times 100\% = 38\%$$

Drat! x is not less than 5% of 0.050 M and so is not small enough to ignore. We could either go back and solve the quadratic equation, or we can use this value for x to make a new estimate for the concentration of H_3PO_4.

$$[H_3PO_4] = 0.050 - 0.019 = .031$$

Our estimate of x is:

$$\frac{x^2}{0.031} = 7.2 \times 10^{-3}$$

$$x = [H_3O^+] = .015$$

- If we repeat this process, our new estimate of $[H_3PO_4] = 0.050 - 0.015 = .035$

$$\frac{x^2}{0.035} = 7.2 \times 10^{-3}$$

$$x = \left[H_3O^+\right] = 0.016 M$$

x is no longer changing much, so we will take 0.016 as the concentration of hydrogen ions from the first dissociation of H_3PO_4.

- Is the concentration of hydrogen ions from the second dissociation of phosphoric acid significant?

$$H_2PO_4^-(aq) + H_2O(l) \rightleftharpoons HPO_4^{2-}(aq) + H_3O^+(aq)$$

	$H_2PO_4^-$	HPO_4^{2-}	H_3O^+
initial conc.	0.016	0	0
change	- x	x	x
equil. conc.	0.016-x	x	x

$$\frac{\left[HPO_4^{2-}\right]\left[H_3O^+\right]}{\left[H_2PO_4^-\right]} = \frac{x^2}{0.016 - x} = 6.3 \times 10^{-8}$$

- We'll assume that x is small compared to 0.016:

$$\frac{x^2}{0.016} = 6.3 \times 10^{-8}$$

$$x = \left[H_3O^+\right] = 3.17 \times 10^{-5} \, M$$

Was our assumption that x is small compared to 0.016 valid?

$$\frac{3.17 \times 10^{-5} \, M}{0.016 \, M} \times 100\% = 0.2\%$$

Good! We're OK this time. The hydrogen ion concentration from the second dissociation of phosphoric acid (3.17×10^{-5}) is insignificant compared to the concentration from the first dissociation (0.016), so, as is usually the case with polyprotic acids, we can assume that the first ionization produces almost all the hydrogen ion.

SECTION 17.6: Theories of acids and bases

Acids and base definitions:

Choose words from the list below to complete the following paragraphs:

a) acid
b) base
c) electron pair
d) hydrogen

e) hydroxide
f) proton
g) proton transfer
h) water

Arrhenius proposed the earliest definition for acids and bases, and it is the most restrictive definition. In the Arrhenius model, an acid is a substance which in water produces an excess of _____(1) ions; a base forms an excess of _____(2) ions.

The Bronsted-Lowry theory extends this concept of acids and bases. A Bronsted-Lowry acid-base reaction involves a _____(3) from one species to another. The species which donates the proton is the _____(4), and the species which accepts the proton is the _____(5). Often the proton acceptor is _____(6) in which case the hydronium forms, and water acts as a _____(7). Water can act as both an acid and a base according the the Bronsted-Lowry approach. Every acid has a conjugate _____(8), and vice versa. In order to be a Bronsted-Lowry acid, the species must contain an ionizable _____(9).

The Lewis acid-base concept, extends the definition of an acid further. An acid must be a species which can accept an _____(10). Examples of Lewis acids are protons, metal ions, and molecules with an incomplete octet of electrons, such as boron trifluoride, BF_3. Generally "Lewis acids" refer to a species such as BF_3 which is a powerful electron pair acceptor.

answers: 1-d, 2-e, 3-g, 4-a, 5-b, 6-h, 7-b, 8-b, 9-f, 10-c.

Chapter 17 Exercises:

17.1) Fill in the empty blocks in the following table.

[H+]	[OH-]	pH	pOH
4.3×10^{-4} M			
		8.90	
	1.5×10^{-11} M		
			4.22

17.2) The freezing point depression of meta-hydroxybenzoic acid, $HC_7H_5O_3$, in water indicates it is a weak electrolyte that ionizes 4.0% in a 0.050 M solution. Calculate K_a for the acid.

17.3) The pH of a 0.10 M adipic acid solution is 5.43. Calculate K_a for the weak acid.

17.4) The pH of a 0.010 M strychnine solution is 6.00. Calculate K_b for the weak base.

17.5) Calculate pK_a for the following weak acids:
   a. trans-cinnamic acid, $HC_9H_7O_2$; $K_a = 3.7 \times 10^{-5}$
   b. diethylbarbituric acid, $HC_8H_{11}O_3N_2$; $K_a = 3.7 \times 10^{-8}$
   c. saccharin, $HC_7H_4O_3NS$; $K_a = 2.5 \times 10^{-2}$

17.6) Calculate pK_b for the following weak bases:
   a. Novocain, $C_{13}H_{20}O_2N_2$; $K_b = 7 \times 10^{-6}$
   b. piperdine, $C_5H_{11}N$; $K_b = 1.6 \times 10^{-3}$
   c. methyl red, $(CH_3)_2N \cdot C_{13}H_9O_2N_2$; $K_b = 3 \times 10^{-12}$

17.7) Calculate the pH and the percent ionization for each of the following weak acid solutions:

   a. 0.10 M HF
   b. 0.10 M H_3BO_3
   c. 0.10 M HCN

17.8) What is the relationship, if any, between the percent ionization of a weak acid and the size of K_a for the acid?

17.9) Calculate the pH and percent ionization for each of the following solutions of acetic acid:

   a. 1.00 M CH_3COOH
   b. 0.10 M CH_3COOH
   c. 0.010 M CH_3COOH

17.10) What is the relationship, if any, between the percent ionization of a weak acid and the concentration of the acid solution?

17.11) Calculate the pH and the percent ionization for each of the following weak base solutions:

     a. 0.10 M CH_3NH_2; pK_a = 10.657

     b. 0.020 M $C_6H_5NH_2$; pK_a = 4.63

17.12) Calculate K_b and pK_b for the salts of the following weak acids.

     a. sodium pelargonate, $NaC_9H_{17}O_2$; K_a = 1.1 x 10^{-5}

     b. potassium hydrogen phosphate, K_2HPO_4; K_a = 6.2 x 10^{-8}

17.13) Calculate the pH of the following solutions:

     a. 0.020 M potassium acid phthalate ($KHC_8H_4O_4$) solution; K_a = 1.3 x 10^{-3}

     b. 0.15 M potassium glutarate ($KHC_5H_8O_4$) solution; K_a = 4.5 x 10^{-5}

     c. 0.050 M morphine hydrochloride ($C_{17}H_{19}O_3N \cdot HCl$) solution; K_b = 7.4 x 10^{-7}

17.14) Fumaric acid, $H_2C_4H_2O_4$, is one of the intermediates of the citric acid cycle of glucose and fatty acid metabolism. It may be represented as H_2Fum. Calculate the pH and the concentrations of the [HFum⁻] and [Fum²⁻] ions in a 0.040 M H_2Fum solution (pK_{a1} = 3.03; pK_{a2} = 4.44).

17.15) Citric acid, $H_3C_6H_5O_7$, can be expressed as H_3Cit (pK_{a1} = 3.14; pK_{a2} = 4.77; pK_{a3} = 6.39). For a 0.10 M citric acid solution, determine the concentration of each of the following species: [H⁺], [H₃Cit], [H₂Cit⁻], [HCit²⁻], and [Cit³⁻].

17.16) For each of the aqueous solutions of the following substances, indicate whether it is acidic, neutral or basic, and indicate whether the substance is a strong acid, weak acid, strong base, weak base, acidic salt, basic salt, or neutral salt.

     a. NaOCl   e. K_2SO_4 i. H_3AsO_4

     b. HNO_3 f. NH_2OH j. $(NH_4)_2SO_4$

     c. $CaCO_3$ g. $HC_7H_5O_2$ k. $AgNO_3$

     d. $Mg(OH)_2$ h. $(CH_3)_2NH$ l. NH_3

Chapter 17 Answers:

17.1)

	[H+]	[OH-]	pH	pOH
	4.3×10^{-4} M	2.3×10^{-11} M	3.37	10.63
	1.3×10^{-9} M	7.9×10^{-6} M	8.90	5.10
	6.7×10^{-4} M	1.5×10^{-11} M	3.18	10.82
	1.7×10^{10} M	6.0×10^{-5} M	9.78	4.22

17.2) $K_a = 8.3 \times 10^{-5}$

17.3) $K_a = 1.4 \times 10^{-10}$

17.4) $K_b = 1.0 \times 10^{-4}$

17.5) a. 4.43; b. 7.43; c. 1.60

17.6) a. 5.15; b. 2.80; c. 11.52

17.7) a. pH = 2.08; 8.25% ionized
b. pH = 5.12; 7.6×10^{-3}% ionized
c. pH = 5.10; 7.9×10^{-3}% ionized

17.8) For a given concentration, the smaller the K_a of the acid, the smaller the percent ionization.

17.9) a. pH = 2.37; 0.42% ionized; b. pH = 2.87; 1.34% ionized; c. pH = 3.37; 4.24%

17.10) The smaller the concentration, the greater the percent ionization of the acid.

17.11) a. pH = 11.83; 6.74% ionized
b. pH = 8.46; 1.46×10^{-2}% ionized

17.12) a. $K_b = 9.1 \times 10^{-10}$; $pK_b = 9.04$
b. $K_b = 1.6 \times 10^{-7}$; $pK_b = 6.79$

17.13) a. pH = 7.59; b. pH = 8.76; c. pH = 4.59

17.14) pH = 2.21; $[HFum^-] = 5.7 \times 10^{-3}$ M; $[Fum^{2-}] = 4.4 \times 10^{-5}$ M

17.15) $[H^+] = 8.53 \times 10^{-3}$ M; $[H_3Cit] = 9.2 \times 10^{-2}$ M; $[H_2Cit^-] = 8.2 \times 10^{-3}$ M;
$[HCit^{2-}] = 3.7 \times 10^{-4}$ M; $[Cit^{3-}] = 1.2 \times 10^{-5}$ M

17.16)
a. basic; basic salt
b. acidic; strong acid
c. basic; basic salt
d. basic; strong base
e. neutral; salt of a strong acid and base
f. basic; weak base
g. acidic; weak acid
h. basic; weak base
i. acidic; strong acid
j. acidic; acidic salt
k. neutral; salt of a strong acid and base
l. acidic; weak acid

CHAPTER 18

Ionic Equilibria in Aqueous Systems

TOPICS:

- Acid-base buffers
- Acid-base titration curves
    - strong acid and strong base
    - weak acid and strong base
    - weak base and strong acid
- Equilibria of slightly soluble ionic compounds
- Equilibria involving complex ions
- Applications to chemical analysis

SECTION 18.1: Acid-base buffers

Buffers temper the change in pH of a solution. A lake has buffering capacity if acid rain does not significantly change its pH. A buffer usually contains a weak acid which reacts with OH^- ions, and the conjugate base of the weak acid which reacts with H^+ ions. A buffer has maximum buffering capacity when the acid/conjugate base ratio is 1:1.

Acetic acid (CH_3COOH) mixed with acetate ion (CH_3COO^-) is a buffer. The buffer in blood is carbonic acid (H_2CO_3) and bicarbonate ion (HCO_3^-). Lactic acid, produced by exercise, could lower pH to a point where it would kill body cells if blood were not buffered.

---

pH change of a buffered solution:

A liter of a buffered solution contains 0.10 mole CH_3COOH and 0.10 mole $NaCH_3COO$.

a) What is the pH of this solution?

b) What is the pH after addition of 100 mL of 0.10 M NaOH?

c) What is the pH after addition of 100 mL of 0.10 M HCl?

d) What is the pH of 100 mL of 0.10 M NaOH added to 1.0 L of water?

e) What is the pH of 100 mL of 0.10 M HCl added to 1.0 L of water?

<u>solutions</u>:

a) CH_3COOH (acetic acid, a weak acid) and CH_3COO^- (acetate ion, its conjugate base) form a buffer solution. To find pH, we must determine the concentration of H^+.

The equilibrium is:

$$CH_3COOH + H_2O \rightleftharpoons CH_3COO^- + H_3O^+$$

initial conc.	0.10	0.10	0
change	-x	+x	+x
equil. conc.	0.10-x	0.10+x	x

$$K_a = \frac{\left[CH_3COO^-\right]\left[H_3O^+\right]}{\left[CH_3COOH\right]} = \frac{(0.1+x)(x)}{(0.1-x)} = 1.8 \times 10^{-5}$$

We look for approximations to make equations such as these easier to solve. Since acetic acid is a weak acid, we can assume that not much dissociates, and $0.1-x \approx 0.1$. Then,

$$\frac{(0.1)(x)}{0.1} = 1.8 \times 10^{-5}$$

$$x = 1.8 \times 10^{-5} M = \left[H_3O^+\right]$$

(notice that 1.8×10^{-5} is much smaller than 0.1, so our assumption was OK.)

$$pH = -\log(1.8 \times 10^{-5}) = 4.74$$

b) NaOH dissociates completely in water producing sodium and hydroxide ions. The initial concentrations are:

$$\left[OH^-\right] = \frac{(0.1\,L)(0.10\,mol/L)}{1.1\,L} = 0.0091\,M$$

$$\left[CH_3COOH\right] = \frac{0.10\,mol/L}{1.1\,L} = 0.091\,M$$

$$\left[CH_3COO^-\right] = \frac{0.10\,mol/L}{1.1\,L} = 0.091\,M$$

Hydroxide ion, a strong base, deprotonates acetic acid to form acetate anion and water. The reaction goes essentially to completion:

$$CH_3COOH + OH^- \rightarrow CH_3COO^- + H_2O$$

initial conc.	0.091	0.0091	0.091
change	-0.0091	-0.0091	+0.0091
equil. conc.	0.0819	0	0.1001

We find pH from the K_a expression for the equilibrium:

$$CH_3COOH + H_2O \rightleftharpoons CH_3COO^- + H_3O^+$$

$$K_a = \frac{[CH_3COO^-][H_3O^+]}{[CH_3COOH]} \quad \text{or} \quad [H_3O^+] = \frac{K_a[CH_3COOH]}{[CH_3COO^-]}$$

$$[H_3O^+] = \frac{(1.8 \times 10^{-5})(0.0819)}{(0.1001)} = 1.47 \times 10^{-5} \, M$$

pH = 4.83

The pH change of the buffered solution after addition of a strong base (NaOH) is 0.09 pH unit.

c) HCl is a strong acid which completely dissociates in water to give H_3O^+ and Cl^- ions. H_3O^+ protonates acetate ion to form acetic acid and water. The reaction goes essentially to completion:

$$CH_3COO^- + H_3O^+ \rightarrow CH_3COOH + H_2O$$

initial conc.	0.091	0.0091	0.091
change	-0.0091	-0.0091	+0.0091
equil. conc.	0.0819	0	0.1001

We find pH from the K_a expression for the equilibrium:

$$CH_3COOH + H_2O \rightleftharpoons CH_3COO^- + H_3O^+$$

$$[H_3O^+] = \frac{(1.8 \times 10^{-5})(0.1001)}{(0.0819)} = 2.20 \times 10^{-5} \, M$$

pH = 4.65

The pH change of the buffered solution due to addition of a strong acid (HCl) is 0.08 pH unit.

d) The pH of pure water is 7.0. To find the pH of 100 mL of 0.10 M NaOH added to 1.0 L of water, calculate the concentration of NaOH, and use the relationship that pH = 14 - pOH:

$$[OH] = \frac{(0.10 \, L)(0.10 \, M)}{1.1L} = 0.0091 M$$

pOH = -log(0.0091) = 2.04

pH = 14 - pOH = 14 - 2.04 = 12.0

The pH change from pure water is 12-7 = 5 pH units.

e) Analogous to part "d", we calculate the concentration of HCl to find pH:

$$[H_3O^+] = \frac{(0.10\ L)(0.10\ M)}{1.1L} = 0.0091\,M$$

$$pH = -\log(0.0091) = 2.04$$

In the buffer solution, the strong acid and base were consumed by acetate ion and acetic acid respectively. The pH change in these solutions was less than 0.1 pH unit. The same amount of acid and base added to pure water changed the pH by 5 pH units.

When we solved for the pH of acetic acid in the above problem, we used the expression for K_a and then solved for H_3O^+ to find the pH:

$$CH_3COOH + H_2O \rightleftharpoons CH_3COO^- + H_3O^+$$

$$K_a = \frac{[CH_3COO^-][H_3O^+]}{[CH_3COOH]} \qquad \text{or} \qquad [H_3O^+] = \frac{K_a[CH_3COOH]}{[CH_3COO^-]}$$

We can solve directly for pH by taking the negative logarithm (base 10) of both sides of the equation:

$$-\log[H_3O^+] = -\log K_a - \log\frac{[CH_3COOH]}{[CH_3COO^-]}$$

$$pH = pK_a + \log\frac{[CH_3COO^-]}{[CH_3COOH]}$$

(the Henderson-Hasselbalch equation).

pH of buffer solutions:

1) What is the pH of a solution containing 0.1 M CH_3COOH and 0.1 M CH_3COO^-?

2) What is the pH of a solution containing 1.0 M CH_3COOH and 1.0 M CH_3COO^-?

3) What is the pH of a solution containing 0.1 M CH_3COOH?

4) What is the pH of one liter of a 0.20 M CH_3COOH solution to which 0.10 mol OH^- has been added?

5) Which of the above solutions has the greatest buffer capacity?

solutions:

1) We can use the Henderson-Hasselbalch equation to solve directly for pH:

$$pH = pK_a + \log \frac{\left[CH_3COO^-\right]}{\left[CH_3COOH\right]}$$

$$pH = pK_a + \log\left(\frac{0.1}{0.1}\right)$$

$$pH = pK_a = 4.74$$

2) Analogous to the solution for "1":

$$pH = pK_a + \log\left(\frac{1.0}{1.0}\right)$$

$$pH = pK_a = 4.74$$

Problems 1 and 2 illustrate that when HA = A⁻, pH = pK_a.

3) In this case, we need to consider the equilibrium:

$$CH_3COOH + H_2O \rightleftharpoons CH_3COO^- + H_3O^+$$

initial conc.	0.10	0	0
change	-x	+x	+x
equil. conc.	0.10-x	x	x

$$K_a = \frac{\left[CH_3COO^-\right]\left[H_3O^+\right]}{\left[CH_3COOH\right]} = \frac{(x)(x)}{(0.1-x)} = 1.8 \times 10^{-5}$$

Since acetic acid is a weak acid, we can assume that only a small amount dissociates, and 0.1-x ≈ 0.1. Therefore,

$$K_a = \frac{x^2}{0.1}$$

$$x^2 = \left(1.8 \times 10^{-5}\right) \times 0.1$$

$$x = \sqrt{1.8 \times 10^{-6}} = 1.34 \times 10^{-3} M = \left[H_3O^+\right]$$

$$pH = -\log(1.34 \times 10^{-3}) = 2.87$$

4) OH⁻, a strong base deprotonates CH_3COOH, a weak acid. The reaction goes essentially to completion:

$$CH_3COOH \;+\; OH^- \;\rightarrow\; CH_3COO^- \;+\; H_2O$$

	CH_3COOH	OH^-	CH_3COO^-
initial conc.	0.20	0.10	0
change	-0.10	-0.10	+0.10
equil. conc.	0.10	0	0.10

We have made a buffer with equal amounts of acetic acid and acetate ions. We have the buffer solution as in part "1" when equal amounts of acetic acid and sodium acetate were mixed.

Again, the Henderson-Hasselbalch equation allows us to solve directly for pH:

$$pH = pK_a + \log\left(\frac{0.1}{0.1}\right)$$

$$pH = pK_a = 4.74$$

5) Solutions 1, 2, and 4 all have the same pH since for these solutions, $[CH_3COOH] = [CH_3COO^-]$. Solution 2 has the greatest buffer capacity since it has the highest concentration of the weak acid and its conjugate base.

A buffer is generally effective if it is within ±1 pH unit of the pK_a. The most effective buffer is one in which its $pH = pK_a$. A buffer may be prepared by mixing an acid and the salt of its conjugate base, or by partially deprotonating a weak acid to form its conjugate base (or vice versa).

Buffer range:

Consider an acetic acid/acetate ion buffer.

a) if pH = pKa, what are the relative concentrations of $[CH_3COOH]$ and $[CH_3COO^-]$?
b) If pH = pKa + 1, what are the relative concentrations of $[CH_3COOH]$ and $[CH_3COO^-]$?
c) If pH = pKa - 1, what are the relative concentrations of $[CH_3COOH]$ and $[CH_3COO^-]$?

solutions:

a) We know from the Henderson-Hasselbalch equation, $pH = pK_a + \log\dfrac{\left[CH_3COO^-\right]}{\left[CH_3COOH\right]}$, that pH=pK$_a$ when $\log \dfrac{[CH_3COO^-]}{[CH_3COOH]} = 0$

The log of 1 = 0, so pH = pKa when $[CH_3COOH] = [CH_3COO^-]$.

b) The log of 10 = 1, so pH = pK$_a$ + 1 when the concentration of acetate ion is 10 times the concentration of acetic acid.

c) The log of 1/10 = -1, so pH = pK$_a$ - 1 when the concentration of acetic acid is 10 times the concentration of acetate ion.

In general, a buffer is effective if a conjugate acid is less than 10 times as concentrated as its conjugate base (or vice versa).

SECTION 18.2: Acid-base titration curves

We can determine the concentration of an acid by titrating with a base or vice versa. We use pH meters or indicators to determine the end point of a titration.

Indicators are weak organic acids whose acid form (HIn) is a different color than its base form (In⁻). The indicator <u>bromthymol blue</u> is yellow when protonated , and blue when deprotonated:

$$HIn(aq) + H_2O(l) \rightleftharpoons In^-(aq) + H_3O^+(aq) \qquad K_a = \dfrac{[In^-][H_3O^+]}{[HIn]}$$

yellow   blue

The acid and/or base forms are intensely colored, so only tiny amounts are required to color a solution, and they therefore do not affect the pH of the test solution. An indicator changes color over a pH range of about two units. You may be wondering how a system which changes color over two pH units (or a <u>100-fold</u> change in [H⁺]) can be used to accurately determine the end point of a titration. As you will see, the pH changes <u>more than</u> 2 pH units with addition of only a tiny amount of titrant at the equivalence point of many acid-base titrations.

Strong acid-strong base titrations:

The **equivalence point** of a titration is the point when the number of moles of added OH⁻ equals the number moles of H$_3$O⁺ originally present. The **end point** of a titration occurs when the indicator changes color. We choose an indicator so that the end point is close to the equivalence point.

Equivalence point:

What is the pH at the equivalence point for each of the following:

a) the titration of a strong acid with a strong base.

b) the titration of a weak acid with a strong base.

answers:

a) The reaction of a strong acid with a strong base is:

$$H_3O^+ + OH^- \rightarrow 2H_2O$$

At the equivalence point, all the acid is neutralized, and the pH =7.

b) the reaction of a weak acid with a strong base is:

$$HA + OH^- \rightarrow A^- + H_2O$$

At the equivalence point, the pH is greater than 7 because the solution contains the weak base, A^-. The pH is determined by the pK_b and the concentration of A^-.

Titration of HCl with NaOH:

40.00 mL of 0.1000 M HCl is titrated with 0.1000 M NaOH. What is the pH of the solution after the following volumes of 0.1000 M NaOH have been added?

a) 10.00 mL NaOH   d) 40.01 mL NaOH

b) 39.99 mL NaOH   e) 50.00 mL NaOH

c) 40.00 mL NaOH

solutions:

a) The reaction is: $H_3O^+ + OH^- \rightarrow 2H_2O$

initial moles H_3O^+ = 0.04000 L x 0.1000 mol/L = 0.004000 mol

moles added OH^- = 0.01000 L x 0.1000 mol/L = 0.001000 mol

moles H_3O^+ not neutralized = 0.004000 mol - 0.001000 mol = 0.003000 mol

$$\left[H_3O^+\right] = \frac{0.003000 \text{ mol}}{\left(0.04000 \text{ L} + 0.01000 \text{ L}\right)} = 0.06000 \text{ M}$$

pH = -log (0.06)= 1.222

b) initial moles H₃O⁺ = 0.04000 L x 0.1000 mol/L = 0.004000 mol

moles added OH⁻ = 0.03999 L x 0.1000 mol/L = 0.003999 mol

moles H₃O⁺ not neutralized = 0.004000 mol - 0.003999 mol = 1 x 10⁻⁶ mol

$$\left[H_3O^+\right] = \frac{1\times10^{-6}\ mol}{\left(0.04000\ L +0.03000\ L\right)} = 1.429\times10^{-5}\ M$$

pH = 4.845

c) The 40.00 mL of 0.1 M NaOH exactly neutralizes the 40.00 mL of 0.1 M HCl. The pH is that of pure water, 7. (The presence of the neutral species Na⁺ and Cl⁻ does not affect the pH).

d) The additional 0.01 mL of base takes us past the equivalence point, and base is in excess.

moles added OH⁻ = 0.04001 L x 0.1000 mol/L = 0.004001 mol

moles of OH⁻ not neutralized = 0.004001 mol - 0.004000 mol = 0.0001 mol

$$\left[OH^-\right] = \frac{0.0001\ mol}{\left(0.04000\ L +0.04001L\right)} = 1.250\times10^{-3}\ M$$

pOH = -log (1.250 x 10⁻³) = 2.903

pH = 14 - pOH = 14 - 2.903 = 11.10

e) We are farther past the equivalence point, and the amount of excess base determines the pH:

moles added OH⁻ = 0.05000 L x 0.1000 mol/L = 0.005000 mol

end moles OH⁻ = 0.005000 mol - 0.004000 mol = 0.001000 mol

$$\left[OH^-\right] = \frac{0.001\ mol}{\left(0.04000\ L +0.05001L\right)} = 1.111\times10^{-2}\ M$$

pOH = -log (1.111 x 10⁻²) = 1.954

pH = 14 - pOH = 14 - 1.954 = 12.04

Near the equivalence point, addition of only 0.02 mL of base (less than half a drop) changes the pH from 4.845 to 11.10. Phenolphthalein is often used as an indicator for these titrations. It changes from colorless at pH 8.3 to pink at pH 10.0. Although the color change does not occur at the equivalence point, it changes within a partial drop of the equivalence point.

The titration curve for this titration is shown on p. 796 of your text. What other indicators would be appropriate for this titration? (Answer.: Any which change color between pH 4 and 10; see the list of indicators on p. 795 of your text).

Weak acid-strong base titrations:

Titration of acetic acid with sodium hydroxide:

Consider the titration of acetic acid (a weak acid) with sodium hydroxide (a strong base):

$$CH_3COOH(aq) + OH^-(aq) \rightleftharpoons CH_3COO^-(aq) + H_2O$$

Choose words from the list below to complete the following paragraph:

a) buffered
b) equals
c) higher
d) methyl red (end point pH 5)
e) phenolphthalein (end point pH 9)

f) rapidly
g) shorter
h) slowly
i) acetate ion

If we compare the titration of acetic acid (a weak acid) to a strong acid, the pH at the beginning of the titration of acetic acid will be _____(1) than that of a strong acid of the same molarity. Initially the pH will rise more _____(2) than it did with a strong acid. When enough base is added to make the acetate concentration 1/10th of the acetic acid concentration, the solution is _____(3). In the buffered region, the pH changes _____(4). At the midpoint of the buffer region, the concentration of acetic acid _____(5) the concentration of acetate ion, and pH = pK$_a$. At the equivalence point, the solution contains only _____(6), which makes the solution basic. The steep portion of the curve is _____(7) than when a strong acid is titrated, so the choice of indicator is more crucial. The pH of a dilute sodium acetate solution is 9, so an appropriate indicator for the titration would be _____(8). An indicator which would change color too early, and therefore be inappropriate would be _____(9).

answers: 1-c, 2-f, 3-a, 4-h, 5-b, 6-i, 7-g, 8-e, 9-d.

Weak base-strong acid titrations

Consider the titration of the weak base, ammonia with HCl:

$$NH_3(aq) + H_3O^+(aq) \rightleftharpoons NH_4^+(aq) + H_2O$$

When we titrate ammonia (a weak base) with a hydrochloric acid (a strong acid), the shape of the titration curve is inverted from the titration curve for acetic acid (a weak acid) and sodium hydroxide (a strong base). The pH starts out high, drops quickly initially, and then drops slowly as a buffered solution forms. At the midpoint of the buffer region, the pH equals the pK$_b$ of ammonia. The pH at the equivalence point is below 7 because the solution contains ammonium ions (NH_4^+) which are weak acids. The indicator must change color in an acidic pH range.

SECTION 18.3: Equilibria of slightly soluble ionic compounds

In Chapter 12, we saw that most solutes do not dissolve completely, and that in a saturated solution, equilibrium exists between the undissolved solid and the dissolved solute.

$$CaF_2(s) \rightleftharpoons Ca^{2+}(aq) + 2F^-(aq)$$

We will assume in this chapter that any salt which dissolves dissociates completely into ions. In reality, some of the dissolved calcium is in the form of CaF^+. When we assume that all the dissolved calcium exists as the Ca^{2+} ion (as we will do in this chapter), the calculated value for the solubility of calcium fluoride will be lower than the observed solubility of calcium fluoride. For this equilibrium we write:

$$K_c = \frac{\left[Ca^{2+}\right]\left[F^-\right]^2}{\left[CaF_2\right]}$$

We assume the concentration of a solid is constant, and so its concentration is combined with the constant K_c to obtain K_{sp} (the solubility-product constant):

$$K_c[CaF_2] = [Ca^{2+}][F^-]^2 = K_{sp}$$

K_{sp} for CaF_2 at 25°C is 3.2×10^{-11}. K_{sp} depends on temperature.

Writing K_{sp} for slightly soluble ionic compounds:

Write the solubility product constant expression for:
- a) calcium fluoride
- b) barium carbonate
- c) lead(II) iodate
- d) mercury(II) sulfide

Which of these compounds is the least soluble in water?

solutions:

a) The equilibrium expression is:

$$CaF_2(s) \rightleftharpoons Ca^{2+}(aq) + 2F^-(aq)$$

$$K_{sp} = [Ca^{2+}][F^-]^2$$

b) The equilibrium expression is:

$$BaCO_3(s) \rightleftharpoons Ba^{2+}(aq) + CO_3^{2-}(aq)$$

$$K_{sp} = [Ba^{2+}][CO_3^{2-}]$$

c) The equilibrium expression is:

$$Pb(IO_3)_2(s) \rightleftharpoons Pb^{2+}(aq) + 2IO_3^-(aq)$$

$$K_{sp} = [Pb^{2+}][IO_3^-]^2$$

d) The equilibrium expression is:

$$HgS(s) + H_2O(l) \rightleftharpoons Hg^{2+}(aq) + HS^-(aq) + OH^-(aq)$$

$$K_{sp} = [Hg^{2+}][HS^-][OH^-]$$

S^{2-} is so basic that it reacts completely with water to form hydrogen sulfide ion and hydroxide ion. The least soluble compound is HgS with the smallest K_{sp} value (2×10^{-53}).

Before working problems, let's review from Chapter 4 the solubility rules for ionic compounds in water.

SOLUBILITY RULES FOR IONIC COMPOUNDS IN WATER

1. **Anions of strong acids are soluble:**

   Most nitrate (NO_3^-), sulfate (SO_4^{2-}) and perchlorate (ClO_4^-) salts are soluble; exceptions are $CaSO_4$, $BaSO_4$, and $PbSO_4$ which are insoluble.

2. **Anions of weak acids are only slightly soluble**

   Most sulfide (S^{2-}), carbonate (CO_3^{2-}), and phosphate (PO_4^{3-}) salts are only slightly soluble.

3. **Most hydroxide salts are only slightly soluble**

   exceptions are NaOH, KOH, and $Ca(OH)_2$ which are soluble.

4. **Most salts of Na^+, K^+, and NH_4^+ are soluble**

5. **Most chloride, bromide and iodide salts are soluble**

   exceptions are AgCl, $PbCl_2$, and Hg_2Cl_2 which are insoluble.

Determining K_{sp} from solubility:

CaF$_2$ is used in the manufacture of steel and glass. When powdered fluorite is shaken with water at 18°C, 1.5 x 10^{-4} g dissolves in 10.0 mL of solution. Calculate K_{sp} of CaF$_2$ at 18°C.

solution:

- First write the equilibrium and K_{sp} expressions:

$$CaF_2(s) \rightleftharpoons Ca^{2+}(aq) + 2F^-(aq); \quad K_{sp} = [Ca^{2+}][F^-]^2$$

- Calculate the molar solubility of CaF$_2$:

$$\frac{1.5 \times 10^{-4} \text{ g}}{0.010 \text{ L}} \times \frac{1 \text{ mol}}{77.074 \text{ g}} = 1.95 \times 10^{-4} \text{ mol / L}$$

- When one mole of CaF$_2$ dissolves, 1 mol of Ca^{2+} and 2 mol of F$^-$ form.

$[Ca^{2+}] = 1.95 \times 10^{-4}$ mol/L
$[F^-] = 2 \times (1.95 \times 10^{-4}$ mol/L$) = 3.89 \times 10^{-4}$ mol/L

- Now we can calculate K_{sp}:

$$K_{sp} = [Ca^{2+}][F^-]^2 = (1.95 \times 10^{-4} \text{ mol/L}) \times (3.89 \times 10^{-4} \text{ mol/L})^2$$
$$= 3.0 \times 10^{-11}$$

Compare this value for K_{sp} to the one at 25°C in your text on p. 806 (3.2 x 10^{-11}). K_{sp} is smaller at 18°C; less CaF$_2$ dissolves at the lower temperature.

Determining solubility from K_{sp}:

The K_{sp} for CaF$_2$ at 25°C is 3.2 x 10^{-11}. Estimate the solubility, in mol/L of this mineral.

solution:

- We can see from the equation:

$$CaF_2(s) \rightleftharpoons Ca^{2+}(aq) + 2F^-(aq)$$

that 1 mol of Ca^{2+} and 2 mol of F$^-$ form for every mol of CaF$_2$ which dissolves. If S is the solubility of CaF$_2$ in mol/L, then:

$$[Ca^{2+}] = S, \text{ and } [F^-] = 2S.$$

• We can substitute into the expression for K_{sp}:

$$K_{sp} = [Ca^{2+}][F^-]^2 = S \times (2S)^2 = 3.2 \times 10^{-11}$$

$$4S^3 = 3.2 \times 10^{-11}$$

$$S = 2.0 \times 10^{-4} \, M$$

The solubility of calcium fluoride changes if a common ion is present. Would the solubility of CaF_2 in 0.10 M $Ca(NO_3)_2$ (a soluble salt) be greater or lower than in water? (Answer: The dissolved $Ca(NO_3)_2$ increases the concentration of Ca^{2+} ions, so the concentration of F^- must <u>decrease</u> to maintain a constant K_{sp} value. The equilibrium shifts to the left, making calcium fluoride less soluble in the 0.10 M $Ca(NO_3)_2$ solution).

Solubility of CaF_2 in the presence of a common ion:

What is the solubility of CaF_2 in 0.10 M $Ca(NO_3)_2$? K_{sp} of $CaF_2 = 3.2 \times 10^{-11}$.

<u>solution</u>:

• The main thing to keep in mind when working this problem is that <u>most</u> of the solid calcium fluoride does not go into solution. The tiny amount which does dissolve produces a concentration of calcium ion, Ca^{2+} which is insignificant compared to the amount which comes from the soluble salt, $Ca(NO_3)_2$.

	$CaF_2(s) \rightleftharpoons$	$Ca^{2+}(aq)$ +	$2F^-(aq)$
initial conc.	--	0.10	0
change	--	+S	+2S
equilibrium conc.	--	0.1+S	2S

$$K_{sp} = [Ca^{2+}][F^-]^2 = (0.1+S) \times (2S)^2 = 3.2 \times 10^{-11}$$

• We assume that 0.1 >> S, so (0.1 + S) $\approx$ S. Then:

$$(0.10)(2S)^2 = 3.2 \times 10^{-11}$$

$$4S^2 = 3.2 \times 10^{-10}$$

$$S = 8.94 \times 10^{-6}$$

Compare this solubility to that in pure water worked in the earlier example problem (S = 3.17×10^{-4}). The presence of the common ion, Ca^{2+}, decreased solubility, as predicted.

Predicting whether a precipitate will form:

Does a precipitate form when 0.500 L of 0.50 M $Ca(NO_3)_2$ is mixed with 0.500 L of 0.050 M NaF?

solution:

- Nitrate and sodium salts are soluble, so the ions present in solution are:

$$Ca^{2+}, NO_3^-, Na^+, \text{ and } F^-$$

$NaNO_3$ is soluble, so the only possibility for a precipitate is CaF_2.

- Calculate the initial concentrations of calcium and fluoride ions:

$$\left[Ca^{2+}\right] = \frac{0.500\ L \times 0.50\ mol/L}{1L} = 0.025\,M$$

$$\left[F^-\right] = \frac{0.500\ L \times 0.050\ mol/L}{1L} = 0.025\,M$$

- Substituting into Q_{sp} gives:

$$Q_{sp} = [Ca^{2+}][F^-]^2 = (0.025) \times (0.025)^2 = 1.56 \times 10^{-5}$$

Since $Q_{sp} > K_{sp}$, CaF_2 will precipitate until $Q_{sp} = 3.2 \times 10^{-11}$.

SECTION 18.4: Equilibria involving complex ions

A complex ion is a metal ion (the central atom) covalently bonded to two or more anions or molecules (called ligands). The metal ion acts as a Lewis acid which accepts an electron pair from each ligand. Compounds containing complex ions are called coordination compounds. Coordination compounds will be discussed in more detail in chapter 22.

Coordination compounds:

For each of the following coordination compounds, give the charge of the complex ion, the charge of the central atom, and list the ligands.

a) $[Pt(NH_3)_4]Cl_2$

b) $[Cu(NH_3)_4]SO_4$

c) $K_2[PtCl_4]$

d) $[Zn(H_2O)_3(OH)]Cl$

answers:

Ligands are written within the brackets; species written outside the brackets are not bonded directly to the central metal atom, and are present as free ions.

a) The NH_3 molecules are neutral and the chloride ions have a charge of -1.

    charge of complex = +2; charge of metal (Pt) = +2

    ligands: 4 NH_3 molecules

b) The NH_3 molecules are neutral and the sulfate ion has a charges of -2.

    charge of complex = +2; charge of metal (Cu) = +2

    ligands: 4 NH_3 molecules

c) The potassium ions have a charge of +1, and chloride ions have a charge of -1.

    charge of complex = -2; charge of metal (Pt) = +2

    ligands: 4 Cl^- ions

d) The water molecules are neutral. The hydroxide and chloride ions have -1 charges.

    charge of complex = +1; charge of metal (Zn) = +2

    ligands = 3 H_2O molecules, and 1 OH^- ion

The formation constant, K_f, for a complex ion determines the extent of ligand substitution:

$$Zn(H_2O)_4^{2+}(aq) + 4NH_3(aq) \rightleftharpoons Zn(NH_3)_4^{2+}(aq) + 4H_2O(l)$$

$$K_f = \frac{\left[Zn(NH_3)_4^{2+}\right]}{\left[Zn(H_2O)_4^{2+}\right]\left[NH_3\right]^4} = 7.8 \times 10^8$$

Notice that this formation constant is large, so the complex ion forms readily and is stable.

Calculating the concentration of complex ions:

If 25.5 mL of 3.10 x 10^{-2} M $Fe(H_2O)_6^{3+}$ is mixed with 35.0 mL of 1.50 M NaCN, what is the final $[Fe(H_2O)_6^{3+}]$? K_f of $Fe(CN)_6^{3-}$ = 4.0 x 10^{43}.

solution:

• The equation and K_f expressions are:

$$Fe(H_2O)_6^{3+}(aq) + 6CN^-(aq) \rightleftharpoons Fe(CN)_6^{3-}(aq) + 6H_2O(l)$$

$$K_f = \frac{\left[Fe(CN)_6^{3-}\right]}{\left[Fe(H_2O)_6^{3+}\right]\left[CN^-\right]^6}$$

- The initial reactant concentrations are:

$$\left[Fe(H_2O)_6^{3+}\right] = \frac{(0.0255\,L) \times (3.1 \times 10^{-2}\,mol/L)}{(0.0255\,L + 0.0350\,L)} = 0.013\,M$$

$$\left[CN^-\right] = \frac{(0.0350\,L) \times (1.5\,mol/L)}{(0.0255\,L + 0.0350\,L)} = 0.87\,M$$

- We set up a reaction table with $x = [Fe(H_2O)_6^{3+}]$ at equilibrium. Since K_f is so large, we can assume that essentially all the $Fe(H_2O)_6^{3+}$ is converted to $Fe(CN)_6^{3-}$.

$$Fe(H_2O)_6^{3+} + 6CN^- \rightleftharpoons Fe(CN)_6^{3-} + 6H_2O(l)$$

initial conc.	0.013	0.87	0
change	$-\approx 0.013$	$-6(0.013)$	$+0.013$
equil. conc.	x	0.792	0.013

- Solving for x gives:

$$K_f = \frac{\left[Fe(CN)_6^{3-}\right]}{\left[Fe(H_2O)_6^{3+}\right]\left[CN^-\right]^6} = \frac{(0.013)}{(x)(0.792)^6} = 4.0 \times 10^{43}$$

$$x = 1.32 \times 10^{-45}\,M$$

x, the amount of $Fe(H_2O)_6^{3+}$ left at equilibrium, is very small, as we would expect since the reaction goes nearly to completion (K_f is a large number).

SECTION 18.5: Applications of ionic equilibria to chemical analysis

Ions in mixtures may be separated by selectively precipitating them. The less soluble compounds (ones with the smallest K_{sp} values) precipitate first.

Separating ions:

100 mL of a 0.50 M $Sn(NO_3)_2$ solution is mixed with 100 mL of a 0.20 M $Cd(NO_3)_2$ solution. If aqueous NaOH is added to the mixture, which ion precipitates first? (K_{sp} of $Sn(OH)_2 = 5.4 \times 10^{-27}$ and K_{sp} of $Cd(OH)_2 = 8.1 \times 10^{-13}$). Calculate the concentration of OH^- that will separate Sn from Cd.

solution:

$Sn(NO_3)_2$ and $Cd(NO_3)_2$ are soluble, so the ions in solution are Sn^{2+}, Cd^{2+} and NO_3^-. Hydroxide ion added to the solution will precipitate Sn^{2+} and Cd^{2+}:

$$Sn(OH)_2(s) \rightleftharpoons Sn^{2+}(aq) + 2OH^-(aq) \qquad K_{sp} = [Sn^{2+}][OH^-]^2 = 5.4 \times 10^{-27}$$

$$Cd(OH)_2(s) \rightleftharpoons Cd^{2+}(aq) + 2OH^-(aq) \qquad K_{sp} = [Cd^{2+}][OH^-]^2 = 8.1 \times 10^{-13}$$

We want to add enough hydroxide ion to precipitate as much of the least soluble compound as we can, but not enough to start precipitating the more soluble compound. $Sn(OH)_2$ has the smallest K_{sp} and is therefore the least soluble compound. If we add hydroxide ion until we have a saturated solution of $Cd(OH)_2$, that will be the most Sn^{2+} that can be removed from the mixture without precipitating any Cd^{2+}. We can then calculate the remaining concentration of Sn^{2+} to see if we accomplished the separation.

• The initial concentration of Cd^{2+} is:

$$[Cd^{2+}] = \frac{(0.1L \times 0.20 \, mol/L)}{0.2L} = 0.10 M$$

• The $[OH^-]$ that gives a saturated $Cd(OH)_2$ solution is:

$$[OH^-] = \sqrt{\frac{K_{sp}}{[Cd^{2+}]}} = \sqrt{\frac{8.1 \times 10^{-13}}{0.10}} = 2.8 \times 10^{-6} M$$

• If we add any more hydroxide ion than 2.84×10^{-6} M, Cd^{2+} ion will begin to precipitate out. How much of Sn^{2+} ion remains at this hydroxide ion concentration?

$$[Sn^{2+}] = \frac{K_{sp}}{[OH^-]^2} = \frac{5.4 \times 10^{-27}}{(2.84 \times 10^{-6})^2} = 6.7 \times 10^{-16} M$$

The initial concentration of Sn^{2+} was 0.25 M, so if we add $2.84 \times 10^{-6} M$ hydroxide ion, essentially all the Sn^{2+} will precipitate leaving Cd^{2+} in solution.

Chapter 18 Exercises:

18.1) a. How would you prepare 1 L of a lactic acid/sodium lactate buffer solution with a pH of 4.00 from a 0.10 M $HC_3H_5O_3$ lactic acid solution and solid sodium lactate, $NaC_3H_5O_3$? (pK_a of $HC_3H_5O_3$ = 3.08).

b. How would you prepare a buffer solution with a pH of 4.50?

18.2) a. How would you prepare 500 mL of a buffer with a pH of 9.00 from 0.50 M aqueous NH_3 and a bottle of reagent grade NH_4Cl? A NH_3/NH_4Cl buffer containing equal molar amounts of base and salt has a pH of 9.25. (K_b of NH_3 = 1.77 x 10^{-5}).

b. Could you prepare a buffer with a pH of 8.00 using the same salt and base?

18.3) a. What effect does the addition of the salt of the weak acid have on the pH of a weak acid/salt buffer?

b. If the salt of a weak base is added to a weak base/salt buffer, is the effect on the pH the same?

18.4) What volume of 0.50 M NaClO must be added to 0.20 M HClO in order to prepare 500 mL of a HClO/NaClO buffer with a pH of 7.00? (K_a HClO = 3.5 x 10^{-8}).

18.5) The average adult has 5.0 L of blood of which 3.0 L is blood plasma with a pH of 7.4. The pH of the blood must be maintained within a very narrow region of pH, 7.35-7.45. Thus, the pH of the blood can vary only 0.05 pH units before a serious health risk results. What is the buffer capacity, with respect to excess acid, for the blood supply of an average adult?

18.6) 20.0 mL of 0.10 M KH_2PO_4 solution is titrated with 0.10 M KOH. Calculate the pH of the solution after the following amounts of KOH solution are added:

a. 0 mL   b. 10.0 mL   c. 19.8 mL   d. 20.0 mL   e. 25.0 mL

18.7) Calcium carbonate, $CaCO_3$, is an active ingredient in many antacids such as Tums. What is the molar solubility of $CaCO_3$ in g/L? How does the calcium carbonate work if it does not dissolve in our stomach? (K_{sp} = 8.7 x 10^{-9}).

18.8) Calculate the K_{sp} for each of the salts below:

a. 0.35 g TlCl/ 100 mL solution

b. 1.35 g $Cu(IO_3)_2$/L solution

18.9) A primary treatment of municipal sewage involves the addition of calcium hydroxide and aluminum sulfate to a sedimentation tank to produce aluminum hydroxide, $Al(OH)_3$. As the $Al(OH)_3$ slowly settles to the bottom, it traps suspended particles and bacteria. What is the $[Al^{3+}]$ if the pH in the sedimentation tank is 8.00? (K_{sp} of $Al(OH)_3$ = 2 x 10^{-32} at 25°C).

18.10) Hard water produces soap scum on clothes and in bathtubs. One way to remove the Ca ions and Mg ions that causes hard water is to add hydrated lime, $Ca(OH)_2$. The Mg ions are removed in the form of $Mg(OH)_2$. If the metal ion concentration is less than 65 mg/gal, the water is considered soft. If the target $[Mg^{2+}]$ is only 30 mg/gal, what is the pH needed to cause the remaining Mg^{2+} to precipitate?

18.11) A nuclear chemist, working with a 5.0×10^{-3} M $Ra(NO_3)_2$ solution, precipitates the Ra^{2+} ion as $RaSO_4$ ($K_{sp} = 2.0 \times 10^{-11}$) in an effort to reduce the amount of radium in the waste. What concentration of Na_2SO_4 must be used in order to reduce the free $[Ra^{2+}]$ to 0.1 parts per trillion (1 ppt = 1 nanogram/L)?

18.12) The concentrations of chloride ion and bromide ion in sea water are 0.55 M and 8.4×10^{-4} M respectively. 1.0 M $AgNO_3$ is added to the sea water by a micropump. Which will precipitate first, AgCl or AgBr? The change in volume is negligible. What percent of the first ion has precipitated when the second ion begins to precipitate?

18.13) The ferrous ion concentration, $[Fe^{2+}]$, in the ground water of a Midwest aquifer is 3.6×10^{-4} M. What is the minimum concentration of NH_3 needed to initiate the precipitation of $Fe(OH)_2$?

18.14) Gold is extracted from low-grade ores by washing the ore with a potassium cyanide solution. The gold is leached out of the ore as a $K[Au(CN)_2]$ solution ($K_d = 5.0 \times 10^{-39}$). What is the concentration of the free Au^{3+} ion in a solution which is 0.40 M $K[Au(CN)_2]$ and 1.0×10^{-3} M KCN?

Chapter 18 Answers:

18.1)   a. Dissolve 93 g of $NaC_3H_5O_3$ in about 500 mL of 0.10 M $HC_3H_5O_3$, dilute to 1.00 L with 0.100 M $HC_3H_5O_3$, and mix.

           b. Dissolve 295 g of $NaC_3H_5O_3$ in about 500 mL of 0.10 M $HC_3H_5O_3$, dilute to 1.00 L with 0.100 M $HC_3H_5O_3$, and mix.

18.2)   a. Dissolve 96.3 g NH_4Cl in about 250 mL of the 0.50 M NH_3 solution, dilute to 500 mL with the 0.20 M NH_3 solution, and mix.

           b. No, the buffer would require 963 g of NH_4Cl which would not completely dissolve in the 500 mL of 0.50 M NH_3.

18.3)   a. The addition of the salt of the weak acid will increase the pH of the buffer.
           b. The addition of the salt of a weak base will decrease the pH of the buffer.

18.4)   61 mL of NaClO must be added to 439 mL of 0.20 M HClO.

18.5) 1.4×10^{-8} moles of H^+

18.6)   a. pH = 4.10;   b. pH = 7.20;   c. pH = 9.20;   d. pH = 9.95;   e. pH = 12.18

18.7) 9.3×10^{-3} g/L
The excess acid consumes the soluble carbonate ion and by LeChatlier's Principle, the lost carbonate ion is replaced by the dissolution of more calcium carbonate.

18.8)   a. $K_{sp} = 2.1 \times 10^{-4}$;   b. $K_{sp} = 2.02 \times 10^{-7}$

18.9) $[Al^{3+}] = 2.0 \times 10^{-14}$ M

18.10)   pH = 10.41

18.11) $[Na_2SO_4] = 4.5$ M

18.12)   Since the K_{sp} for AgBr is smaller than that for silver chloride, we might expect AgBr to precipitate first. However, because there is such a large excess of chloride ion compared to bromide ion, silver chloride precipitates until $[Ag^+] = 5.95 \times 10^{-10}$ at which point, when more silver is added, silver bromide begins to precipitate. The chloride ion concentration remaining at this point is $[Cl^-] = 0.27$ M (51% of the chloride has precipitated).

18.13)   pH = 8.37

18.14) $[Au^{3+}] = 2.00 \times 10^{-36}$ M

CHAPTER 19
Thermodynamics

TOPICS:

- The Second Law of Thermodynamics
- Entropy, free energy, and work
- Gibb's Free Energy, Equilibrium, and Reaction direction

SECTION 19.1: The Second Law of Thermodynamics

The first law:

In Chapter 6 we saw that the total energy of the universe is constant; energy cannot be created or destroyed (the first law of thermodynamics). The first law of thermodynamics recasts thermochemical "heats of reaction" into ΔH values. It does not tells us why a process occurs in a given direction. The second law of thermodynamics answers the why question.

Exothermic and endothermic reactions:

Determine which of the following reactions are endothermic and which are exothermic. Which reactions are spontaneous (ones that occur by themselves)?

a) ice at room temperature melting

b) water below 0°C freezing

c) water vaporizing

d) methane burning

e) sodium chloride dissolving in water

f) iron rusting

g) diamond converting to graphite

answers:

The processes which absorb energy from their surroundings (endothermic reactions) are:

a) $\Delta H^o_{rxn} = 6.02$ kJ/mol

c) $\Delta H^o_{rxn} = 44.0$ kJ/mol

e) $\Delta H^o_{rxn} = 3.90$ kJ/mol

The processes which release energy (exothermic reactions) are:

b) $\Delta H^o_{rxn} = -6.02$ kJ/mol

d) $\Delta H^o_{rxn} = -802$ kJ/mol

f) $\Delta H^o_{rxn} = -826$ kJ/mol

g) $\Delta H^o_{rxn} = -1.90$ kJ/mol

All of the reactions are spontaneous. Spontaneous reactions may be exothermic or endothermic, and may be fast or slow. Diamond changes to graphite spontaneously, but hardly instantaneously.

We can not use the standard heat of formation for a reaction to determine if a reaction is spontaneous.

The second law:

Picture a ball rolling uphill; a rusty nail changing back to a shiny one; carbon dioxide and water coming together to form wood; water freezing at room temperature. These processes never happen. Experience tells us you can't go back. **All spontaneous processes increase the entropy of the universe.** This is the second law of thermodynamics. **Entropy (S)** is the measure of a system's disorder; entropy increases as systems become more disordered.

$$S_{disordered\ system} > S_{ordered\ system}$$

The entropy of the universe is always increasing. A spontaneous process may occur in which the system becomes more ordered, but the entropy of the system + the surroundings must be greater than zero:

$$\Delta S_{universe} = \Delta S_{sys} + \Delta S_{surr} > 0$$

We will show how we can calculate both ΔS_{sys} and ΔS_{surr}.

Entropy of a familiar system:

Water freezes spontaneously when the temperature is below 0°C. Does the entropy of the water molecules increase or decrease as they change from the liquid to the solid state. How can one understand the fact that water freezing below 0°C is a spontaneous process?

answer:

The entropy of water molecules decreases as they become more ordered in a solid crystal lattice. Since water freezes spontaneously below 0°C, the increase in entropy of the surroundings must be larger than this decrease in entropy of the system, making the total entropy change of the system + surroundings > 0.

The heat released to the surroundings as water freezes increases the random motion of particles in the surroundings causing their disorder (entropy) to increase.

Entropy of the system (ΔS_{sys}):

Entropy is a state function, so it depends only on the initial and final states. $\Delta S_{sys} = S_{final} - S_{initial}$. Unlike enthalpies which are all relative values, we can determine absolute entropies. The third law of thermodynamics states that a perfect crystal has zero entropy at absolute zero. At absolute zero, the crystal has no thermal motion, and so has only one arrangement of atoms. Entropy is an extensive property, it depends on the amount of a substance. The entropy of a mole of a substance in its standard state is its standard molar entropy (S^o in J/K mol).

Predicting relative S^o values:

Which of the following would you predict would increase the entropy of a system?

    a) temperature increase

    b) solid changing to a liquid

    c) liquid changing to a gas

    d) solid or liquid dissolving in a liquid

    e) gas dissolving in a solid or liquid

    f) number of atoms increases (same physical state)

answers: a,b,c,(d),f

All of the above would increase the disorder of a system except for a gas dissolving in a liquid or solid. Gas molecules are so disordered, that their disorder decreases as they incorporate into a liquid or solid matrix. The entropy of some solids may decrease when dissolved because, although the entropy of their ions is much greater in solution than in the crystal, the entropy of the solvent may decrease as water molecules become highly ordered around hydrated ions.

Predicting relative entropy values:

Choose the member of each pair with the higher entropy.

a) 1 mol liquid nitrogen; 1 mol gaseous nitrogen

b) 1 mol $O_2(g)$; 1 mol $O_3(g)$

c) water at room temperature; water at 90°C

d) 1 mol dry ice ($CO_2(s)$); 1 mol $CO_2(g)$

e) 1 mol $CO_2(g)$; 1 mol $CO_2(g)$ dissolved in a cola

solutions:

a) 1 mol gaseous nitrogen; for a given substance, entropy increases in the sequence solid < liquid < gas

b) 1 mol O_3; both oxygen and ozone are gaseous molecules which contain oxygen atoms. Ozone has more atoms in a molecule, so there are more types of motion available to it and its entropy is higher.

c) water at 90°C; entropy increases with an increase in temperature

d) 1 mol gaseous CO_2; see solution "a"

e) 1 mol gaseous CO_2; the entropy of a gas decreases as it dissolves into a liquid or solid matrix.

Predicting the sign of $\Delta S°$:

Predict the sign of $\Delta S°$ for each of the following reactions:

a) $4NH_3(g) + 5O_2(g) \rightarrow 4NO(g) + 6H_2O(g)$

b) $CH_4(g) + 2O_2(g) \rightarrow CO_2(g) + 2H_2O(g)$

c) $CaCO_3(s) \rightarrow CaO(s) + CO_2(g)$

answers:

a) In general, when a reaction involves gaseous molecules, the change in the number of molecules of gaseous reactants and products dominates the entropy change. In reaction (a), 9 gaseous molecules produce 10 gaseous molecules. The number of gaseous molecules increases, and entropy increases; ΔS is positive.

b) 3 gaseous molecules produce 3 gaseous products, so the entropy change for this reaction is approximately 0.

c) A gas is produced from a solid reactant, so entropy increases, and $\Delta S°$ is positive.

We can add standard molar entropies to find the standard entropy of reaction:

$$\Delta S^\circ_{rxn} = \sum m S^\circ_{(products)} - \sum n S^\circ_{(reactants)}$$

where m and n are the coefficients of the balanced equation.

Calculating the standard entropy of reaction (ΔS_{sys}):

Calculate $\Delta S°$ at 25°C for the reaction

$$CH_4(g) + 2O_2(g) \rightarrow CO_2(g) + 2H_2O(g)$$

given the standard entropy values:

$CH_4(g)$	186 J/K mol
$O_2(g)$	205 J/K mol
$CO_2(g)$	214 J/K mol
$H_2O(g)$	189 J/K mol

<u>solution:</u>

$$\Delta S°_{rxn} = \sum m S°_{(products)} - \sum n S°_{(reactants)}$$

$$= (1 \text{ mol} \times 214 \text{ J/K mol}) + (2 \text{ mol} \times 189 \text{ J/K mol})$$

$$-(1 \text{mol} \times 186 \text{ J/K mol}) - (2 \text{mol} \times 205 \text{ J/K mol})$$

$\Delta S° = 214 \text{ J/K} + 378 \text{J/K} - 186 \text{ J/K} - 410 \text{ J/K}$

$\Delta S° = -4 \text{ J/K}$

Notice that in the previous example we predicted that the entropy change for this reaction would be approximately 0 because moles of gaseous reactants = moles of gaseous products. When we calculate $\Delta S°$, we find that it is close to 0.

Entropy of the surroundings (ΔS_{surr}):

Knowing the entropy change of a system will not tell us if a reaction is spontaneous. We must know the entropy change of a system <u>and</u> the surroundings ($\Delta S_{universe} = \Delta S_{system} + \Delta S_{surroundings}$). If the surroundings add heat to the system (endothermic reaction), the entropy of the surroundings decreases; if the system gives off heat to the surroundings (exothermic reaction), the entropy of the surroundings increases. The sign of ΔS_{surr} depends on the direction of heat flow, and the magnitude of ΔS_{surr} depends on the temperature. The change in entropy of the surroundings is greater when heat is added at a lower temperature:

$$\Delta S_{surr} = -\frac{q_{sys}}{T}$$

$$\Delta S_{surr} = -\frac{\Delta H_{sys}}{T} \quad (\text{const. P})$$

We can calculate ΔS_{surr} by measuring ΔH_{sys} and the temperature at which the change takes place.

Entropy of solution formation:

We saw in Chapter 12 that the entropy of solution formation (ΔS_{sys}) is positive, and favorable, but that ΔS_{surr} depends on the heat of solution. Explain why one of the following processes is spontaneous, and the other is not.

        a) oil mixing with water

        b) ethanol mixing with water

solution:

ΔH for the mixing of polar water molecules and nonpolar oil molecules is large and positive. Since $\Delta S_{surr} = -\dfrac{\Delta H_{sys}}{T}$, a large positive ΔH makes ΔS_{surr} a large negative number. Although ΔS_{sys} is positive, the large negative ΔS_{surr} makes $\Delta S_{universe}$ negative, and solution formation is nonspontaneous.

ΔH for the mixing of polar molecules is small, and the positive value for ΔS_{sys} predominates. $\Delta S_{universe}$ is positive and so solution formation is spontaneous.

Determining if a reaction is spontaneous (calculation of $\Delta S_{universe}$):

Does the combustion of methane with oxygen occur spontaneously at 298K?

$$CH_4(g) + 2O_2(g) \rightarrow CO_2(g) + 2H_2O(g) \qquad \Delta S^{o}_{sys} = -4 \text{ J/K}$$

$$\Delta H^{o}_{f} (CH_4) = -75 \text{ kJ}$$

$$\Delta H^{o}_{f} (CO_2) = -393.5 \text{ kJ}$$

$$\Delta H^{o}_{f} (H_2O) = -242 \text{ kJ}$$

solution:

• For this reaction to be spontaneous, $\Delta S_{universe}$ must be > 0.

  Since $\Delta S_{universe} = \Delta S_{sys} + \Delta S_{sys}$, and since $\Delta S^{o}_{sys} = -4$ J/K (above),

  ΔS^{o}_{surr} must be greater than +4 J/K in order for $\Delta S_{universe}$ to be > 0.

  We calculate ΔS^{o}_{surr} from ΔH^{o}_{sys} :

$$\Delta S^{o}_{surr} = -\frac{\Delta H^{o}_{sys}}{T}$$

$$\Delta H^o_{sys} = \Delta H^o_{rxn} - \sum m\Delta H^o_{f(products)} - \sum n\Delta H^o_{f(reactants)}$$

$$= (2 \text{ mol } H_2O \times -242 \text{ kJ/mol}) + (1 \text{ mol } CO_2 \times -393.5 \text{ kJ/mol})$$
$$- (1 \text{ mol } CH_4 \times -75 \text{ kJ/mol})$$

$$= (-484 \text{ kJ/mol}) - (-393.5 \text{ kJ/mol}) + (75 \text{ kJ/mol})$$

$$= -802.5 \text{ kJ/mol}$$

- $$\Delta S^o_{surr} = -\frac{\Delta H^o_{sys}}{T} = -\frac{-802.5 \text{ kJ/mol}}{298 \text{ K}} \times \frac{1000 \text{ J}}{kJ} = 2693 \text{ J/K}$$

(note: don't forget that the units of ΔS^o are J/K, not kJ/K)

- $\Delta S^o_{universe} = \Delta S^o_{surr} + \Delta S^o_{sys} = 2693 \text{ J/K} - 4 \text{ J/K} = 2689 \text{ J/K}$

$\Delta S^o_{universe} > 0$, so the reaction is spontaneous at 298 K. The entropy change of the system is almost zero, but the heat that this reaction gives off to its surroundings greatly increases the entropy of the surroundings and makes $\Delta S^o_{universe} > 0$.

At equilibrium, neither the forward nor the reverse reaction is spontaneous:

$$\Delta S_{sys} = -\Delta S_{surr}$$

$$\Delta S_{universe} = \Delta S_{surr} + \Delta S_{surr} = 0$$

SECTION 19.2: Entropy, Free Energy, and Work

We can predict the spontaneity of a process from ΔS_{univ}. Another thermodynamic function which is related to spontaneity is the Gibb's Free Energy (G):

$$G = H - TS$$

For a process at constant temperature and pressure, the change in free energy (ΔG) is:

$$\Delta G = \Delta H - T\Delta S$$

All quantities in this equation refer to the system. (When no subscript is used, assume the quantity refers to the system). When ΔG is negative, the process is spontaneous.

Usually the enthalpy contribution (ΔH) to the free energy change (ΔG) is much larger than the entropy contribution ($T\Delta S$), especially at lower temperatures. The entropy contribution is more likely to become critical to the spontaneity of a process if the temperature is high. Most exothermic reactions, which have a negative ΔH, are spontaneous. In order for an exothermic reaction to be nonspontaneous (ΔG positive), there would need to be an increase in the order of the system (a negative ΔS term) and a temperature that is high enough to make the $T\Delta S$ term larger than ΔH.

Temperature and spontaneity:

The formation of solid sodium chloride from elemental sodium and chlorine is an exothermic reaction ($\Delta H° = -822.2$ kJ). The increase in order of the sodium and chlorine atoms in the sodium chloride lattice makes the entropy of the system negative ($\Delta S° = -181.7$ J/K). At what temperatures is this exothermic reaction nonspontaneous?

<u>solution</u>:

- The reaction is not spontaneous when $\Delta G°$ is positive. Since both $\Delta H°$ and $\Delta S°$ are negative, $\Delta G°$ will be positive when the $T\Delta S°$ term is larger than the $\Delta H°$ term.

- Calculate when $\Delta H° = T\Delta S°$:

$$-822.2 \text{ kJ} = T \times (-181.7 \text{ J/K})$$

$$T = \frac{-822,200 \text{ J}}{-181.7 \text{ J/K}} = 4225 \text{ K}$$

The reaction is not spontaneous at temperatures above 4525 K (4252°C).

Gibb's Free Energy (G):

A process is spontaneous if $\Delta S_{univ} > 0$. To see how ΔG relates to spontaneity, relate ΔG to ΔS_{univ}.

<u>solution</u>:

- We know that: $\Delta S_{univ} = \Delta S_{surr} + \Delta S_{sys}$, and

$$\Delta S_{surr} = -\frac{\Delta H_{sys}}{T} \quad (\text{const. } P)$$

- Substituting $-\Delta H_{sys}/T$ for ΔS_{surr} in the first equation gives:

$$\Delta S_{univ} = -\frac{\Delta H_{sys}}{T} + \Delta S_{sys}$$

- Multiply through by $-T$:

$$-T\Delta S_{univ} = \Delta H_{sys} - T\Delta S_{sys}$$

Comparing this equation to the one for ΔG above ($\Delta G = \Delta H - T\Delta S$), we see that the right hand sides of the two equations are the same, and therefore:

$$\Delta G = -T\Delta S_{univ} \quad (\text{const. T,P})$$

$$\text{or,} \quad \Delta S_{univ} = -\frac{\Delta G_{sys}}{T}$$

A process is spontaneous if $\Delta S_{univ} > 0$. Since temperature in Kelvin is always > 0, a process carried out at constant temperature and pressure will be spontaneous when ΔG is negative.

When:

$\Delta G < 0$, the process is spontaneous

$\Delta G > 0$, the process is nonspontaneous

$\Delta G = 0$, the system is at equilibrium

ΔS_{univ} and ΔG:

When any process is spontaneous, the entropy of the universe increases ($\Delta S_{univ} > 0$). When a process at constant temperature and pressure is spontaneous, the change in free energy is less than 0 ($\Delta G < 0$).

Spontaneity of melting ice:

Use the change in free energy ($\Delta G°$) and the change in the entropy of the universe (ΔS_{univ}) to predict the spontaneity of ice melting at:

a) -10°C, b) 0°C, c) 10°C

For the process: $H_2O(s) \rightarrow H_2O(l)$, $\Delta H° = 6.03 \times 10^3$ J/mol

$\Delta S° = 22.1$ J/K mol

(The superscript degree symbol (°) indicates all substances are in their standard states).

solution:

a) $\Delta G°$: substitute -10°C (263 K), and the values for $\Delta H°$ and $\Delta S°$ into the free energy equation:

$$\Delta G° = \Delta H° - T\Delta S°$$

$$\Delta G° = 6.03 \times 10^3 \text{ J/mol} - (263 \text{ K} \times 22.1 \text{ J/K mol})$$

$$\Delta G° = 2.17 \times 10^2 \text{ J/mol}$$

ΔS_{univ}:

$$\Delta S_{univ} = \Delta S° + \Delta S_{surr}$$

$$\Delta S°_{surr} = -\frac{\Delta H°_{sys}}{T} = \frac{-6.03 \times 10^3 \text{ J / mol}}{263} = -22.9 \text{ J / K mol}$$

$$\Delta S_{univ} = (22.1 \text{ J/K mol}) + (-22.9 \text{ J/K mol}) = -0.800 \text{ J/K mol}$$

$\Delta G°$ is positive, and ΔS_{univ} is negative, so ice does not melt spontaneously at -10°C.

b) $\Delta G°$: substitute 0°C (273 K), and the values for $\Delta H°$ and $\Delta S°$ into the free energy equation:

$$\Delta G° = \Delta H° - T\Delta S°$$

$$\Delta G° = 6.03 \times 10^3 \text{ J/mol} - (273 \text{ K} \times 22.1 \text{ J/K mol})$$

$$\Delta G° = 0$$

ΔS_{univ}:

$$\Delta S_{univ} = \Delta S° + \Delta S_{surr}$$

$$\Delta S°_{surr} = -\frac{\Delta H°_{sys}}{T} = \frac{-6.03 \times 10^3 \text{ J / mol}}{273} = -22.1 \text{ J / K mol}$$

$$\Delta S_{univ} = (22.1 \text{ J/K mol}) + (-22.1 \text{ J/K mol}) = 0 \text{ J/K mol}$$

$\Delta G°$ and ΔS_{univ} equal 0, so water and ice are at equilibrium at 0°C.

c) $\Delta G°$: substitute 10°C (283 K), and the values for $\Delta H°$ and $\Delta S°$ into the free energy equation:

$$\Delta G° = \Delta H° - T\Delta S°$$

$$\Delta G° = 6.03 \times 10^3 \text{ J/mol} - (283 \text{ K} \times 22.1 \text{ J/K mol})$$

$$\Delta G° = -2.24 \times 10^2 \text{ J/mol}$$

ΔS_{univ}:

$$\Delta S_{univ} = \Delta S° + \Delta S_{surr}$$

$$\Delta S°_{surr} = -\frac{\Delta H°_{sys}}{T} = \frac{-6.03 \times 10^3 \text{ J / mol}}{283} = -21.3 \text{ J / K mol}$$

$$\Delta S_{univ} = (22.1 \text{ J/K mol}) + (-21.3 \text{ J/K mol}) = 0.793 \text{ J/K mol}$$

$\Delta G°$ is negative, and ΔS_{univ} is positive so ice melts spontaneously at 10°C.

When ice melts:

H_2O molecules become more disordered and their entropy increases; ΔS_{sys} (ΔS°) favors the melting of ice.

Heat is absorbed from the surroundings and the entropy of the surroundings decreases; ΔS_{surr} favors the freezing of ice.

Below 0°C, $\Delta S_{surr} > \Delta S_{sys}$, and ice freezes.
Above 0°C, $\Delta S_{sys} > \Delta S_{surr}$, and ice melts.
At 0°C, $\Delta S_{surr} = \Delta S_{sys}$, and ice and water coexist.

When mercury boils:

Mercury has an enthalpy of vaporization of 58.51 kJ/mol, and an entropy of vaporization of 92.92 J/K mol. What is the normal boiling point of mercury?

$$Hg(l) \rightarrow Hg(g) \qquad \Delta H^\circ = 58.51 \text{ kJ/mol}$$
$$\Delta S^\circ = 92.92 \text{ J/K mol}$$

solution:

At the boiling point of mercury, liquid and gaseous mercury are in equilibrium ($\Delta G^\circ = 0$).

$$\Delta G^\circ = \Delta H^\circ - T\Delta S^\circ = 0$$
$$\Delta H^\circ = T\Delta S^\circ$$

$$T = \frac{\Delta H^\circ}{\Delta S^\circ} = \frac{58,510 \text{ J / mol}}{92.92 \text{ J / k mol}} = 629.7 \text{ K } (356.7°C)$$

ΔG°_{rxn} can be calculated using values for the standard free energy of formation:

$$\Delta G^\circ_{rxn} = \sum m \, \Delta G^\circ_{f(products)} - \sum n \, \Delta G^\circ_{f(reactants)}$$

where $\sum$ means "the sum of", ΔG°_f is the standard free energy of formation, and m and n are the coefficients of the reactants and products from the balanced equation.

Calculation of ΔG°_{rxn}:

Calculate ΔG°_{rxn} at 25°C for the following reactions:

a) $CH_4(g) + 2O_2(g) \rightarrow CO_2(g) + 2H_2O(g)$

b) $2H_2S(g) + SO_2(g) \rightleftharpoons 3S(s) + 2H_2O(g)$

given the values for the standard free energies of formation:

$CH_4(g)$	-51 kJ/mol	$H_2S(g)$	-34 kJ/mol
$CO_2(g)$	-394 kJ/mol	$SO_2(g)$	-300 kJ/mol
$H_2O(g)$	-229 kJ/mol		

<u>solution:</u>

a) $\Delta G^\circ_{rxn} = \sum m\Delta G^\circ_{f(products)} - \sum n\,\Delta G^\circ_{f(reactants)}$

$= (1 \text{ mol} \times -394 \text{ kJ/mol}) + (2 \text{ mol} \times -229 \text{ kJ/mol})$

$\quad - (1 \text{ mol} \times -51 \text{ kJ/mol}) - (2 \text{ mol} \times 0)$

$= -394 \text{ kJ} - 458 \text{ kJ} + 51 \text{ kJ}$

$= -801 \text{ kJ}$

b) $\Delta G^\circ_{rxn} = \sum m\Delta G^\circ_{f(products)} - \sum n\,\Delta G^\circ_{f(reactants)}$

$= (3 \text{ mol} \times 0) + (2 \text{ mol} \times -229 \text{ kJ/mol})$

$\quad - (2 \text{ mol} \times -34 \text{ kJ/mol}) - (1 \text{ mol} \times -300 \text{ kJ/mol})$

$= -458 \text{ kJ} + 68 \text{ kJ} + 300 \text{ kJ}$

$= -90 \text{ kJ}$

Both reactions are spontaneous in the forward direction at 25°C.

Free energy and work:

For a spontaneous process, ΔG represents the energy that is free to do useful work.

$$\Delta G = w_{max}$$

For a nonspontaneous process, ΔG represents the minimum amount of work that must be done to make the process occur. The amount of work we obtain from any real process is always less than the maximum possible amount because some of the work is changed to heat the surroundings. We can determine the work <u>potential</u> of a process, but we can never achieve this potential. Thermodynamicist Henry Bent sums up these thoughts as follows:

> First law: You can't win, you can only break even.
>
> Second law: You can't break even.

SECTION 19.3: Free Energy, Equilibrium and Reaction Direction

In Chapter 15 on kinetics, we saw that at equilibrium, the forward and reverse reaction rates are equal. From a thermodynamic point of view, equilibrium is the lowest value of free energy available to the system. The free energy is related to the equilibrium constant by:

$$\Delta G = \Delta G° + RT \ln Q$$

At equilibrium, $\Delta G = 0$, and $Q = K$, so:

$$\Delta G° = - RT \ln K$$

$\Delta G°$ for the autoionization of water:

What is $\Delta G°$ for the autoionization of water at 25°C?

solution:

For the process, $H_2O(l) \rightleftharpoons H^+(aq) + OH^-(aq)$, $K = 1.0 \times 10^{-14}$.

We can use the relationship between $\Delta G°$ and K to calculate $\Delta G°$:

$$\Delta G° = - RT \ln K$$

$$\Delta G° = -(8.3148 \text{ J/K mol} \times 298 \text{ K}) \ln (1.0 \times 10^{-14})$$

$$\Delta G° = 80 \text{ KJ}$$

$\Delta G°$ is positive, so the reverse reaction is favored; only a tiny percent of water molecules exist in their ionized form.

Chapter 19 Exercises:

19.1) Without consulting any tables, indicate which of the two substances has a greater amount of entropy. Explain your answer.

a. $CH_4(g)$ (natural gas) or $CH_3CH_2CH_3(g)$ (propane gas)
b. $N_2(g)$ (atmospheric nitrogen) or $N_2(l)$ (liquid nitrogen)
c. $Fe(OH)_2$ (Iron(II) hydroxide) or $Fe(OH)_3$ (Iron(III) hydroxide)
d. $Br(g)$ (atomic bromine) or $Br_2(g)$ (molecular bromine)

19.2) Calculate: ΔS^o_{sys}, ΔS^o_{surr}, and ΔS^o_{univ} for the reaction below (25°C) and indicate if the reaction is spontaneous.

Partially hydrogenated fats or oils are used in junk food to help eliminate the greasy appearance in the food. The reaction below illustrates the principle of hydrogenation.

$$CH_3CH=CH_2(g) + H_2(g) \rightleftharpoons CH_3CH_2CH_3(g)$$

19.3) Calculate: ΔS^o_{sys}, ΔS^o_{surr}, and ΔS^o_{univ} for the reaction below (25°C) and indicate if the reaction is spontaneous.

An iron nail placed in a solution of copper(II) sulfate will develop a copper coating on the nail

$$Fe(s) + CuSO_4(aq) \rightleftharpoons Cu(s) + FeSO_4(aq)$$

19.4) Calculate ΔG^o_{rxn} for each reaction at 25°C.

a. Near the earth surface ozone is actually an air pollutant with a tolerance level of 0.12 ppm. Ozone reacts in the presence of ultraviolet light with another air pollutant, NO_2, to produce nitric acid.

$$O_3(g) + H_2O(g) + 2NO_2(g) \rightleftharpoons HNO_3(l) + O_2(g)$$

b. Phosgene gas, $COCl_2$, was first used as a chemical weapon during WWI. It reacts with water in the lungs.

$$COCl_2(g) + H_2O(l) \rightleftharpoons CO_2(g) + 2HCl(aq)$$

19.5) Before development of metal catalysts, SO_2 was converted to SO_3 using NO_2 as a catalyst in a method referred to as the lead chamber process.

$$2SO_2(g) + 2NO_2(g) \rightleftharpoons 2SO_3(g) + 2NO(g)$$
$$\underline{2NO(g) + O_2(g) \rightleftharpoons 2NO_2(g)}$$
$$2SO_2(g) + O_2(g) \rightleftharpoons 2SO_3(g) \qquad \text{(net reaction)}$$

What is ΔG^o_{rxn} for each step at 100°C?

19.6)   At what temperatures are the following reactions spontaneous?

a. An innovative, efficient coal gasification process converts coal into methane which is easier to transport and burns more cleanly than coal. The overall process is:

$$2C(s) + 2H_2O(g) \rightleftharpoons CH_4(g) + CO_2(g)$$

b. The use of ammonia as a fertilizer in agriculture makes it one of the most important chemicals manufactured in the world. Fritz Haber developed a method for the synthesis of ammonia directly from nitrogen and hydrogen.

$$N_2(g) + 3H_2(g) \rightleftharpoons 2NH_3(g)$$

c. The first step in the manufacturing of nitric acid by the Oswalt process is the oxidation of ammonia.

$$4NH_3(g) + 5O_2(g) \rightleftharpoons 4NO(g) + 6H_2O(g)$$

d. Elemental mercury is obtained by roasting cinnabar in the presence of oxygen.

$$HgS(s) + O_2(g) \rightleftharpoons Hg(l) + SO_2(g)$$

19.7)   What are the equilibrium constants for the following reactions?

a. The photochemical decomposition of Freon in the stratosphere leads to the decomposition of the ozone present. The average temperature in the stratosphere is -55°C.

$$2O_3(g) \rightleftharpoons 3O_2(g)$$

b. The slightly acidic nature of unpolluted rainfall is the result of the carbon dioxide in the atmosphere reacting with water.

$$H_2O(l) + CO_2(g) \quad H_2CO_3(aq)$$

Chapter 19 Answers:

19.1)   a. propane gas; more complex molecules contain more types of molecular motion.
b. $N_2(g)$; gases have more random movement than liquids, less intermolecular forces of attraction.
c. $Fe(OH)_3$; more complex molecules contain more types of molecular motion.
d. $Br_2(g)$; more complex molecules contain more types of molecular motion.

19.2) $\Delta S^o_{sys} = -130.2$ J/mol K

$$\Delta S^o_{surr} = -\frac{\Delta H^o_{sys}}{T} = -\frac{-124300 \text{ J / mol}}{298 \text{ K}} = 417 \text{ J/mol K}$$

$\Delta S^o_{univ} = \Delta S^o_{sys} + \Delta S^o_{surr} = -130.2$ J/mol K + 417 J/mol K = 286.8 J/mol K

Since ΔS^o_{univ} is greater than zero, the reaction is spontaneous.

19.3) $\Delta S^o_{sys} = -32.2$ J/mol K

$$\Delta S^o_{surr} = -\frac{\Delta H^o_{sys}}{T} = -\frac{-153800 \text{ J / mol}}{298 \text{ K}} = 516 \text{ J/mol K}$$

$\Delta S^o_{univ} = \Delta S^o_{sys} + \Delta S^o_{surr} = -32.2$ J/mol K + 516 J/mol K = 483.8 J/mol K

Since ΔS^o_{univ} is greater than zero, the reaction is spontaneous.

19.4)   a. $\Delta G^o_{rxn} = -575$ kJ; spontaneous
b. $\Delta G^o_{rxn} = -215$ kJ; spontaneous

19.5)   Step 1: $\Delta G^o_{rxn} = -99.3$ kJ; Step 2: $\Delta G^o_{rxn} = -59.54$ kJ;
Net reaction: $\Delta G^o_{rxn} = -99.3$ kJ - 59.54 = -158.8 kJ

19.6)   a. spontaneous above 1403 K
b. spontaneous below 1380 K
c. The reaction is always spontaneous
d. The reaction is always spontaneous

19.7)   a. $K_p = 1.13 \times 10^{78}$
b. $K_p = 0.034$

CHAPTER 20

Electrochemistry
"Where the electrons go"

TOPICS:

- Half-Reactions

- Voltaic cells
  - Electrode potentials
  - Relative oxidizing and reducing strengths

- Free energy and electrical work
  - Charge and current
  - Nernst equation
  - Equilibrium constants from $E°$
  - Concentration cells

- Electrolytic cells
  - Metal plating
  - Electrolysis of water
  - Overvoltage

SECTION 20.1: Half-reactions

Remember from Chapter 4 that electron transfer reactions are called **oxidation-reduction** (or redox) reactions. **Oxidation** is the loss of electrons, **reduction** is the gain of electrons. "Oil rig" is a good mnemonic device for remembering this. Now we will see how to harness a redox reaction to produce an electrical current, or to use an electrical current to drive a redox reaction.

Balancing a redox equation (review):

In Chapter 4, we balanced the redox reaction which occurs between chloride and permanganate ions in acidic solution by the oxidation-number method. Here we will balance the same equation using the half-reaction method. The equation is:

$$MnO_4^-(aq) + H^+(aq) + Cl^-(aq) \rightarrow Mn^{2+}(aq) + Cl_2(g) + H_2O(l)$$

- Split the overall reaction into two half-reactions containing only species which are oxidized or reduced.

   oxidation: $Cl^- \rightarrow Cl_2$

   reduction: $MnO_4^- \rightarrow Mn^{2+}$

Cl^- ions lose electrons (are oxidized) to obtain an oxidation number of 0 in elemental Cl_2. The oxygen atoms in MnO_4^- (oxidation number -2) contribute a negative charge of 4 x -2= -8. The charge on the polyanion is -1, so manganese contributes a +7 charge. Mn gains electrons (is reduced) to obtain a +2 charge.

- Balance the atoms and charges in each half-reaction.

$$\text{oxidation: } 2Cl^- \rightarrow Cl_2 + 2e^-$$

$$\text{reduction: } 5e^- + MnO_4^- + 8H^+ \rightarrow Mn^{2+} + 4H_2O$$

In the oxidation reaction, we simply balance the atoms by adding a coefficient of 2 in front of Cl^-, and balance the charge by adding two electrons to the right hand side of the equation.

In the reduction reaction, we must add 4 waters to the right hand side of the equation to make oxygen atoms balance. Then, since the reaction is in acidic solution, we balance the hydrogen atoms with protons on the left side of the equation. The charge balances by adding $5e^-$ to the left side of the equation. The $5e^-$ reduce Mn in MnO_4^- from +7 to +2. (Each side of the reaction has an overall charge of +2).

- Multiply each half-reaction by an integer to make e^- gained in reduction equal electrons lost in oxidation.

$$\text{oxidation: } 5[2Cl^- \rightarrow Cl_2 + 2e^-]$$

$$10Cl^- \rightarrow 5Cl_2 + 10e^-$$

$$\text{reduction: } 2[5e^- + MnO_4^- + 8H^+ \rightarrow Mn^{2+} + 4H_2O]$$

$$10e^- + 2MnO_4^- + 16H^+ \rightarrow 2Mn^{2+} + 8H_2O$$

This step is easy to check: the 10e- on the right side of the oxidation equation equal the 10e- on the left side of the reduction equation. Make sure to multiply each member of the equation by the multiplication integer.

- Add the balanced half-reactions and include states of matter.

$$10Cl^- + 10e^- + 2MnO_4^- + 16H^+ \rightarrow 2Mn^{2+} + 8H_2O + 5Cl_2 + 10e^-$$

$$10Cl^-(aq) + 2MnO_4^-(aq) + 16H^+(aq) \rightarrow 2Mn^{2+}(aq) + 8H_2O(l) + 5Cl_2(g)$$

- Check that the atoms and charges balance.

Note:
    For a reaction in basic solution, balance the equation as if it were in acidic solution. Add an OH^- ion to both sides of the equation for every H^+ ion present. The OH^- and H^+ ions on one side of the equation make water, and OH^- ions appear on the other side of the equation. Cancel excess water molecules that appear on both sides of the equation.

SECTION 20.2:   Voltaic Cells

A spontaneous redox reaction occurs on its own. For example, a spontaneous redox reaction occurs when we add zinc to a blue copper sulfate solution. The blue color disappears, and a reddish brown deposit forms on the surface of the zinc. What is the redox reaction?

$$Zn(s) + Cu^{2+}(aq) \rightarrow Zn^{2+}(aq) + Cu(s)$$

Zinc metal reduces aquated Cu^{2+} ions (blue colored) to Cu metal (the reddish brown deposit). Zn is oxidzed to Zn^{2+}. What color are aquated Zn^{2+} ions? (answer: colorless).

We can use this spontaneous redox reaction to do electrical work by constructing a voltaic cell. We <u>separate</u> the two half reactions and connect them with a wire.

$$oxidation: Zn \rightarrow Zn^{2+} + 2e^-$$
$$reduction: Cu^{2+} + 2e^- \rightarrow Cu$$

The electrons given up by zinc atoms travel through the wire and do electrical work before they reduce the Cu^{2+} ions. Oxidation occurs at the **anode**; reduction occurs at the **cathode**. The electrode which "pumps" electrons into the external circuit (the anode) is considered to be the negative pole of the cell. The electrode which "pulls" electrons from the external circuit (the cathode) is considered to be the positive pole of the cell. Which are the negative and positive electrodes for the Zn/Cu^{2+} cell? (Answer: the zinc half-reaction produces electrons which flow to the copper half-reaction. Zinc is the negative electrode; copper is the positive electrode).

An excess of positive Zn^{2+} ions builds up around the zinc electrode, and the region around the copper electrode becomes deficient in positive ions as Cu^{2+} ions are reduced. A **salt bridge** connection allows ions to flow from one half-reaction to the other to eliminate this charge build up. A salt-bridge contains (guess what?) a salt solution which takes no part in the electrode reactions. Ions diffuse through the salt bridge from one half-reaction to the other keeping both solutions neutral.

The voltaic cell containing this reaction is shown on p. 883 of your text.

Memory devices:

Oxidation occurs at the **anode** (both begin with vowels)

Reduction occurs at the **cathode** (both begin with consonants)

A way to remember more information about the cathode and anode for a <u>galvanic</u> cell is:

The **anode** is drawn on the **left** and is where **oxidation** takes place; it is the **negative** electrode for a **galvanic** cell.

<u>a</u>node = <u>l</u>eft = <u>o</u>xidation = <u>n</u>egative = <u>g</u>alvanic (**ALONG**)

Conversely then, the cathode is drawn on the right and is where reduction takes place; it is the positive electrode.

A Voltaic cell:

Consider the voltaic cell on p. 883 of your text containing the reaction between zinc and Cu^{2+} ion. The container on the left contains a zinc rod in a 1 M solution of $ZnSO_4$ (a Zn^{2+} electrolyte). The container on the right contains a copper rod in a 1 M solution of $CuSO_4$ (a Cu^{2+} electrolyte). The electrodes are connected by a wire containing a voltmeter, and the solutions are connected by a salt bridge.

Choose words from the list below to complete the following paragraphs.

a) active
b) anode
c) cathode
d) decreases
e) electrons
f) inactive

g) increases
h) left to right
i) negative
j) positive
k) reduce
l) right to left

When zinc atoms are oxidized to Zn^{2+}, _____(1) are generated which enter the wire. The zinc electrode, which generates electrons, is considered to have a _____(2) charge. The cell containing the oxidation reaction is called the _____(3).

Electrons flow through the wire to the solution containing Cu^{2+} ions where they _____(4) Cu^{2+} to Cu^0. The electrode on the copper side of the cell has a _____(5) charge. The cell containing the reduction reaction is called the _____(6).

It is the convention to draw the anode on the left, and the cathode on the right. In this setup, electrons move through the wire from _____(7). In order to maintain neutrality in the two solutions, negatively charged anions (in this case, SO_4^{2-} ions) move through the salt bridge from _____(8).

For this voltaic cell, the electrodes are components of the reaction, and so are _____(9). As the reaction progresses, the mass of the zinc electrode _____(10) as zinc atoms are oxidized to Zn^{2+} ions, and the mass of the copper electrode _____(11) as Cu^{2+} ions are reduced to Cu^0 and "plate out" on the electrode. In some redox reactions, there are no reactants or products which can serve as electrodes. In these cases, electrodes are used which allow electrons to flow, but are not part of the reaction chemistry. These electrodes are called _____(12). For the example problem showing the redox reaction between chloride and permanganate ions, a graphite or platinum electrode might be used.

answers: 1-e, 2-i, 3-b, 4-k, 5-j, 6-c, 7-h, 8-l, 9-a, 10-d, 11-g, 12-f

Notation for a Voltaic cell:

A shorthand notation is useful for describing a voltaic cell. For the cell described above, and shown on p. 883 of your text, the shorthand notation is:

$$Zn(s)\,|\,Zn^{2+}(aq)\,||\,Cu^{2+}(aq)\,|\,Cu(s)$$

The convention is:

- The anode is written on the left
- The double lines, $||$, separate the two half reactions
- The single lines, $|$, represent a phase boundary
- A comma separates members of the same phase

Shorthand notation for a voltaic cell:

Write the shorthand notation for a galvanic cell containing the reaction described in the first example of this section. Use graphite (inactive) electrodes.

$$MnO_4^-(aq) + H^+(aq) + Cl^-(aq) \rightarrow Mn^{2+}(aq) + Cl_2(g) + H_2O(l)$$

solution:

$$Graphite\,|\,Cl^-(aq)\,|\,Cl_2(g)\,||\,H^+(aq),\,MnO_4^-(aq),\,Mn^{2+}(aq)\,|\,Graphite$$

Check:

- The anode (oxidation) is written on the left. Cl^- is oxidized to Cl_2
- A double line separates the anode half-reaction and the cathode half-reaction
- A single line indicates the phase boundaries between the solid graphite electrodes and the aqueous solutions; and between aqueous Cl^- and gaseous Cl_2.
- Commas separate aqueous members of the same phase.

Electrode potentials:

Which metal, zinc or copper, gives up its electrons more easily? (Answer: zinc; it gives up its electrons to reduce Cu^{2+} ions to Cu^0). The difference in electrical potential between two electrodes is the voltage, or cell potential (E_{cell}). The Zn/Cu^{2+} cell described above measures 1.1 volts. This voltage is the sum of the voltage of the two half reactions:

$$E^o_{cell} = E^\circ(Cu^{2+} \rightarrow Cu) + E^\circ(Zn \rightarrow Zn^{2+})$$

The superscript indicates the reactants are in their standard states. When a reaction is spontaneous, $E_{cell} > 0$.

We cannot measure the voltage of an electrode by itself; we can only measure the difference in voltage between two electrodes. We arbitrarily choose an electrode to have a zero voltage so that all other electrodes can be measured relative to it. The half reaction of 1.0 M aqueous protons reduced to hydrogen gas (1 atm.) is defined to be exactly 0.00 volts:

$$2H^+(aq) + 2e^- \rightarrow H_2(g)$$

Let's set up a cell to measure the voltage for the Zn/Zn^{2+} half-reaction. The anode compartment contains a zinc metal electrode in a 1 M solution of Zn^{2+} and SO_4^{2-} ions. The cathode contains a platinum electrode in a 1 M solution of H^+ ions, and bathed with hydrogen gas at 1 atm. (the standard hydrogen electrode).

$$Zn(s) | Zn^{2+}(aq) || H^+(aq) | H_2(g) | Pt$$

Note: All components are in their standard states: the aqueous solutions are 1 M; H_2 gas is at 1 atm. pressure.

• Which is the active electrode, and which is the inactive electrode? (Answer: Zn(s) is the active electrode; Pt is the inactive electrode)

• The voltage reads 0.76 volts. What is E° for the two half reactions?
  We cannot measure the potential of the individual electrode processes, but since the H_2/H^+ half reaction is assigned a potential of 0 volts, the reaction
  $Zn \rightarrow Zn^{2+} + 2e^-$ has a potential of 0.76 volt:

$$E^°_{cell} = E°(H^+ \rightarrow H_2) + E°(Zn \rightarrow Zn^{2+})$$

$$0.76 = 0 + E°(Zn \rightarrow Zn^{2+})$$

$$0.76 = E°(Zn \rightarrow Zn^{2+})$$

Voltage of Zn and Cu half-reactions:

Assign E° values for the two half reactions in the Zn/Cu^{2+} cell. (The voltmeter reads 1.1 V).

$$Zn(s) + Cu^{2+}(aq) \rightarrow Zn^{2+}(aq) + Cu(s)$$

solution:

The half reactions are:

oxidation: $Zn \rightarrow Zn^{2+} + 2e^-$

reduction: $Cu^{2+} + 2e^- \rightarrow Cu$

$$E^{\circ}_{cell} = E^{\circ}(Cu^{2+} \rightarrow Cu) + E^{\circ}(Zn \rightarrow Zn^{2+})$$

$$E^{\circ}(Zn \rightarrow Zn^{2+}) = 0.76 \text{ V (above)}$$

$$1.1 \text{ V} = E^{\circ}(Cu^{2+} \rightarrow Cu) + 0.76 \text{ V}$$

$$E^{\circ}(Cu^{2+} \rightarrow Cu) = 0.34 \text{ V}$$

In standard tables, all half-reactions are given as <u>reduction</u> processes. E° values for these half reactions are called **standard reduction potentials.**

Standard reduction potentials for zinc and copper:

What are the standard reduction potentials for zinc and copper?

<u>solution:</u>

• In the previous exercise, we found that:

$$E^{\circ}(Cu^{2+} \rightarrow Cu) = 0.34 \text{ V}$$
$$E^{\circ}(Zn \rightarrow Zn^{2+}) = 0.76 \text{ V}$$

• We must write both half-reactions as reductions.

$E^{\circ}(Cu^{2+} \rightarrow Cu)$ is a reduction half-reaction, so the standard reduction potential of copper is 0.34 V.

$E^{\circ}(Zn \rightarrow Zn^{2+})$ is an oxidation half-reaction, so the reduction reaction equals
$$-E^{\circ}(Zn \rightarrow Zn^{2+}) = -0.76 \text{ V}$$

Table 20.2 in your text lists the standard reduction potentials at 298 K for many reactions. The more positive the E° value, the more readily the reduction reaction occurs.

Relative oxidizing and reducing strengths:

Consider the following reduction reactions:

$$2H^+(aq) + 2e^- \rightarrow H_2(g) \qquad E^{\circ} = 0 \text{ V}$$
$$Zn^{2+} + 2e^- \rightarrow Zn \qquad E^{\circ} = -0.76 \text{ V}$$
$$Cu^{2+} + 2e^- \rightarrow Cu \qquad E^{\circ} = +0.34 \text{ V}$$

1) Which reaction has the greatest tendency to take place? The least?

2) Which ion is the strongest oxidizing agent?

3) Which is the strongest reducing agent, H_2, Zn, or Cu?

answers:

1) Cu^{2+} reduction has the largest E° value, and so has the most tendency to take place; Zn^{2+} reduction has the smallest E° value, and so has the least tendency to take place.

2) Cu^{2+} is reduced most easily and so is the strongest oxidizing agent.

3) If we reverse the direction of the half reactions, the signs of E° change. Zn oxidation has the largest E° value, and so Zn is the strongest reducing agent.

All this oxidizing and reducing can be confusing. Keep the following in mind:

• Standard electrode potentials are written as reductions.

• If the table shows a large E° value, the reaction has a high tendency to occur.

• An atom, ion, or molecule that is easily reduced is a strong oxidizing agent; it will take electrons from another species in order to be reduced.

• An atom, ion, or molecule that is easily oxidized is a strong reducing agent; it will donate electrons to another species in order to be oxidized.

• E° is an intensive property and does not depend on amount; do not change an E° value when coefficients of a balanced equation are changed.

The extreme:

Fluoride ion is the most electronegative element. Would you predict the reduction of $F_2(g)$ to $2F^-(aq)$ reaction to have a large or small E° value? Is F_2 a strong or weak oxidizing agent? Is F^- a strong or weak reducing agent?

answers:

Fluorine molecules have a very high tendency to pick up electrons to become fluoride ions, so the reduction of fluorine to fluoride should have a large E° value. (In fact, E°=2.87, the largest value in the table).

Fluorine is easily reduced so it is a strong oxidizing agent. F^- is not easily oxidized to F_2 (E°=-2.87), so it is a weak reducing agent.

Comparing oxidizing agents:

Use the data below from table 20.2 in your text to answer the following questions:

$$Sn^{4+}(aq) + 2e^- \rightarrow Sn^{2+}(aq) \qquad E° = +0.13$$
$$Sn^{2+}(aq) + 2e^- \rightarrow Sn(s) \qquad E° = -0.14$$

1) Which is the stronger oxidizing agent, $Sn^{2+}(aq)$ or $Sn^{4+}(aq)$?

2) Which is the stronger reducing agent, $Sn^{2+}(aq)$ or $Sn(s)$?

3) Is $Sn^{2+}(aq)$ a stronger oxidizng or reducing agent?

answers:

1) The strongest oxidizing agent is the most easily reduced. $Sn^{4+}(aq)$ is the stronger oxidizing agent because the reduction potential is more positive.

2) The strongest reducing agent is the most easily oxidized.

$$Sn^{2+}(aq)^- \rightarrow Sn^{4+}(aq) + 2e^- \qquad E° = -0.13$$
$$Sn(s) \rightarrow Sn^{2+}(aq) + 2e^- \qquad E° = +0.14$$

Sn(s) is the stronger reducing agent because the E° value is more positive.

3) Compare oxidation and reduction reactions:

$$Sn^{2+}(aq)^- \rightarrow Sn^{4+}(aq) + 2e^- \qquad E° = -0.13$$
$$Sn^{2+}(aq) + 2e^- \rightarrow Sn(s) \qquad E° = -0.14$$

It is slightly easier to oxidize $Sn^{2+}(aq)$ than to reduce it. $Sn^{2+}(aq)$ is a slightly stronger reducing agent than oxidizing agent.

Calculating cell voltages:

We can use table 20.2 in your text containing standard reduction potentials to calculate the standard cell voltage for a cell. Consider the cell:

$$Zn(s) \,|\, Zn^{2+}(aq) \,\|\, Fe^{3+}(aq), Fe^{2+}(aq) \,|\, Pt \qquad \text{all solns. 1 } M$$

What voltage will the cell develop?

solution:

• First write both half reactions as reductions:

$$Zn^{2+} + 2e^- \rightarrow Zn \qquad E° = -0.76 \text{ V}$$

$$Fe^{3+} + e^- \rightarrow Fe^{2+} \qquad E° = -0.44 \text{ V}$$

Fe^{3+} is the most easily reduced; Fe^{3+} will be reduced and Zn will be oxidized.

• Combine the two half-reactions to give the spontaneous cell reaction:

$$2(Fe^{3+} + e^- \rightarrow Fe^{2+}) \qquad\qquad E° = -0.44 \text{ V}$$

$$Zn \rightarrow Zn^{2+} + 2e^- \qquad\qquad E° = +0.76 \text{ V}$$

Sum: $Zn + 2Fe^{3+} \rightarrow Zn^{2+} + 2Fe^{2+} \qquad E°_{cell} = +0.32 \text{ V}$

(Remember not to multiply the Fe^{3+}/Fe^{2+} half-reaction potential by 2 even though the half-reaction is multiplied by 2 to give an overall balanced equation with no electrons).

Reactivity of metals with acid and water:

In Chapter 13 we learned that the alkali metal, sodium, violently decomposes water to sodium hydroxide and hydrogen gas, while the alkaline earth metal, magnesium reacts only slowly with water at ordinary temperatures. Iron does not react with water at all, but reacts with acid. Explain these observations using table 20.2 in your text of standard electrode potentials.

<u>solution:</u>

• Sodium reacts with water to form a basic solution:

$$2Na(s) + 2H_2O(l) \rightarrow 2NaOH(aq) + H_2(g)$$

Sodium is oxidized from 0 to +1. It <u>reduces</u> hydrogen in H_2O from +1 to 0. The half reactions are:

$$2H_2O + 2e^- \rightarrow 2H_2(g) + 2OH^-(aq) \qquad E° = -0.83 \text{ V}$$

$$2Na(s) \rightarrow 2Na^+(aq) + 2e^- \qquad E° = +2.71 \text{ V}$$

$$E°_{cell} = E°(H_2O \rightarrow H_2) + E°(Na \rightarrow Na^+)$$

$$E°_{cell} = -0.83 + 2.71 = 1.88 \text{ V}$$

A positive $E°_{cell}$ means the reaction is spontaneous.

• The half reaction for the oxidation of magnesium is:

$$Mg(s) \rightarrow Mg^{2+}(aq) + 2e^- \qquad\qquad E° = +2.37 \text{ V}$$

$$E^o_{cell} = E^°(H_2O \rightarrow H_2) + E^°(Mg \rightarrow Mg^{2+})$$

$$E^o_{cell} = -0.83 + 2.37 = 1.54 \text{ V}$$

Again, the positive value for $E^°_{cell}$ means the reaction is spontaneous. The difference in E^o_{cell} for sodium and magnesium does not account for the large difference in their reactivity in water. Magnesium quickly forms an oxide coating that prevents further reaction with water.

Metals that lie below the water half-reaction in Table 20.2 oxidize spontaneously as H^+ in water is reduced. Iron lies above the reduction of water, but below the standard hydrogen (reference) half reaction. Therefore, E^o_{cell} for reaction of iron with water is negative, but for reaction with acid is positive (spontaneous).

Copper in nitric acid:

A copper penny dissolves in dilute nitric acid. How can you explain this observation using the standard electrode potentials? Are you safe (from your spouse) if you put your gold wedding ring in acid?

solution:

The oxidation of $Cu(s)$ to $Cu^{2+}(aq)$ lies above the standard hydrogen half-reaction, so reaction of copper metal with acid is nonspontaneous:

$$2H^+(aq) + 2e^- \rightarrow H_2(g) \qquad\qquad E^°= 0.00 \text{ V}$$

$$Cu(s) \rightarrow Cu^{2+}(aq) + 2e^- \qquad\qquad E^°= -0.34 \text{ V}$$

$$E^o_{cell} = E^°(H^+ \rightarrow H_2) + E^°(Cu \rightarrow Cu^{2+})$$

$$E^o_{cell} = 0 - 0.34 = -0.34 \text{ V}$$

Non-oxidizing acids (i.e. HCl) have little effect on copper metal, but copper dissolves in the oxidizing acid, HNO_3. Gold does not dissolve in nitric acid, but it dissolves in concentrated HCl if a strong oxidizing agent is present. A 3:1 mixture of conc. HCl and conc. HNO_3 is named *Aqua regia* by alchemists because it dissolves gold, the king of metals.

SECTION 20.3: Free Energy and Electrical Work

Charge:

The charge on one electron = 1.60219 x 10⁻¹⁹ C. What is the charge of one mole of electrons? What is the charge on n moles of electrons?

solution:

- The charge on one mole of electrons is:

$$1.60219 \times 10^{-19} \frac{C}{e^-} \times 6.02205 \times 10^{23} \frac{e^-}{mol} = 96{,}487 \frac{C}{mol\ e^-}$$

The charge on one mole of electrons is the **faraday** (F).

$$1\ F = 96{,}487 \frac{C}{mol\ e^-}$$

- The charge on n moles of electrons is:

$$charge\ (q) = 1.60219 \times 10^{-19} \frac{C}{e^-} \times 6.02205 \times 10^{23} \frac{e^-}{mol} \times n$$

$$or,\ charge\ (q) = 96{,}487 \frac{C}{e^-} \times n$$

$$q = nF$$

Current:

Current is the rate of flow of electric charge; it is measured in amperes (or amps).

1 ampere = 1 coulomb of charge transferred per second.

$$i = \frac{q}{t},\ or$$

$$q = i \times t$$

Electrical work:

We can harness a current flow to do work, just as we can harness a water flow to do work. The maximum amount of work we can harness depends on the difference in potential for two points in a circuit, and the amount of charge transferred:

$$w_{max} = -E_{cell} \times \text{charge}$$

where work is in Joules, potential is in volts, and charge is in Coulombs. How much work is done when one Coulomb of charge is transferred over a potential difference of 1 volt? (answer: 1 Joule).

Charge = nF, so $\qquad w_{max} = -E_{cell} \times nF$

Units of a Joule:

Thermochemistry (Chapter 6): $\quad 1 \text{ J} = 1 \text{ kgm}^2/\text{s}^2$

Electrochemistry (Chapter 20): $\quad 1 \text{ J} = 1 \text{ V} \times 1 \text{ C}$

Note: $\text{C} = \text{J/V}$

$F = 96,487 \text{ J/V mol e}^-$

Remember from Chapter 19 that the free energy change, ΔG represents the maximum work we can obtain from a spontaneous process.

$$w_{max} = -E_{cell} \times nF = \Delta G$$

When all components of a cell are in their standard states:

$$\Delta G° = -nFE°_{cell}$$

Dependence of cell voltage on concentrations:

For problems in this chapter so far, we have calculated $E°_{cel}$ for a very specific set of conditions: all constituents of the cells have been in their standard states (1 M concentration for aqueous solutions, and 1 atm. pressure for gases). How does changing the concentration of reactants affect the cell voltage?

Recall from Chapter 19 that $\Delta G = \Delta G° + RT \ln Q$. From above, we have that $\Delta G = -nFE$, and $\Delta G° = -nFE°$; therefore:

$$-nFE = -nFE°_{cell} + RT \ln Q$$

Dividing through by -nF gives the **Nernst equation**:

$$E = E° - \frac{RT}{nF} \ln Q$$

At 25°C, this equation becomes:

$$E = E° - \frac{0.0592}{n} \log Q$$

At equilibrium, Q=K and E=0:

$$E° = \frac{RT}{nF} \ln K$$

At 25°C, this equation becomes:

$$E° = \frac{0.0592}{n} \log K$$

Equilibrium constant from E°:

1) When copper foil is placed in a silver nitrate solution, the solution turns light blue, and silver precipitates. What is the equilibrium constant for the following reaction:

$$Cu(s) + 2AgNO_3(aq) \rightarrow Cu(NO_3)_2(aq) + 2 Ag(s)$$

solution:

- First we'll write the half-reactions and calculate E° for the overall reaction:

$$Cu(s) \rightarrow Cu^{2+}(aq) \qquad E° = -0.34 \text{ V}$$
$$Ag^+(aq) \rightarrow Ag(s) \qquad E° = +0.8 \text{ V}$$

$$E°_{cell} = -0.34 + 0.8 = 0.46 \text{ V}$$

- Now we can calculate K:

$$E° = \frac{0.0592}{n} \log K \;;\; n = 2 \text{ since 2e}^- \text{ are transferred in the reaction}$$

$$0.46 \text{ V} = \frac{0.0592}{2} \log K$$

$$\log K = \frac{2 \times 0.46}{0.0592} = 15.54$$

$$K = 3.47 \times 10^{15}$$

2) How many faradays of electricity are given off in this spontaneous reaction when 190.5 g of copper metal are used up?

solution:

- moles of copper = $190.5 \text{ g Cu} \times \dfrac{1 \text{ mol}}{63.5 \text{ g Cu}}$ = 3 mol Cu

- For every mole of copper, 2 moles of electrons are used, so

$$3 \text{ mol Cu} \times \dfrac{2 \text{ mol } e^-}{1 \text{ mol Cu}} \times \dfrac{1 \text{ faraday}}{\text{mol } e^-} = 6 \text{ faradays}$$

Concentration cell:

What is the cell voltage of the following cell?

$$Zn(s) \mid Zn^{2+}(aq, 0.001 \text{ M}) \parallel Zn^{2+}(aq, 1.0 \text{ M}) \mid Zn(s)$$

solution:

- $E^\circ_{cell} = 0$, since the two half-reactions are the same:

$$Zn^{2+}(aq) + 2e^- \rightarrow Zn(s) \qquad\qquad E^\circ = -0.76 \text{ V}$$

$$Zn(s) \rightarrow Zn^{2+}(aq) + 2e^- \qquad\qquad E^\circ = +0.76 \text{ V}$$

- Q for the reaction is:

$$Q = \dfrac{[Zn^{2+}]_{dilute}}{[Zn^{2+}]_{conc.}} = \dfrac{0.001}{1} = 0.001$$

- Calculate E:

$$E = E^\circ - \dfrac{0.0592}{n} \log Q = 0 - \dfrac{0.0592}{2} \log 0.001$$

$$E = 0 - \dfrac{0.0592}{2} \times (-3)$$

$$E = 0.0888 \text{ V}$$

This cell is called a concentration cell. Spontaneous reaction occurs in the direction which reduces the amount of the concentrated reactant; the 1 M zinc solution is reduced to zinc metal.

SECTION 20.4: Electrolytic cells

Galvanic cells harness spontaneous reactions to do work. An electrolytic cell uses electrical energy to drive a nonspontaneous reaction. We can convert voltaic cells into electrolytic cells by supplying an electric potential greater than the cell potential of the voltaic cell. Electrolysis has many practical applications; for example, producing aluminum metal, purifying metals, charging a battery, and chrome plating an object are all done electrolytically.

Electrolytic cells:

Oxidation occurs at the **anode** (both begin with vowels)

Reduction occurs at the **cathode** (both begin with consonants)

(same as galvanic cells)

For an electrolytic cell, the cathode, where electrons are supplied, is the negative electrode; the anode is the positive electrode.

Metal plating:

Plating a metal means depositing a neutral metal on an electrode by reducing the metal ions. When a spoon is electroplated with silver, a silver bar is the anode where oxidation to Ag^+ occurs. A power source "pumps" the electrons to the cathode, a spoon, where silver plates out:

$$Ag^+ + e^- \rightarrow Ag$$

Silver plating:

A 7.00 amp current is applied to a Ag^+ solution to produce silver metal. How long must the current be applied to produce 15.5 g of silver metal?

solution:

• We need to know how many electrons are required to reduce enough Ag^+ ions to produce 15.5 g of silver metal. We know that each Ag^+ ion requires one electron to become a silver atom:

$$Ag^+ + e^- \rightarrow Ag$$

Once we know the number of electrons required, we can calculate the charge, since we know the charge on one electron = 1.60219×10^{-19} C (or the charge of a mole of electrons is 96,487 C). If we know q (charge) and i (current, 7.00 A, or 7.00 C/s), we can calculate time, t: $i = q/t \Rightarrow t = q/i$

- moles of e- required to produce 15.5 g Ag:

$$15.5 \text{ g Ag} \times \frac{1 \text{ mol Ag}}{107.9 \text{ g Ag}} \times \frac{1 \text{ mol e}^-}{1 \text{ mol Ag}} = 0.144 \text{ mol e}^-$$

- Charge on 0.144 mol e⁻:

$$0.144 \text{ mol e}^- \times \frac{96,487 \text{ C}}{\text{mol e}^-} = 13,900 \text{ C}$$

- time that 7.00 A current must be applied to generate 13,860 C:

$$t = \frac{13,900 \text{ C}}{7.00 \text{ C} / \text{s}} = 1986 \text{ s, or } 33.1 \text{ minutes}$$

Tin plating:

Tin cans are actually steel cans with a thin coating of tin. How much tin is plated out when a 5.00 amp current is passed through a solution containing Sn^{2+} for 40.0 minutes?

solution:

- We first calculate charge from current and time:

$$q = i \times t$$

$$q = 5.00 \text{ C/s} \times 2400 \text{ s} = 12,000 \text{ C}$$

- 12,000 coulombs is how many electrons?

$$12,000 \text{ C} \times \frac{1 \text{ mol e}^-}{96,487 \text{ C}} = 0.124 \text{ mol e}^-$$

- Each Sn^{2+} ion requires two electrons to become a tin atom, so the amount of tin that 0.124 mol e⁻ plates out is:

$$0.124 \text{ mol e}^- \times \frac{1 \text{ mol Sn}}{2 \text{ mol e}^-} \times \frac{118.7 \text{ g Sn}}{\text{mol Sn}} = 7.36 \text{ g Sn}$$

We worked examples of silver and tin plating out at a cathode. If an electrolytic cell contains both Ag^+ and Sn^{2+} ions, which order will the metals plate out onto the cathode if the voltage starts low, and is gradually increased? Compare the standard reduction potentials of Ag^+ and Sn^{2+} ions:

$$Ag^+ + e^- \rightarrow Ag \qquad\qquad E° = 0.80 \text{ V}$$
$$Sn^{2+} + 2e^- \rightarrow Sn \qquad\qquad E° = -0.14 \text{ V}$$

The more positive the E° value, the more the reaction has a tendency to proceed. Ag^+ is reduced more easily than Sn^{2+}, so silver will plate out first.

Why can't aluminum be plated out of an aqueous solution of Al^{3+}? The answer is that water itself is reduced before Al^{3+}:

$$2H_2O + 2e^- \rightarrow H_2 + 2OH^- \qquad\qquad E° = -0.83 \text{ V}$$
$$Al^{3+} + 3e^- \rightarrow Al \qquad\qquad E° = -1.66 \text{ V}$$

Until a practical electrolytic process for producing aluminum was discovered, aluminum was a more precious metal than gold or silver. Ions must be mobile for electrolysis to occur, such as when they are in a water solution, or when a salt is melted. The high melting point of aluminum oxide (2050°C) however, makes melting solid Al_2O_3 an impractical route to electrolysis. In 1854, it was discovered that a mixture of aluminum oxide and Na_3AlF_6 has a melting point of 1000°C, and aluminum metal could be obtained electrolytically from the molten mixture. After this discovery, the price of aluminum dropped by a factor of one thousand.

Electrolysis of water:

Electrolysis of water produces oxygen and hydrogen:

$$2H_2O \rightarrow O_2 + 2H_2$$

The anode reaction is:

$$2H_2O \rightarrow O_2 + 4H^+ + 4e^- \qquad -E° = -1.23 \text{ V}$$

The cathode reaction is:

$$4H_2O + 4e^- \rightarrow 2H_2 + 4OH^- \qquad E° = -0.83 \text{ V}$$

The net reaction is:

$$6H_2O \rightarrow 2H_2 + O_2 + 4H^+ + 4OH^-$$
$$(4H_2O)$$
$$2H_2O \rightarrow O_2 + 2H_2 \qquad E° = -2.06 \text{ V}$$

The potentials assume 1 M H^+ and 1 M OH^-. In pure water, $[H^+]=[OH^-]=10^{-7}$ M, and the potential for the process is -1.24 V.
($-E_{oxidation}$ = -0.82 V; $E_{reduction}$ = -0.42 V; E_{cell}=-1.24 V)

Overvoltage:

When an aqueous sodium chloride solution is electrolyzed, hydrogen gas forms at the cathode, and chlorine gas forms at the anode. Can you explain these observations using the standard electrode potentials?

solution:

• The possible oxidizing agents are Na^+ and H_2O:

$$2H_2O + 2e^- \rightarrow H_2 + 2OH^- \qquad E = -0.42 \text{ V}$$

$$Na^+ + e^- \rightarrow Na \qquad E° = -2.71 \text{ V}$$

Water has the less negative electrode potential, so it is easier to reduce than Na^+; H_2 forms at the cathode, as observed.

• The possible reducing agents are Cl^- and H_2O:

$$2H_2O \rightarrow O_2 + 4H^+ + 4e^- \qquad -E = -0.82 \text{ V}$$

$$2Cl^- \rightarrow Cl_2 + 2e^- \qquad -E° = -1.36 \text{ V}$$

Water has the more positive electrode potential, so water is easier to oxidize than Cl^- and we expect to see O_2 produced at the anode. But, instead, chlorine forms. We cannot explain this observation using the standard electrode potentials. An additional 0.4 to 0.6 volts (the **overvoltage**) is required to produce oxygen at metal electrodes. The causes of overvoltage are complex, and make life less certain when predicting the order of oxidation or reduction of species in an electrolytic cell.

Chapter 20 Exercises:

20.1) Calculate the standard cell potential, E^0_{cell}, and draw the voltaic cell for the following reactions:

      a. $2Cr(s) + 3Pb^{2+}(aq) \rightarrow 3Pb(s) + 2Cr^{3+}(aq)$

      b. $Cl_2(g) + Fe^{2+}(aq) \rightarrow 2Cl^-(aq) + Fe^{3+}(aq)$

20.2) Electrical energy and drinking water are generated from hydrogen-oxygen fuel cells while a space shuttle is in flight.

$$2H_2(g) + O_2(g) \rightarrow 2H_2O(l)$$

The fuel cell reaction involves a hot potassium hydroxide electrolyte solution and uses Ni/NiO embedded carbon electrodes instead of platinum. What is the cell potential produced by a hydrogen-oxygen cell?

20.3) Calculate the cell potential, E_{cell}, for the following reactions.

      a. $2Al(s) + 3Ni^{2+}(aq, 0.50M) \rightarrow 2Al^{3+}(aq, 0.20\ M) + 3Ni(s)$

      b. $2VO_2^+(aq) + 4H^+(aq) + Zn(s) \rightarrow 2VO^{2+}(aq) + 2H_2O(l) + Zn^{2+}(aq)$ $E^0_{cell} = 1.76\ V$

      $[VO_2^+] = 1.5\ M;\ [H^+] = 0.60\ M;\ [VO^{2+}] = 1.3 \times 10^{-2}\ M;\ [Zn^{2+}] = 2.0 \times 10^{-1}\ M$

20.4) In the following reaction, the pH of the MnO_2/Mn^{2+} half cell can be changed. Calculate the cell potential, E_{cell} for the conditions listed below.

$$2Ag(s) + 2Cl^-(aq, 0.10M) + MnO_2(s) + 4H^+(aq) \rightarrow$$
$$2AgCl(s) + Mn^{2+}(aq, 0.10\ M), + 2H_2O(l)$$

      a. pH = 3.00; b. pH = 7.00; c. What effect does pH have on E_{cell} for the reaction?

20.5) Aluminum is prepared by the electrolysis of aluminum oxide in the Hall process. The overall reaction is:

$$2Al_2O_3(s) \rightarrow 4Al(s) + 3O_2(g)$$

From the standard Gibb's free energy of the reaction, calculate the electromotive force for this reaction.

20.6) Chrome plated bumpers and grills on automobiles were the style during the 1950's. How long would it take to plate out 50.0 g of Cr metal by passing a 5.00 ampere current through a $Cr(NO_3)_3$ solution?

20.7) The final step in the preparation of copper is the electrolytic refining of "blister" copper to pure copper. 590 kg slabs of the crude copper are placed in a vat containing copper(II) sulfate. The pure copper plates out at the cathode on copper starter sheets. What current (amperes) is required to plate out 500 kg of copper in a 24 hour period?

20.8) A lab technician came across a bottle in the stockroom labeled only "Cobalt Chloride". In an effort to determine the oxidation number of the cobalt, she decided to carry out an analytical electrolysis of a sample of the solution. The mass of the cathode before electrolysis was 38.4124 g. After 20.00 minutes of electrolysis with a 0.500 amp current, the mass of the cathode had increased to 38.5346 g. What is the oxidation number of the cobalt ion? How should the lab technician relabel the bottle?

Chapter 20 Answers:

20.1). (a)

$E^0_{cell} = +0.61$ V

(b)

$E^0_{cell} = +0.59$ V

20.2) $E^0_{cell} = +1.23$ V

20.3) a. $E_{cell} = 1.43$ V; b. $E_{cell} = 1.88$ V

20.4) a. $E_{cell} = 0.90$ V
b. $E_{cell} = 0.78$ V
c. Increasing the pH decreases the cell potential

20.5) $E_{cell} = 1.37$ V

20.6) 15.5 hours

20.7) 1.76×10^4 amperes, or 17.6 kA

20.8) 3.0 eq/mole; cobalt(III) chloride

CHAPTER 21

Nuclear chemistry
"The chemistry of the nucleus"

TOPICS:

- Radioactive decay
- Kinetics of nuclear change
- Bombardment reactions
- Fission and fusion

INTRODUCTION:

In Chapters 1-20 we studied chemical changes that occur as valence electrons are transferred or shared among atoms to form new compounds. In Chapter 2 we learned that <u>mass is always conserved in chemical transformations</u>; atoms are indestructible particles which are never transformed by chemical reactions, but which merely rearrange (Law of Mass Conservation). Now we consider reactions of the nucleus (nuclear reactions) in which elements are transformed, and mass is not conserved (the total quantity of mass-energy <u>is</u> conserved). The energy changes of nuclear reactions are so large that changes in mass ($E=mc^2$) become detectable.

SECTION 21.1: Radioactive decay

The nucleus is tiny and heavy - it is extremely dense - about 10^{14} g/mL. It contains protons (p^+) with a positive charge, and neutrons (n^0) with no charge. The protons and neutrons account for essentially all the mass of the atom. Many nuclei are unstable (**radioactive**). Over time, they decay as they emit radiation at a characteristic rate.

Terminology:

nucleons: The particles that make up the nucleus, the protons and neutrons.

mass number (A): The total number of protons and neutrons in an atom.

notation: In Chapter 2, we used the notation for an atom: $^A_Z X$

where X is the symbol for an element, A is the mass number, and Z is the atomic number (the number of protons in the nucleus). We extend this notation to elementary particles (protons, electrons, and neutrons) where Z is the charge of the particle:

electron: $^0_{-1}e$ (electrons have essentially no mass)

proton: $^1_1 p$

neutron: $^1_0 n$ (neutrons have no charge)

---

nuclide: An atom with a specific nuclear composition.

Nuclides are often designated with the element name followed by the mass number:

carbon-12, or carbon-14, for example.

isotopes: Atoms with the same number of protons (the same Z), but different numbers of neutrons, and therefore different mass numbers (different A).

alpha particle (α): a helium nuclei, $^4_2$He

beta particle (β or $^0_{-1}\beta$): fast electrons

gamma rays $\left(\gamma\right)$: high-energy photons

parent: the decaying nuclide

daughter: the product nuclide

Review exercises:

1) How many protons and neutrons are in the nucleus of a carbon-12 atom and a carbon-14 atom?

answer:

All carbon atoms contain 6 protons. The protons and neutrons together make up the mass of the nucleus.

Carbon-12 contains: 12 - 6 = 6 neutrons.

Carbon-14 contains: 14 - 6 = 8 neutrons.

2) When does atomic mass equal mass number?

answer:

Atomic mass equals the mass number when all atoms under consideration are the same isotope. In naturally occurring samples of an element where there is a mixture of isotopes, the atomic mass is the weighted average of the masses of all the isotopes.

Types of radioactive decay:

1) **Alpha decay:** The loss of an alpha particle $\left(^{4}_{2}He\right)$ from a nucleus. All elements heavier than lead (Pb; Z=82), and a few lighter ones exhibit alpha decay.

2) **Beta decay:** The ejection of a beta particle from the nucleus. A neutron converts into a proton and a ß particle which is expelled:

$$^{1}_{0}n \rightarrow {}^{1}_{1}p + {}^{0}_{-1}\beta$$

3) **Positron decay:** The ejection of a positron from the nucleus. A proton converts into a neutron and a positron. Positron decay is the opposite of beta decay. A positron is the antiparticle of an electron. :

$$^{1}_{1}p \rightarrow {}^{1}_{0}n + {}^{0}_{1}\beta$$

Positron decay is not observed in natural radioactivity, but is a common mode of decay for radioisotopes produced in the laboratory.

4) **Electron capture:** The nucleus "captures" an electron from the lowest energy level. The net effect is the same as positron decay; a proton is converted into a neutron:

$$^{1}_{1}p + {}^{0}_{-1}e \rightarrow {}^{1}_{0}n$$

When an electron moves down from a higher energy level to fill the vacancy left by the captured electron, an x-ray photon is generated.

5) **Gamma emission:** The radiation of high-energy gamma photons from an excited nucleus. Gamma emission accompanies most other types of decay. Gamma rays have no mass or charge and so do not change A or Z.

Radioactive decay exercise:

Choose words from the list below to complete the following paragraphs:

a) -1
b) +1
c) 2
d) 4
e) alpha
f) beta
g) charge
h) electron

i) excited
j) gamma
k) higher
l) lead
m) lower
n) mass
o) neutron
p) proton

A helium nuclei contains _____(1) protons and _____(2) neutrons in its nucleus giving it a mass number (A) of _____(3) and a charge (Z) of _____(4). A helium nuclei is called an _____(5) particle. An atom which undergoes alpha decay reduces its mass number by _____(6) and its charge by _____(7). Every element heavier than _____(8) undergoes alpha decay.

A beta particle is a "fast electron" that is generated when a neutron converts into a _____(9). Protons and neutrons have the same _____(10), but a proton has a charge of _____(11) while a neutron is neutral. Like a valence electron in an atom, a beta particle has no _____(12) and has a charge of _____(13). In beta decay, because the product nuclide has one more proton than the parent, the element with the next _____(14) atomic number forms.

A positron is the "Antiparticle" of an _____(15). Like an electron, a positron has no mass, but instead of a -1 charge, it has a charge of _____(16). Positron decay is the opposite of _____(17) decay. In positron decay, a positron is generated when a proton converts into a _____(18). Because the product nuclide has one less proton than the parent, the element with the next _____(19) atomic number forms.

Electron capture transforms a proton into a _____(20) just as in positron decay. However, instead of emitting a positron, a nucleus draws in a surrounding _____(21), usually one from the lowest energy level. When a higher energy electron fills the orbital vacancy, an x-ray photon is generated.

Many nuclear changes leave the nucleus in an _____(22) state. These excited nuclei radiate high-energy _____(23) photons. Gamma rays have no _____ or _____(24, 25), so gamma emission does not change A or Z.

answers:

1-c, 2-c, 3-d, 4-c, 5-e, 6-d, 7-c, 8-l, 9-p, 10-n, 11-b, 12-n, 13-a, 14-k, 15-h, 16-b, 17-f, 18-o, 19-m, 20-o, 21-h, 22-i, 23-j, (24,25)-g,n

Classifying radioactive decay:

Determine the type of radioactive decay for each of the following:

1) $^{214}_{84}Po \rightarrow ^{4}_{2}He + ^{210}_{82}Pb$

2) $^{63}_{28}Ni \rightarrow ^{63}_{29}Cu + ^{0}_{-1}\beta$

3) $^{14}_{6}C \rightarrow ^{14}_{7}N + ^{0}_{-1}\beta$

4) $^{11}_{6}C \rightarrow ^{11}_{5}B + ^{0}_{1}\beta$

5) $^{55}_{26}Fe + ^{0}_{-1}e \rightarrow ^{55}_{25}Mn + X\text{-ray}$

6) $^{238}_{92}U \rightarrow ^{234}_{90}Th + ^{4}_{2}He + 2\gamma$

7) $^{36}_{17}Cl + ^{0}_{-1}e \rightarrow ^{36}_{16}S$

8) $^{232}_{90}Th \rightarrow ^{228}_{88}Ra + ^{4}_{2}He$

answers: 1-alpha; 2-beta; 3-beta; 4-positron emission; 5-electron capture; 6-alpha, gamma emission; 7-electron capture; 8-alpha

Balancing nuclear equations:

Balance the following nuclear reactions:

1) $^{226}_{88}Ra \rightarrow ^{4}_{2}He + \underline{\quad}$ 2) $^{38}_{19}K \rightarrow ^{0}_{-1}\beta + \underline{\quad}$

solutions:

1) The answer is some element, $^{A}_{Z}X$. First we balance subscripts and superscripts:

A: $A + 4 = 226$; $A = 222$

Z: $Z + 2 = 88$; $Z = 86$

Now find the element with atomic number 86: $^{222}_{86}Rn$

2) Balance subscripts and superscripts:

A: $A + 0 = 38$; $A = 38$

Z: $Z - 1 = 19$; $Z = 20$

Now find the element with atomic number 20: $^{38}_{20}Ca$

Nuclear stability:

- All nuclei of atomic number greater than 83 (Z>83) are unstable and decay. Bismuth-209 is the heaviest stable nuclide.

- Light stable nuclei have at least as many neutrons as protons ($_1^1H$ and $_2^3He$ are exceptions). Heavier stable nuclei have more neutrons than protons.

- Nuclei with an even number of protons and neutrons are more stable than nuclei with odd numbers of protons and/or neutrons

- There are magic numbers of N or Z values which make stable nuclei.

If a nuclide is neutron-rich, it will undergo _betadecay_ to convert neutrons into protons. If a nuclide is proton-rich, it will undergo _positron decay_ or _electron capture_ to convert protons into neutrons. If a nuclide is heavy (Z > 83), it undergoes alpha decay which reduces Z and N values by two units per emission.

Predicting nuclear decay:

Predict the mode of decay for the most abundant radioactive isotope of iron, iron-53.

solution:

Neutron rich Iron-59 undergoes beta decay converting a neutron into a proton:

$$_{26}^{59}Fe \rightarrow {_{27}^{59}Co} + {_{-1}^{0}\beta}$$

Iron-59 has a half-life of 45.1 days, and is used as a radiotracer for red blood cells.

SECTION 21.2: Kinetics of nuclear change

The rate of decay of radioactive nuclei is a _first-order process_. The number of nuclei that decay over some period of time, t, is proportional to the number of nuclei present:

$$Rate = \frac{-\Delta N}{\Delta t} = kN$$

The integrated form of the rate law is:

$$\ln\left(\frac{N_t}{N_o}\right) = -kt$$

The half-life is the time required for half of the nuclides present to decay. For a first order process, the half-life, $t_{1/2}$ is:

$$t_{\frac{1}{2}} = \frac{-\ln\left(\frac{\frac{1}{2}N_o}{N_o}\right)}{k} = \frac{\ln(2)}{k}$$

$$t_{\frac{1}{2}} = \frac{0.693}{k}$$

The half-lives of radioactive nuclei vary over a wide range: Uranium-235 has a half-life of 7.1×10^6 years; Lithium-8 has a half-life of 842 milliseconds.

Half-life of $^{222}_{86}Rn$:

$^{222}_{86}Rn$ has a half-life of 3.82 days.

    1) How long does it take for half of a 100 g sample of $^{222}_{86}Rn$ to decay?

    2) How long does it take for half of a 1 g sample of $^{222}_{86}Rn$ to decay?

    3) How much of the 100 g sample of $^{222}_{86}Rn$ remains after 10 days?

<u>solutions:</u>

1) 3.82 days

2) 3.82 days; half-life is independent of the number of nuclei present.

3) We can calculate k from the expression for half-life:

$$k = \frac{0.693}{t_{\frac{1}{2}}}$$

$$k = \frac{0.693}{3.82 \text{ days}} = 0.181 \text{days}^{-1}$$

Now we can calculate the amount left after 10 days by using the integrated rate law:

$$\ln\left(\frac{N_t}{N_o}\right) = -kt$$

$N_{10 \text{ days}} = N_o e^{-kt}$

$N_{10 \text{ days}} = 100 \text{ g} \times e^{-(0.181 \text{ days}^{-1})(10 \text{ days})}$

$N_{10 \text{ days}} = 16.4 \text{ g}$

Check: 10 days is about 2.6 half-lives.

after 1 half-life, 50 g remains

after 2 , half-lives, 25 g remains

after 3 half-lives, 12.5 g remains

Our answer falls in between the amount that remains after 2 and 3 half-lives, so it seems reasonable.

We can count the number of disintegrations of radioactive nuclei per second (dps) to measure the concentration of a radioactive substance. **Radiocarbon dating** is used to determine the age of things that were once living. Cosmic rays in the atmosphere produce nearly constant amounts of carbon-14 from ^{14}N atoms.

$$^{14}_{7}N + ^{1}_{0}n \rightarrow ^{14}_{6}C + ^{1}_{1}p$$

The carbon-14 formed by this nuclear reaction eventually incorporates into the carbon dioxide in the air giving our atmosphere a constant concentration of carbon-14 atoms (there is about one atom of carbon-14 for every 10^{12} atoms of carbon-12). Living plants, which take in carbon dioxide, have this same $^{14}C/^{12}C$ ratio, as do plant-eating animals and animals who eat plant-eating animals. When an organism dies, the intake of ^{14}C stops, and the $^{14}C/^{12}C$ ratio decreases as ^{14}C decays. The half-life of ^{14}C is 5730 years.

Radiocarbon dating:

A sample of linen from a funeral shroud taken from an Egyptian tomb has a carbon-14 content of 0.560 times that of a living plant. Estimate the age of the linen.

solution:

• We first calculate the first order rate constant from the expression for half-life. (The half-life of ^{14}C is 5730 years).

$$k = \frac{0.693}{5720 \text{ years}} = 1.21 \times 10^{-4} \text{ yr}^{-1}$$

• Use the integrated rate law to solve for time:

$$\ln\left(\frac{N_t}{N_o}\right) = -kt$$

$$\ln\left(\frac{N_t}{N_o}\right) = -\left(1.21 \times 10^{-4} \text{ yr}^{-1}\right) \times t$$

$$N_t = 0.560(N_o) , \text{ so}$$

$$\ln\left(\frac{0.560\ N_o}{N_o}\right) = \ln(0.560) = -0.580$$

$$-0.580 = -(1.21\times 10^{-4}\ yr^{-1}) \times t$$

$$t = 4790 \text{ years.}$$

Note: This method assumes that the ratio of $^{14}C/^{12}C$ in the atmosphere has not changed over the years.

The first atomic explosion:

On July 16, 1945, the first atomic bomb was detonated in the desert in New Mexico. What fraction of the strontium-90 ($t_{1/2} = 28.8$ y) produced by the explosion remains today?

solution:

- Calculate the first order rate constant from the expression for half-life:

$$k = \frac{0.693}{28.8\ yr} = 2.41\times 10^{-2}\ yr^{-1}$$

- Use the integrated rate law to solve for the fraction of strontium-90 remaining:

$$\ln\left(\frac{N_t}{N_o}\right) = -kt$$

$$t=50.5 \text{ y (on Jan. 16, 1996)}$$

$$\ln\left(\frac{N_t}{N_o}\right) = -(2.41\times 10^{-2}\ yr^{-1}) \times (50.5\ yr)$$

$$\ln\left(\frac{N_t}{N_o}\right) = -1.22$$

$$\frac{N_t}{N_o} = 0.297 , \text{ less than 1/3 remains}$$

Check: After 28.8 y (1 half-life), half of the strontium-90 remained. After another 28.8 y (in the year 2003), 1/4 of the strontium-90 will remain. At 50.5 y, in 1996, more than 1/4 , but less than 1/3 remains.

The age of rocks can be determined using "radioactive clocks" such as the radioactive decomposition of uranium-238 to lead-206, and the beta-decay of rubidium-87 ($t_{1/2} = 5.7 \times 10^{10}$ years. Rocks dated by these and other radioactive metals gives an approximate age of the earth (≈ 4.5 billion years old).

Nuclear decomposition of uranium-235:

Uranium-235 undergoes the following decay series:

$$\alpha, \beta, \alpha, \beta, \alpha, \alpha, \alpha, \alpha, \beta, \alpha, \beta.$$

What is the final product of decay of uranium-235?

solution:

To find the final product of this decay series, we'll show each successive decay with the intermediate nuclides:

$$^{235}_{92}U \rightarrow {}^4_2He + {}^{231}_{90}Th$$

$$^{231}_{90}Th \rightarrow {}^0_{-1}\beta + {}^{231}_{91}Pa$$

$$^{231}_{91}Pa \rightarrow {}^4_2He + {}^{227}_{89}Ac$$

$$^{227}_{89}Ac \rightarrow {}^0_{-1}\beta + {}^{227}_{90}Th$$

$$^{227}_{90}Th \rightarrow {}^4_2He + {}^{223}_{88}Ra$$

$$^{223}_{88}Ra \rightarrow {}^4_2He + {}^{219}_{86}Rn$$

$$^{219}_{86}Rn \rightarrow {}^4_2He + {}^{215}_{84}Po$$

$$^{215}_{84}Po \rightarrow {}^4_2He + {}^{211}_{82}Pb$$

$$^{211}_{82}Pb \rightarrow {}^0_{-1}\beta + {}^{211}_{83}Bi$$

$$^{211}_{83}Bi \rightarrow {}^4_2He + {}^{207}_{81}Tl$$

$$^{207}_{81}Tl \rightarrow {}^0_{-1}\beta + {}^{207}_{82}Pb$$

The final product of decay of uranium-235 is lead-207. The overall reaction is:

$$^{235}_{92}U \rightarrow {}^{207}_{82}Pb + 7\,{}^4_2He + 4\,{}^0_{-1}\beta$$

Uranium-235 makes up only about 0.7% of naturally occurring uranium. The more abundant isotope, uranium-238 decays to lead-206, and it is the $^{238}U/^{206}Pb$ ratio that is used to date rocks. As you will see, uranium-235 undergoes fission while uranium-238 does not.

SECTION 21.3: Bombardment reactions

Chemists can convert stable, nonradioactive nuclei to radioactive nuclei by bombarding them with neutrons or positive ions. The first radioactive isotopes made in the laboratory were achieved by bombarding stable isotopes with alpha particles. Bombardment of aluminum foil with alpha particles produced phosphorus-30 and neutrons:

$$^{27}_{13}Al + ^{4}_{2}He \rightarrow ^{30}_{15}P + ^{1}_{0}n$$

A shorthand notation for the above reaction is $^{27}Al(\alpha,n)$ ^{30}P. Phosphorus-30 is radioactive and decays by positron emission. What are the products of this nuclear reaction? (answer: a positron, the antiparticle of an electron, is emitted as a proton converts into a neutron):

$$^{30}_{15}P \rightarrow ^{30}_{14}Si + ^{0}_{1}\beta$$

Particle accelerators, such as the cyclotron (invented in 1930), use electromagnets to accelerate cations to very high velocities so they can overcome coulombic repulsion of the target nucleus and cause nuclear reactions.

Elements with atomic numbers greater than uranium (atomic number 92), the transuranium elements have been prepared by bombardment reactions.

Synthesis of transuranium elements:

Elements 93-95 can be prepared by neutron bombardment; high-energy positive ions are used to prepare the heavier transuranium elements. Fill in the blanks for the examples below:

1) **Neptunium** and **Plutonium** (atomic numbers 93 and 94)

Uranium-238 is bombarded with neutrons to form uranium-239 which undergoes _____ decay to form neptunium and plutonium:

$$^{238}_{92}U + ^{1}_{0}n \rightarrow ^{239}_{92}U$$

$$^{239}_{92}U \rightarrow ^{239}_{93}Np + \underline{\quad}$$

$$^{239}_{93}Np \rightarrow ^{239}_{94}Pu + \underline{\quad}$$

answers: beta decay; $^{0}_{-1}\beta$ particles emitted

---

2) **Curium** (atomic number 96)

Plutonium-239 is bombarded with alpha particles:

$$^{239}_{94}Pu + \underline{\quad} \rightarrow {}^{242}_{96}Cm + \underline{\quad}$$

answers: $^{4}_{2}He$, $^{1}_{0}n$

3) **Element 104 (Rf/Ku)***

Californium-249 is bombarded with carbon-12:

$$^{249}_{98}Cf + \underline{\quad} \rightarrow {}^{257}_{104}Rf + \underline{\quad}$$

answer: $^{12}_{6}C$, $4\ ^{1}_{0}n$

*Americans have suggested the name Rutherfordium honoring Ernest Rutherford. The Russians prefer the name Kurchatovium after the Russian physicist J.V.Kurchatov.

SECTION 21.4: Fission and fusion

Binding energy:

The mass of a nucleus is always less than the mass of its constituent protons and neutrons. Consider a carbon-12 atom. One carbon-12 atom weighs exactly 12.00000 amu. The mass of the nucleus is 12.00000 amu minus the mass of 6 electrons:

mass of ^{12}C nucleus = 12.00000 - 6(0.000549 amu) = 11.99671 amu

The mass of the constituents of a ^{12}C nucleus, 6 neutrons and 6 protons is:

mass of 6 protons: (6 x 1.00728 amu) = 6.04368 amu
mass of 6 neutrons: (6 x 1.00867 amu) = <u>6.05202 amu</u>
12.09570 amu

A ^{12}C nucleus weighs <u>less</u> than the sum of its parts. Mass is not conserved. The missing mass was converted into <u>energy</u> when the nucleus was formed. In order to break apart the nucleus, we would need to provide this energy (called the binding energy). We can calculate the binding energy using Einstein's equation:

$$\Delta E = \Delta mc^2$$

The binding energy for a carbon-12 nucleus:

How much energy is required to break one mole of carbon-12 nuclei into protons and neutrons?

solution:

- The difference in the mass of a carbon-12 nucleus, and the sum of its parts is called the mass defect, Δm.

$$\Delta m = m_{final} - m_{initial}$$

- In this case, since we are breaking the nucleus apart, $m_{final} > m_{initial}$ and Δm is positive. If we were calculating the binding energy, $m_{initial} > m_{final}$, and Δm would be negative.

$$\Delta m = 12.09570 \text{ amu} - 11.99671 \text{ amu} = 0.09899 \text{ amu}$$

- In one mole of carbon-12 nuclei, 0.09899 grams would be "missing";

$$\Delta m = 0.09899 \text{ g/mol} = 9.899 \times 10^{-5} \text{ kg/mol}$$

- We use Einstein's equation to solve for ΔE:

$$\Delta E = \Delta m \times c^2$$

$$= 9.899 \times 10^{-5} \text{ kg/mol} \times (2.9979 \times 10^8 \text{ m/s})^2$$

$$= 9.899 \times 10^{-5} \text{ kg/mol} \times (8.9874 \times 10^{16} \text{ m}^2/\text{s}^2)$$

$$= 8.897 \times 10^{12} \text{ kg m}^2/\text{s}^2 \text{ mol} = 8.897 \times 10^{12} \text{ J/mol}$$

$$= 8.897 \times 10^9 \text{ kJ/mol}$$

Note: Energy is required to break apart a nucleus, so ΔE is positive; the binding energy is the energy <u>released</u> when a nucleus forms, so it has a negative ΔE value.

A more convenient energy unit for binding energies is a megaelectron volt, MeV (million electron Volts).

$$1 \text{ MeV} = 1.602 \times 10^{-13} \text{ J}$$

A useful conversion factor converts a <u>mass</u> defect (in amu) directly to its <u>energy</u> equivalent in electron volts:

$$1 \text{ amu} = 931.5 \text{ MeV}$$

Conversion of mass defect into its energy equivalent:

Calculate the binding energy of a carbon-12 atom in MeV given:
a) The binding energy = -8.978 x 10^{12} J/mol
b) The mass defect, Δm = -0.09899 amu/nucleus

solution:

a) We use the equality 1 MeV = 1.602 x 10^{-13} J:

$$\left(- 8.897 \times 10^{12} \text{ J / mol} \right) \times \left(\frac{1 \text{ MeV}}{1.602 \times 10^{-13} \text{ J}} \right) = -5.554 \times 10^{25} \text{ MeV / mol}$$

b) We use the equality 1 amu = 931.5 MeV:

$$\frac{- 0.09899 \text{ amu}}{\text{nucleus}} \times \frac{931.5 \text{ MeV}}{\text{amu}} = -92.21 \text{ MeV/nucleus}$$

If we convert this to moles we obtain:

$$\frac{- 92.21 \text{ MeV}}{\text{nucleus}} \times \frac{6.02 \times 10^{23} \text{ nuclei}}{\text{mol}} = - 5.551 \times 10^{25} \text{ MeV/mol}$$

We can compare the stability of nuclides of different elements by comparing their binding energy per nucleon. For ^{12}C, we have:

$$\frac{- 92.21 \text{ MeV}}{\text{nucleus}} \times \frac{\text{nucleus}}{12 \text{ nucleons}} = - 7.684 \text{ MeV/nucleon}$$

The most stable nucleus is $^{56}_{26}$Fe. Its binding energy per nucleon is -8.790 MeV. The binding energy per nucleon increases for elements up to A≈60, and then slowly decreases. Nuclei can gain stability by forming nuclei closer to A≈60. Heavier nuclei can split into lighter nuclei (fission); lighter nuclei can combine to form heavier nuclei (fusion).

Uranium-235 can be split into fragments by neutron bombardment. Uranium-238 (the more abundant isotope) does not undergo fission. During World War II, several processes were studied to separate the two isotopes. About 84,000,000 kJ of energy is given off for every gram of U-235 that reacts. The heat of combustion of coal is about 33 kJ/g.

Fusion reactions produce 3 to 10 times as much energy per gram of starting material as fission processes. Fusion reactions produce stable isotopes rather than hazardous radioactive isotopes formed by fission. The fusion of two deuteriums forms helium:

$$^2_1\text{H} + ^2_1\text{H} \rightarrow ^4_2\text{He}$$

The fusion of only a small per cent of deuterium in seawater would supply the annual energy requirements of the world. Unfortunately, nuclear fusion requires extremely high temperatures and so is not yet a practical energy source.

Chapter 21 Exercises:

21.1) Predict the type of nuclear decay for the following nuclei:

a. $^{102}_{47}Ag$ b. $^{226}_{90}Th$ c. $^{24}_{11}Na$ d. $^{83}_{35}Br$

21.2) Complete the following nuclear equations:

a. $^{231}_{91}Pa \rightarrow ^{227}_{89}Ac + ?$

b. $^{166}_{67}Ho \rightarrow ? + ^{0}_{-1}e$

c. $? \rightarrow ^{239}_{93}Np + ^{4}_{2}He$

d. $^{59}_{29}Cu \rightarrow ? + ^{0}_{+1}e$

21.3) Complete the following nuclear equations:

a. $^{131}_{53}I \rightarrow ? + ^{0}_{-1}e$

b. $? \rightarrow ^{222}_{86}Rn + ^{4}_{2}He$

c. $^{67}_{31}Ga + ^{0}_{-1}e \rightarrow ? + x\text{-ray}$

d. $? \rightarrow ^{32}_{16}S + ^{0}_{-1}e$

21.4) Complete the following nuclear transformations:

a. $^{239}_{94}Pu + ^{4}_{2}He \rightarrow ? + ^{1}_{0}n$

b. $^{249}_{98}Cf + ? \rightarrow ^{260}_{105}Ha + 4^{1}_{0}n$

c. $^{242}_{96}Cm + 17^{1}_{0}n \rightarrow ? + 8^{0}_{-1}e$

d. $? + ^{54}_{24}Cr \rightarrow ^{262}_{107}Uns + ^{1}_{0}n$

21.5) Technetium-99m is the radioisotope of choice in diagnostic medical imaging because it is only a gamma ray emitter and has a half-life of only 6.00 hours. If a young patient was administered 40 mCi, how long would it take for 90% of the ^{99m}Tc to decay?

21.6) Gadolinium-153 is used in the determination of bone density when diagnosing osteoporosis. If a patient is administered 120 μCi of ^{153}Gd, how much ^{153}Gd remains in her system one year later?

21.1)   a. electron capture or positron decay
b. alpha particle decay
c. beta particle decay
d. beta particle decay

21.2)   a. $^{231}_{91}Pa \rightarrow ^{227}_{89}Ac + ^{4}_{2}\alpha$

b. $^{166}_{67}Ho \rightarrow ^{166}_{68}Er + ^{0}_{-1}e$

c. $^{243}_{95}Am \rightarrow ^{239}_{93}Np + ^{4}_{2}He$

d. $^{59}_{29}Cu \rightarrow ^{59}_{28}Ni + ^{0}_{+1}e$

21.3)   a. $^{131}_{53}I \rightarrow ^{131}_{54}Xe + ^{0}_{-1}e$

b. $^{226}_{88}Ra \rightarrow ^{222}_{86}Rn + ^{4}_{2}He$

c. $^{67}_{31}Ga + ^{0}_{-1}e \rightarrow ^{67}_{30}Zn + x\text{-ray}$

d. $^{32}_{15}P \rightarrow ^{32}_{16}S + ^{0}_{-1}e$

21.4)   a. $^{239}_{94}Pu + ^{4}_{2}He \rightarrow ^{242}_{96}Cm + ^{1}_{0}n$

b. $^{249}_{98}Cf + ^{15}_{7}N \rightarrow ^{260}_{105}Ha + 4^{1}_{0}n$

c. $^{242}_{96}Cm + 17^{1}_{0}n \rightarrow ^{259}_{88}Ra + 8^{0}_{-1}e$

d. $^{209}_{83}Bi + ^{54}_{24}Cr \rightarrow ^{262}_{107}Uns + ^{1}_{0}n$

21.5)   19.8 hours

21.6)   42.2 μCi remains

CHAPTER 22

The Transition Elements

TOPICS:

- Properties of Transition Elements
- The Inner Transition Elements
    - Lanthanides
    - Actinides
- Coordination Compounds
    - Geometry
    - Naming
    - Isomerism
- Bonding
    - Valence Bond Theory
    - Crystal Field Theory

REVIEW:

In Chapter 8 we learned rules for filling orbitals with electrons, and saw how electron configurations explain periodic patterns of chemical reactivity. Transition elements are the elements which have atomic d or f orbitals being filled. The d-block metals fill the 3d, 4d, and 5d shells. The Lanthanides fill the 4f shell, and the actinides (radioactive elements) fill the 5f shell. Transition metals form compounds that are often colored and paramagnetic (contain unpaired electrons).

Review exercises:

1) Show the electron configurations and orbital diagrams for Cu, Ag, Au, Cr, and Mo:

answers:

Cu: $[Ar]4s^13d^{10}$

Ag: $[Kr]5s^14d^{10}$

Au: $[Xe]6s^14f^{14}5d^{10}$

Cr: $[Ar]4s^13d^5$

Mo: $[Kr]5s^14d^5$

Remember: half filled and filled shells are unusually stable.
Generally orbitals fill in the order: ns, (n-1)d, np. However, the energies of the ns and (n-1)d sublevels are so close, deviations in filling patterns occur. Irregularities are found particularly in the 5d transition row, and in the lanthanides and actinides where the orbitals are very close in energy.

2) Write electron configurations for the Zn^{2+}, Cu^{2+}, and Cr^{3+}

   answers:

   Zn^{2+}: $[Ar]3d^{10}$ $[Ar]$ __   ↿⇂ ↿⇂ ↿⇂ ↿⇂ ↿⇂
                                           4s   3d

   Cu^{2+}: $[Ar]3d^{9}$ $[Ar]$ __   ↿⇂ ↿⇂ ↿⇂ ↿⇂ ↿
                                           4s   3d

   Cr^{3+}: $[Ar]3d^{3}$ $[Ar]$ __   ↿ ↿ ↿ __ __
                                           4s   3d

The 4s orbitals are lower in energy than the 3d orbitals when they are filled by the aufbau process, but when they form ions, transition metals appear to lose the valence shell electrons first. Except for the lanthanides, electrons with the highest n value are lost first.

Zinc loses its two 4s electrons to form Zn^{2+} with an electron configuration of $[Ar]3d^{10}$. This filled inner 3d level is a **pseudo-noble gas configuration**.

SECTION 22.1:   Properties of the Transition Metals

   Size:

   Atomic size of the transition metals in period 4 decreases slightly from scandium to iron, and then remains fairly constant (see figure 22.3, A in your text). Remember from Chapter 8 that electrons which add to the same shell (approximately the same distance from the nucleus) are not effective at shielding one another from the nuclear charge. As we move across a period then, each additional proton added to the nucleus increases the effective charge on the outer electrons and pulls them in more tightly.

   Transition metals in period 5 are larger than in period 4, but transition metals in period 6 are nearly the same size as those in period 5 (See exercise below).

Lanthanide contraction:

   The atomic radius (in Å) of the transition metals in group 4B are:

                titanium   1.44

                zirconium: 1.59

                hafnium: 1.56

Use their electron configurations to explain why zirconium in period 5 and hafnium in period 6 are nearly the same size.

solution:

Ti: $[Ar]4s^2 3d^2$

Zr: $[Kr]5s^2 4d^2$

Hf: $[Xe]6s^2 4f^{14} 5d^2$

Titanium and zirconium are the 4th elements in their periods. Hafnium is the 18th member of its period because the 15 elements La-Lu (the lanthanides) fall before it. The lanthanides steadily decrease in size with increasing nuclear charge as electrons fill the 4f sublevel. This "**lanthanide contraction**" offsets the normal increase in atomic size between periods, so transition elements in periods 5 and 6 are about the same size.

Electronegativity, Ionization Energy, and Density:

Consistent with the small changes in the size of transition metals in a period, electronegativity and first ionization energies show only small differences. Density increases with atomic mass since volume across a period is approximately constant.

Densities of gold and lead:

The density of lead is 11.3 g/mL; the density of gold is 18.9 g/mL. Why is gold denser than lead?

answer:

A lead atom (M=82 amu/atom) weighs more than a gold atom (M=70 amu/atom), so in order for the density of gold to be greater than lead, the volume of a gold atom must be smaller than the volume of a lead atom (D=m/V). The valence electrons in gold occupy the 5d and 6s sublevels. Its atomic radius is small (1.44 Å) due to lanthanide contraction. The valence electrons in lead are in the 6p sublevel giving it a larger atomic radius (1.75Å).

Oxidation States:

Transition metals often form +2 and +3 ions, but they may exist in many oxidation states. The ns and (n-1)d electrons are close enough in energy that most (or all) of these electrons can be used in bonding. The highest oxidation state of elements in groups 3B through 7B is equal to the group number. The higher oxidation states cannot exist as hydrated ions because the highly charged metal ion "pulls" electron density away from the water molecules causing them to lose protons. Chromium +3 forms the complex, $Cr(H_2O)_6^{3+}$; chromium +6 forms the oxides CrO_4^{2-} and $Cr_2O_7^{2-}$.

Oxidation Potentials:

Metals tend to act as reducing agents. All the transition metals in period 4 except for copper reduce H^+ from aqueous acid to hydrogen gas:

$$Ni(s) + 2H^+(aq) \rightarrow H_2(g) + Ni^{2+}(aq)$$

Reducing strength decreases from left to right across the table, although chromium and zinc are exceptions.

Oxidation states of transition metals:

What is the oxidation state for the transition metal in each of the following compounds?

a) CrO_4^{2-}

b) $Cr_2O_7^{2-}$

c) VO_4^{3-}

d) MnO_4^-

e) $CrCl_3$

f) $VOCl$

g) VI_2

h) Mn_2O_7

answers:

a) Cr^{6+}, b) Cr^{6+}, c) V^{5+}, d) Mn^{7+}, e) Cr^{3+}, f) V^{3+}, g) V^{2+}, h) Mn^{7+}

Notice that vanadium, in group 5B can exist in an oxidation state of +5; chromium, in group 6B can exist in an oxidation state of +6, and manganese, in group 7B can exist in an oxidation state of +7.

Colors:

Many transition metal compounds have striking colors making them fun to work with in the lab. Main group chemists, with their colorless, white (or if they're impure, perhaps yellowish-colored) compounds have been known to peer enviously at the brilliantly colored compounds in the hoods of transition metal chemists. The energy difference between partially filled d sublevels and unfilled d sublevels is the same as the energy of visible light. Ions with partially filled d shells are therefore often colored. The color depends on the metal and the groups surrounding the metal.

The many colors of transition metals:

Which of the following compounds would you expect to be colored, and which would you expect to be colorless?

a) ZnO

b) $Cu(H_2O)_6^{2+}$

c) $TiCl_4$

f) PCl_3

g) TiO_2

h) $MnSiO_3$

d) Cr_2O_3 i) Sc_2O_3

e) $CoSO_4$ j) $CoCr_2O_4$

answers:

Colored compounds (partially filled "d" orbitals):

b) Cu^{2+} has the electron configuration $[Ar]3d^9$
   $Cu(H_2O)_6^{2+}$ is blue.

d) Cr^{3+} has the electron configuration $[Ar]3d^3$
   Chromic oxide is light to dark green. It is used as a pigment in coloring glass and printing fabrics and bank notes.

e) Co^{2+} has the electron configuration $[Ar]3d^7$
   Cobaltous sulfate is red to lavender. It is used in Co pigments for decorating porcelain.

h) Mn^{2+} has the electron configuration $[Ar]3d^5$
   Manganese silicate is red. It is used as a color for special glass and for producing red glazes on pottery.

j) Co^{2+} has the electron configuration $[Ar]3d^7$
   Cobaltous chromate is a brilliant greenish-blue. It is used as a green pigment for ceramics.

Colorless compounds (full or empty "d" orbitals):

a) Zn^{2+} has a filled d shell (electron configuration $[Ar]3d^{10}$).
   Zinc oxide is white. It is used as a pigment in white paint, and on the noses of life guards to protect them from the sun.

c) Ti^{4+} has no d electrons (electron configuration $[Ar]$).
   Titanium tetrachloride is a colorless liquid. It is used in the manufacture of iridescent glass and artificial pearls.

f) P^{3+} has no d electrons (electron configuration $[Ne]3s^2$
   Phosphorus trichloride is a colorless fuming liquid.

g) Ti^{4+} has no d electrons (electron configuration $[Ar]$).
   Titanium dioxide is white. It is used as a whitener in the food industry, and as a white pigment in paints and films.

i) Sc^{3+} has no d electrons (electron configuration $[Ar]$).
   Scandium oxide is a white powder.

Note: Chromium comes from the Greek word *chroma*, meaning color.
   "Cobalt blue", the dark blue color seen in blue glass gets its color from mainly cobalt oxide and alumina.

SECTION 22.2: The Inner Transition Elements

Lanthanides:

The lanthanides, or rare earths, fill the 4f sublevel. All the lanthanides are relatively active metals. They tend to exist in the +3 oxidation state. They react with chlorine to form chloride salts (MCl_3), and with water to form hydroxides ($M(OH)_3$) and evolve hydrogen.

$$2M(s) + 3Cl_2(g) \rightarrow 2MCl_3(s)$$

$$2M(s) + 6H_2O(l) \rightarrow 2M(OH)_3(s) + 3H_2(g)$$

Lanthanides are difficult to separate from one another because of their similar properties. Until recently, the only pure sample that was commercially available was cerium, the most abundant lanthanide.

Actinides:

The actinides fill the 5f sublevel and are all radioactive. The metals are silvery and chemically reactive forming highly colored compounds. Thorium and uranium occur in nature, but the transuranium elements (elements with a molar mass greater than uranium, Z >92) have been produced in controlled nuclear reactions.

SECTION 22.3: Coordination compounds

We can think of **coordination compounds** as Lewis acid-base adducts. Remember from Chapter 17 that a Lewis acid is a species that accepts an electron pair from a Lewis base. Boron trifluoride, with an incomplete octet of electrons around boron, acts as a Lewis acid when it accepts an electron pair from ammonia:

$$:NH_3 + BF_3 \longrightarrow H_3N \rightarrow BF_3$$

When ammonia donates an electron pair to a metal ion such as Ni^{2+}, it forms a **complex ion**, and we call the Lewis base a **ligand**. The Ni^{2+} metal ion acts as a Lewis acid which accepts an electron pair from each ligand.

$$6NH_3 + Ni^{2+} \rightarrow Ni(NH_3)_6^{2+}$$

Ligands may be molecules, like ammonia, or they may be anions such as Cl⁻. Some molecules with two separate lone pairs of electrons can occupy two positions on the metal ion. Ethylenediamine (abbreviated "en") is a common **bi-dentate** ("two teeth") ligand.

$$:NH_2-CH_2-CH_2-H_2N:$$

It forms a five-membered ring around the metal ion.

EDTA (ethylenediaminetetraacetate) can occupy six positions on a metal ion. The ligands stay attached to the central metal in water. Counter ions maintain electrical neutrality and separate from the complex ion in water as an ionic salt would.

$$\overset{H_2O}{Ni(NH_3)_6Cl_2 \longrightarrow Ni(NH_3)_6^{2+} + 2Cl^-}$$

The **coordination number** is the number of ligands bonded directly to the central metal ion. The coordination number of $Ni(NH_3)_6^{2+}$ is 6.

Coordination compounds:

For each of the following coordination compounds, give the charge of the complex ion, the charge of the central atom, and list the ligands and counter ions.

a) $[Co(NH_3)_6]Cl_3$

b) $[Co(NH_3)_5Cl]Cl_2$

c) $[Co(NH_3)_4Cl_2]Cl$

d) $[Co(NH_3)_3Cl_3]$

answers:

Ligands are written within the brackets; species written outside the brackets are not bonded directly to the central metal atom, and are present as free ions.

a) The NH_3 molecules are neutral and the chloride ions have a charge of -1.

   charge of complex = +3; charge of Co = +3
   ligands: 6 NH_3 molecules; counter ions: 3 Cl^- ions

b) charge of complex = +2; charge of Co = +3
   ligands: 5 NH_3 molecules and 1 Cl^- ion; counter ions: 2 Cl^- ions

c) charge of complex = +1; charge of Co = +3
   ligands: 4 NH_3 molecules and 2 Cl^- ions; counter ion: 1 Cl^- ion

d) charge of complex = 0; charge of Co = +3
   ligands: 3 NH_3 molecules and 3 Cl^- ions; counter ions: none

The "discovery" of coordination compounds:

When the above cobalt compounds were discovered they were written:

$CoCl_3 \cdot 6\ NH_3$ (orange-yellow) $CoCl_3 \cdot 4\ NH_3$ (green)

$CoCl_3 \cdot 5\ NH_3$ (violet) $CoCl_3 \cdot 3\ NH_3$ (green)

Ammonia was apparently tightly bound to cobalt since reaction with HCl did not remove the ammonia. Reaction with silver nitrate precipitated all the chlorine from compound 1 as AgCl, 2/3 of the chlorine from compound 2, and only 1/3 of the chlorine from compound 3. When compound 4 was treated with silver nitrate, no precipitate formed.

Alfred Werner proposed in 1893 the idea of coordination compounds. He suggested that all of the compounds above have six ligands attached directly to the cobalt ion (written within square brackets). Only the Cl^- ions outside the coordination sphere (the counter ions) are precipitated by silver nitrate.

He tested his hypothesis by studying the conductivities of water solutions of the four compounds. As predicted, he found that compounds 1-3 behaved as 3:1, 2:1, and 1:1 electrolytes respectively, and compound 4 behaved as a nonelectrolyte.

Naming Coordination Compounds:

The name for $[Co(NH_3)_5Cl]Cl_2$ is pentaamminemonochlorocobalt(III) chloride. OK, it's not short, but it does the job. Break the name up into parts, and it won't seem so cumbersome:

1) First name the cation, then the anion.

   The same rule is used to name sodium chloride. In the example , the cation is not sodium but, pentaamminemonochlorocobalt(III)

2) Name the ligands (in alphabetical order) before the metal in the complex ion.

   The ligands are named before cobalt.

3) To name the ligands:

   a) Use the name of the molecule for neutral ligands.
      Exceptions are: <u>aquo</u> for H_2O; <u>ammine</u> for NH_3; <u>carbonyl</u> for CO; <u>nitrosyl</u> for NO

      We use <u>ammine</u> for the ammonia ligands

   b) Add "o" to the root name of anions.

      We use <u>chloro</u> for the Cl^- ligand

c) Prefixes denote the number of a particular ligand.
   mono-, di-, tri-, tetra-, penta-, and hexa- for simple ligands
   bis-, tris-, tetrakis- for ligands which already contain a numerical prefix
   The prefixes do not affect the alphabetical order.

The ligands are pentaammine (5 ammonia molecules) and monochloro (one Cl^- ion). Ammine is listed alphabetically before chloro.

4) Use a Roman numeral in parentheses to designate the oxidation state of the central metal ion.

Cobalt is in the +3 oxidation state, so (III) appears after the name cobalt.

5) If the complex ion is an anion, add the suffix -ate to the name of the metal.

The complex ion is a cation, so the metal name appears unchanged as cobalt.

Naming Coordination Compounds:

Name the following compounds:

1) $[Zn(H_2O)_3(OH)]NO_3$ 3) $[Ni(en)_3]Cl_2$

2) $K[Pt(NH_3)_3Cl_3]$ 4) $[Cu(NH_3)_4]SO_4$

answers:

1) Water is a neutral ligand; hydroxide, OH^- and NO_3^- each have -1 charges, so Zn has a +2 oxidation state. The name is:

   Triaquomonohydroxozinc(II) nitrate

2) Potassium ion has a +1 charge; the ammonia ligands are neutral, and the Cl^- ions have -1 charges, so platinum has a +2 oxidation state. The name is:

   Potassium triamminetrichloroplatinate (II)

3) Ethylenediammine is neutral; the Cl^- ions have -1 charges, so nickel has a +2 oxidation state. The name is:

   Tris-ethylenediamminenickel(II) chloride

4) The ammonia ligands are neutral, and the sulfate ion has a -2 charge, so copper has a +2 oxidation state. The name is:

   Tetraamminecopper(II) sulfate

Geometry and isomerism:

The geometry of coordination compounds is summarized in the table below.

Coordination number	Geometry
2	Linear
4	square planar or tetrahedral
6	octahedral

The complexes of platinum are square planar; the complexes of zinc with four ligands are tetrahedral. **Geometrical isomers** can exist for square planar complexes but not for tetrahedral complexes. Consider the square planar complex $[Pt(NH_3)_2Cl_2]$. Two forms of this compound exist:

Cis   Trans

Cis-$Pt(NH_3)_2Cl_2$ ("cisplatin") has identical ligands next to each other. It is a highly effective antitumor agent. *Trans*-$Pt(NH_3)_2Cl_2$ has identical ligands across from each other and is an ineffective antitumor agent.

Recall from Chapter 10 that six groups around a central atom form an octahedral arrangement. All 6 positions are equivalent and can be visualized as a derivative of a square planar complex:

Geometrical isomerism can occur in octahedral complexes. The compound, $[Co(NH_3)_4Cl_2]Cl$ can exist as *cis* and *trans* isomers. The *trans* isomer has chloro ligands opposite each other and is green as described earlier. The *cis* isomers has chloro ligands next to each other and is a violet color.

Cl
$H_3N\ _{////,}$ |,,,,,,NH_3
H_3N | Cl
|
NH_3

Cis

Cl
$H_3N\ _{////,}$ |,,,,,,NH_3
H_3N | NH_3
|
Cl

Trans

Alfred Werner, who recognized the possibility of isomers, was able to isolate the *cis* isomer.

The number of isomers:

How many isomers are possible for the complex $[Co(NH_3)_3Cl_3]$?

<u>solution</u>:

We can simplify the drawing of isomers by using the general form XA_3B_3 and consider only the A ligands. First, if we make two A ligands *trans* to each other, all the four remaining positions are equivalent, and we have our first isomer:

If we move one of the A ligands that are *trans* to each other so that all three ligands are *cis*, we come up with a different isomer:

With a little "fiddling", you should be able to convince yourself that these are the only two isomers. Remember that isomers drawn on paper, may look different, but may actually be equivalent. The two compounds drawn below are the same isomer (Two groups *trans* with the third group *cis* to each):

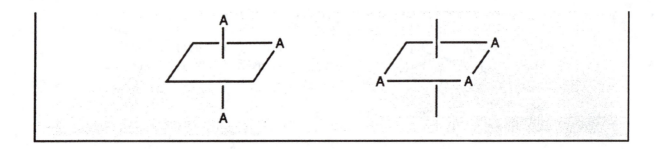

Optical isomers occur when a molecule and its mirror image cannot be superimposed. Optical isomers are identical except for the direction which they rotate the plane of polarized light. If a molecule is superimposable on its mirror image, it is not optically active. This is often difficult to see without making models and trying to superimpose them. If you can find a plane of symmetry in a molecule (one that divides a molecule into two equal parts), the molecule is not optically active. Figure 22.13 in your book shows the *cis* and *trans* isomers of $[Co(en)_2Cl_2]^+$. The *cis*-isomer is optically active; the *trans*-isomer is not. can you find the plane of symmetry in *trans*-$[Co(en)_2Cl_2]^+$?

SECTION 22.4: Bonding in Complex Ions

The valence bond and molecular orbital theories discussed in Chapter 10 can describe the electronic structures and geometries of complex ions. Another model called **crystal-field theory** is more successful at predicting magnetic properties and color of complex ions.

Valence bond model:

Remember that ligands in complex ions act as Lewis bases and donate electron pairs to metal ions (Lewis acids). In the valence bond model, the electron pairs on the ligands enter hybrid orbitals on the central atom. The hybrid orbitals have the correct geometry to form linear, tetrahedral, square planar, and octahedral complexes.

Hybrid orbital	Coordination Number	Shape	Example
sp	2	linear	$[Ag(NH_3)_2]^+$
sp^3	4	tetrahedral	$[Pt(NH_3)_4]^{2+}$
dsp^2	4	square planar	$[Zn(NH_3)_4]^{2+}$
d^2sp^3	6	octahedral	$[Cr(NH_3)_4Cl_2]^+$

You have seen all these hybrid orbitals before in Chapter 10 except for the square planar, dsp^2 orbital. Let's look at an example where the valence bond model successfully predicts the magnetic behavior of a complex ion, $Fe(CN)_6^{4-}$.

The coordination number of $Fe(CN)_6^{4-}$ is 6, so the the bonding electrons from the CN^- ligands enter octahedral d^2sp^3 orbitals. How many electrons does the iron atom contribute? The charge on the complex is -4, and each CN^- ligand contributes a -1 charge, so the charge on iron is +2 (electron configuration $[Ar]3d^6$). The 6 electrons from iron remain in the three unhybridized d orbitals:

metal e^- ligand e^-

3d   4s   4p

equivalent d^2sp^3 orbitals

Since all electrons are paired, the valence bond theory would predict that this complex is diamagnetic, which experiments confirm.

Orbital diagram for $Fe(H_2O)_6^{2+}$:

What is the orbital diagram for the complex ion, $Fe(H_2O)_6^{2+}$?

solution:

The orbital diagram should be identical to $Fe(CN)_6^{4+}$ since $Fe(H_2O)_6^{2+}$ has six ligands bonded to a Fe^{2+} ion. We would predict, based on the valence bond model, that $Fe(H_2O)_6^{2+}$ would have all electrons paired and be diamagnetic.

Now we have a problem. Experiments show $Fe(H_2O)_6^{2+}$ to be paramagnetic with four unpaired electrons. The valence bond theory does not hold up for this complex ion. We need a theory which will predict the magnetic properties of complex ions consistently, and one which can explain the most striking visual feature of complex ions, their color.

Crystal Field Theory:

The crystal field model assumes that electrostatic forces between the ligands and metal atom hold complex ions together. In this model, when ligands approach the central metal ion, their electrons do not fill hybrid orbitals on the metal ion. Instead, the electrostatic interaction of ligand electrons with electrons on the metal ion, split the d orbitals on the metal ion into different energy levels.

In Chapter 7 we showed the five orientations of the d-orbitals. Two of the orientations ($d_{x^2-y^2}$ and d_{z^2}) lie along the x,y, and z axes:

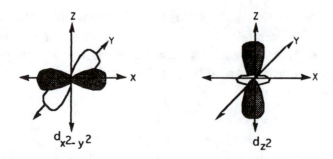

The other three orientations (d_{xy}, d_{yz}, and d_{xz}) lie <u>between</u> the x,y, and z axes:

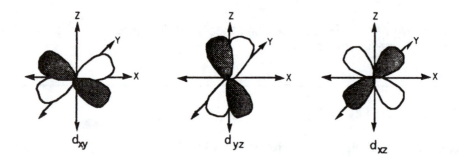

    When ligands approach a metal ion <u>along</u> x,y, and z axes, as they do for an octahedral complex, they interact most strongly with electrons in metal orbitals that lie along the axes (the $d_{x^2-y^2}$ and d_{z^2} orbitals). The ligands interact less strongly with electrons in metal orbitals which lie between x, y, and z axes, (d_{xy}, d_{yz}, and d_{xz} orbitals), so these orbitals are lower in energy.

    The amount that the d orbitals split (called the **crystal field splitting energy, Δ**) depends on how strongly the ligands interact with the metal ion electrons. Some ligands, such as CN^-, interact strongly with metal ion electrons; others, such as water, interact less strongly, and split the d orbitals less. Now we have a way to explain the color differences of complex ions.

The extent of splitting determines the wavelength of light absorbed, and therefore the color that we see when we look at a complex ion. A large split in the d orbitals absorbs a shorter wavelength of light (higher energy); a smaller split absorbs a longer wavelength of light (lower energy).

The order of ligands in their increasing ability to split the d orbitals of a metal ion is called the spectrochemical series:

$$Cl^- < F^- < H_2O < SCN^- < NH_3 < en < NO_2^- < CN^- < CO$$

weak field ligands   strong field ligands

Let's see how the spectrochemical series can explain why $Fe(CN)_6^{4+}$ is diamagnetic and $Fe(H_2O)_6^{2+}$ is paramagnetic with four unpaired electrons. CN^- is a strong field ligand and splits the d orbitals farther apart than H_2O which is a weaker field ligand. When the split in d orbitals is large, it takes less energy to pair electrons in the lower energy d orbitals than to fill the higher energy orbitals with unpaired electrons:

3d orbitals with ligands interacting (average energy)

d_{z^2} $d_{x^2-y^2}$

large Δ

d_{xy}, d_{xz}, d_{yz}

3d orbitals split by six strong field ligands

LOW-SPIN COMPLEX

The six Fe^{2+} electrons are paired in the lower energy d orbitals. This type of complex is called a **low-spin complex**. Notice that in this complex, Hund's rule is violated because electrons pair before all d orbitals are filled with unpaired electrons.

H_2O is a weaker field ligand than CN^- and splits the d orbitals to a smaller extent. In this case, it requires more energy to pair electrons than to fill a higher energy d orbital with an unpaired electron:

3d orbitals with
ligands interacting
(average energy)

d_{z^2} $d_{x^2-y^2}$

small Δ

d_{xy}, d_{xz}, d_{yz}

3d orbitals split
by six weak field
ligands

HIGH-SPIN COMPLEX

Only two of the six Fe^{2+} electrons are paired, and this theory successfully predicts that $Fe(H_2O)_6^{2+}$ has 4 unpaired electrons. This type of complex is called a **high-spin   complex**; Hund's rule is not violated in high-spin complexes.

In tetrahedral complexes, ligands approach the central metal ion <u>between</u> the x, y, and z axes, so the splitting of the d orbitals is the reverse of the octahedral case: the d_{xy}, d_{yz}, and d_{xz} orbitals become higher in energy than the $d_{x^2-y^2}$ and d_{z^2} orbitals. The splitting of the d orbitals is less than in octahedral complexes, so only high-spin tetrahedral complexes are known.

Square planar complexes split the d orbitals into 4 levels as shown on p. 1003 in your text. The orbitals that interact most strongly with ligands in a square planar complex are those in the xy plane (the $d_{x^2-y^2}$ and d_{xy} orbitals).

Crystal field theory successfully predicts the color and magnetic behavior of complex ions, but it cannot explain why, for example, CO, which is only slightly polar, is a strong field ligand if only electrostatic forces are in effect. **Ligand field theory** includes a covalent contribution in the description of bonding in complex ions in order to explain this result.

Chapter 22 Exercises:

22.1)   According to a dictionary, what does the stem of the word "chromium" mean?

22.2)   a. What are the highest oxides of Sc, Ti, V, Cr, and Mn?

b. In terms of the formulas of their highest oxides, Sc, Ti, V, Cr, and Mn are like which elements in the second row of the p block?

c. In what groups are Sc, Ti, V, Cr, and Mn?

22.3)   Match the following:

1. Red rouge   a. TiO_2

2. Chrome oxide green   b. Cr_2O_3

3. White paint pigment   c. Fe_3O_4

22.4)   Before recognition of the f-block, uranium was placed in the same family as what elements?

22.5) OsO_4 corresponds to what compound in the p block?

22.6)   Draw an orbital diagram to describe the crystal field splitting for the Ni^{2+} ion in $[Ni(en)_3]Cl_2$. Would you expect Δ to be relatively large or small? Would you expect the compound to be paramagnetic or diamagnetic?

22.7)   How many isomers are possible for $Co(en)_2(NH_3)(H_2O)$? Are any of the isomers optically active?

Chapter 22 Answers:

22.1) Color. Most compounds of chromium are colored.

22.2) a. Sc_2O_3; TiO_2; V_2O_5; CrO_3; Mn_2O_7
b. Al, Si, P, S, and Cl
c. 3,4,5,6,7

22.3) 1-c; 2-b; 3-a.

22.4) Cr, Mo, and W (based on its maximum oxidation state)

22.5) XeO_4

22.6)

E

d_{z^2} $d_{x^2-y^2}$

Δ

3d orbitals with
ligands interacting
(average energy)

d_{xy}, d_{xz}, d_{yz}

3d orbitals split
by six ligands

Free ion d orbitals

The "en" ligand is a relatively strong field ligand, so Δ is relatively large, and the compound is low spin. It has two unpaired electrons making it paramagnetic.

22.7)

H₃N, Co, N, H₂O, N

N, Co, NH₃, N, OH₂

H₂O, N, Co, N, NH₃

optical isomers

no plane of symmetry

non-superimposable mirror images

CHAPTER 23

Industrial Chemistry: A Case Study

INTRODUCTION:

Most industrial processes rely on combined expertise in chemistry, biochemistry, physics, mathematics, and engineering to successfully develop and manufacture products. Industrial chemists often find themselves as members of interdisciplinary research groups comprised of members with diverse scientific backgrounds (and personalities). As an example of a process which requires expertise from many areas, we will discuss the development and manufacture of photographic film.

A cross section of photographic film has the general structure:

A flexible support (the polyester base) is coated with the light sensitive silver halide crystals suspended in a gelatin-based matrix (the emulsion layer). This "layer" may be made of up to 15 separate layers in specialty color films. The overcoat is often a hardened gelatin-based layer which protects the emulsion layers from scratches and abrasions. It may contain polymer beads as matte particles to facilitate smooth transport of film in camera equipment.

It is an engineering accomplishment to coat emulsion and overcoat layers simultaneously at speeds of 300 ft./min., or faster, without coating defects, and at a uniform coating weight across the width of the coating. Furthermore, continuous drying of the film requires not only engineering skill, but an understanding of the kinetics of the film matrix hardening reactions that occur during the coating and drying processes.

Coating weight reduction:

The silver coating weight of your standard product is 4.2 g/m^2. You discover a way to reduce the thickness of the silver halide grains, and so reduce the coating weight of the silver to 3.8 g/m^2. If the price of silver is \$72 per pound, how much money will you save your company per square foot of film?

<u>solution</u>:

- The cost per <u>gram</u> of silver is:

$$\frac{\$72}{lb} \times \frac{1\ lb}{454\ g} = \$0.16/g$$

- The initial coating weight in g/ft^2 and cost per square foot is:

$$\frac{4.2\ g}{m^2} \times \left(\frac{1\ m}{100\ cm}\right)^2 \times \left(\frac{2.54\ cm}{1\ in}\right)^2 \times \left(\frac{12\ in}{1\ ft}\right)^2 = 0.390\ g/ft^2$$

$$\frac{0.390\ g}{ft^2} \times \frac{\$0.16}{g} = \$.062/ft^2$$

- The reduced coating weight in g/ft^2 and cost per square foot is:

$$\frac{3.8\ g}{m^2} \times \left(\frac{1\ m}{100\ cm}\right)^2 \times \left(\frac{2.54\ cm}{1\ in}\right)^2 \times \left(\frac{12\ in}{1\ ft}\right)^2 = 0.353\ g/ft^2$$

$$\frac{0.353\ g}{ft^2} \times \frac{\$0.16}{g} = \$.057/ft^2$$

- The savings is:

$$\$.062/ft^2 - \$.057/ft^2 = \$.005,\ \text{about half a cents per square foot.}$$

To put this in context, for a product sales volume of 130 million sq. ft. per year, this represents a savings of $650,000.00/year.

BASE:

The polyester support (or base) on which the emulsion is coated is a polymer sheet which is flexible (to allow the film to bend through a processor), strong (to withstand relatively high tensions used during coating), and dimensionally stable. The base may be clear, or it may be tinted with a color. Medical X-ray film base is often tinted blue to help give the final image a blue-black color that radiologists prefer.

The polyester base is a hydrophobic material. An aqueous coating solution will neither wet nor adhere to the base. Typically, a submicron layer of a saran polymer coating (the subbing) provides a bridge between the hydrophobic base and the hydrophilic binder, binding with each.

GELATIN MATRIX:

Gelatin, a familiar biopolymer, has been used as a binder for silver halide emulsions for over a hundred years. Gelatin is produced from collagen (a protein found in the bones and tissue of animals) by either acid or base digestion. Collagen has a helical structure of polypeptide chains which is stabilized by hydrogen bonds. Gelatin retains some of this three dimensional structure. Since gelatin is derived from a natural product, it is a challenge to consistently produce batches of gelatin with identical properties. Gelatin must be highly purified for photographic use since the photographic response of silver halide crystals is highly dependent on the nature of the binder that surrounds the crystals.

No synthetic polymers have been found which can effectively replace gelatin. The ability of gelatin to chill set, or gel upon cooling, is crucial in film manufacture. Gelatin swells in aqueous solution to allow penetration of processing solutions. Gelatin contains reactive side groups on the polymer chains which can be chemically cross-linked to harden the coating, and gelatin acts as a protective colloid of the silver halide microcrystals and prevents their coalescence .

Gelatin has its drawbacks. It exerts extreme pressure as it dries and can cause films to curl. Other binder components are added to counter these effects and adjust the physical properties of the gelatin binder. Latex particles can be added to the emulsion to impart flexibility and shock absorption qualities. Humectants, which prevent the film from drying out and becoming brittle, are often added. Surfactants are used to improve material compatibility and coatability.

Gelatin is hardened by chemically crosslinking reactive side groups. One type of crosslinking is amine-amine crosslinking in which amine groups are linked by a hardening agent such as formaldehyde (example 1 below, where X = an amine-amine crosslinker). Another type of crosslinking, peptide coupling, involves reaction of a gelatin amine group with a carboxyl group (example 2 below, where PC = peptide coupler) to form a peptide bond.

Crosslinking of gelatin:

If there are 0.4 mmol of amine groups in 1 g of gelatin, how many grams of formaldehyde (CH_2O) would you need to add to crosslink 25% of the amine groups in 200 g of gelatin?

It requires 1 molecule of formaldehyde to crosslink 2 amine groups:

$$0.25 \times 200 \text{ g gel} \times \left(\frac{0.4 \text{ mmol amine}}{1 \text{ g gel}}\right) \times \left(\frac{1 \text{ mmol } CH_2O}{2 \text{ mmol amine}}\right) \times \left(\frac{30 \text{ mg } CH_2O}{\text{mmol}}\right) = 300 \text{ mg } CH_2O$$

SILVER HALIDE GRAINS:

Silver halide microcrystals are solid state devices; when the microcrystals are exposed to light, photoelectrons are generated. The valence and conduction bands of the microcrystal allow these photogenerated electrons to move throughout the crystal. If the silver halide grain is properly sensitized, these photogenerated electrons create catalytic centers (called "latent image" sites) which catalyze reduction of the crystal in developer solution.

Silver chloride, silver bromide, and silver iodide are all used in silver halide crystals in film. Often the best crystal for use in photographic film is comprised of a mixture of halides.

Solubility of silver halides:

Given the following K_{sp} values:

$$\text{AgI} \quad K_{sp} = 8.49 \times 10^{-17}$$
$$\text{AgBr} \quad K_{sp} = 5.32 \times 10^{-13}$$
$$\text{AgCl} \quad K_{sp} = 1.76 \times 10^{-10}$$

order silver iodide, silver bromide, and silver chloride from most soluble to least soluble.

<u>solution</u>:

The smallest K_{sp} indicates the least soluble compound:

$$\text{AgCl} > \text{AgBr} > \text{AgI}$$

Silver nitrate reacts with potassium bromide to produce light sensitive silver bromide crystals (or "grains", as they are referred to in the field). The grains are produced in a kettle containing a gelatin/water mixture and excess potassium bromide at around 60°C.

$$Ag^+(aq) + NO_3^-(aq) + K^+(aq) \; Br^-(aq) \;\; \longrightarrow \;\; AgBr(s) + K^+(aq) + NO_3^-(aq)$$

The temperature, the flow rates of silver nitrate and potassium bromide, the bromide ion concentration, and the gelatin concentration all determine the size and shape of the resulting silver bromide grains. In particular, the bromide ion concentration determines the shape (or "morphology") of the grains. Silver halide precipitations are usually carried out under excess halide ion concentration. As halide ion increases, silver halide exists in solution mainly as complexes such as $AgBr_2^-$, and $AgBr_3^{2-}$. These complexes influence the crystal morphology. At lower bromide ion concentration, cubic grains form. As the bromide ion concentration increases, the morphology of the grains changes from cubic to tetradecahedral (cubes with their corners truncated), to octahedral, to tabular.

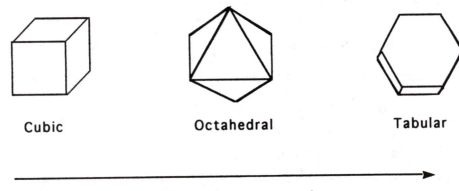

| Cubic | Octahedral | Tabular |

Increasing bromide ion concentration

Ammonia is sometimes used during part, or all, of the precipitation to increase solubility by formation of the complex ion $Ag(NH_3)_2^+$. The increase in solubility increases the size of the silver halide crystals and may make the grain size distribution more uniform as smaller grains dissolve and redeposit onto larger grains.

pAg	Morphology
<8.5	cubes
8.5-9.5	tetradecahedra
9.5-10.5	octahedra
>10.5	tabulars, irregular

pAg and crystal morphology:

The K_{sp} of silver bromide at 60°C is 1.58×10^{-11}. If a patent discloses that their process begins precipitation at a pAg of 8.5, what bromide ion concentration should you have in the kettle at the start of precipitation?

solution:

- For the equilibrium at 60°C,

$$AgBr(s) \rightleftharpoons Ag^+(aq) + Br^-(aq)$$

$$K_{sp} = [Ag^+][Br^-] = 1.58 \times 10^{-11}$$

- We can take -log of each side of the K_{sp} equation to obtain:

$$-\log K_{sp} = -\log[Ag^+] + -\log[Br^-]$$

$$pK_{sp} = pAg + pBr$$

- Substituting in for pK_{sp} and pAg gives:

$$10.8 = 8.5 + pBr$$

$$pBr = 10.8 - 8.5 = 2.3$$

$$[Br^-] = 10^{-2.3} = 5.01 \times 10^{-3} M$$

Silver ions can form water-soluble complex ions with many inorganic and organic compounds. This property explains the ability of the insoluble silver salts to dissolve in some aqueous solutions. As you will see later, the formation of a silver thiocyanate complex allows silver halide the be dissolved from the film during the fixing stage of processing.

Complex ion formation:

The dissociation constant for the silver complex, $Ag(NH_3)_2^+$ at 25°C is 6.2×10^{-8}. What is the formation constant for this complex?

solution:

The formation constant, or "stability" constant for $Ag(NH_3)_2^+$, is the reciprocal of the dissociation constant:

$$Ag(NH_3)_2^+ \rightleftharpoons Ag^+ + 2NH_3 \quad \text{(dissociation)}$$

$$K = \frac{[Ag^+][NH_3]^2}{[Ag(NH_3)_2^+]} = 6.2 \times 10^{-8}$$

$$Ag^+ + 2NH_3 \rightleftharpoons Ag(NH_3)_2^+ \quad \text{(formation)}$$

$$K = \frac{[Ag(NH_3)_2^+]}{[Ag^+][NH_3]^2} = \frac{1}{6.2 \times 10^{-8}} = 1.6 \times 10^7$$

Silver bromide crystallizes in a face-centered cubic structure with alternating silver and halide ions.

The alternating arrangement of ions on the surface of this cubic crystal is a 100 surface, or a "cubic surface". A second important type of crystal surface is the 111, or octahedral surface, which we obtain if we slice diagonally through the cubic structure shown above. 100 surfaces tend to be less reactive than the 111 surface because the ions are bound more symmetrically. The adsorption of sensitizing materials such as sensitizing dyes, discussed below, may be different on the two types of surfaces.

The AgBr unit cell:

1) The ionic radii of silver and bromide are given by Pauling as: rAg = 1.26 Å, rBr = 1.95 Å. What would the dimension of the unit cell for silver bromide be based on these values?

We can show one face of a unit cell for a silver bromide crystal as:

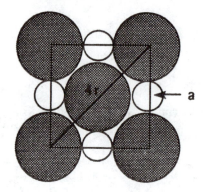

where "a" is the unit cell dimension. "a" is twice the radius of a bromide ion plus the diameter (twice the radius) of a silver ion:

$$a = (2 \times rAg) + (2 \times rBr)$$

$$a = (2 \times 1.26 \text{ Å}) + (2 \times 1.95 \text{ Å})$$

$$a = 6.42 \text{ Å}$$

2) The measured unit cell dimension for silver bromide is 5.77 Å. How can you explain the discrepancy between this measured value and our calculated value above (6.42 Å)?

solution:

The hard sphere model does not apply accurately to the silver halides because the bonding in silver halides is not strictly ionic.

3) Using 5.77 Å as the value for the unit cell dimension for silver bromide, are the bromide ions touching along the diagonal of the face of the face-centered cube (drawn above)?

solution:

• We can find the length of the diagonal using the Pythagorean theorem:

$$a^2 + b^2 = c^2$$

where a = b = 5.77.

• Solving for c gives:

$$c = 4r = \sqrt{(5.77)^2 + (5.77)^2}$$

$$c = 4r = 8.16$$

$$r = 2.04$$

• The radius of a bromide ion is 1.95 Å, so the bromide ions do not touch along the diagonal face of the unit cell.

4) If you measure a tabular silver halide grain to have a thickness of 0.15 μ, how many unit cells thick is the grain?

solution:

$$0.15 \; \mu x \; \frac{10^4 \text{ Å}}{1 \, \mu} \; x \; \frac{\text{unit cell}}{5.77\text{Å}} = 260 \text{ unit cells}$$

Crystal defects such as twin planes and dislocations, which are interruptions in the regular crystal lattice, and surface kink sites, which are surface irregularities, are very important in determining the response of the crystal to sensitization. The number of "interstitial silver ions" - ions which leave lattice positions and move interstitially through the crystal - also play an important role in the ability of a silver halide crystal to be sensitized.

SENSITIZATION:

Chemical sensitization:

Photographic sensitivity can be increased with chemical treatment of the silver halide grains. Sulfur and gold are common sensitizing agents. Sulfur, often in the form of thiosulfate, adsorbs to silver bromide and reacts to produce a silver sulfide speck on the grain surface.

$$S_2O_3^{2-} + 2AgBr + H_2O \rightarrow Ag_2S + 2Br^- + SO_4^{2-} + 2H^+$$

The Ag_2S acts as a trap for electrons and is the site of latent image formation discussed below. If sensitization is carried too far, the sulfide particles may catalyze development without exposure.

Oxidation states of sulfur:

For the reaction of thiosulfate with silver bromide to produce silver sulfide shown above, what are the oxidation states of sulfur in each of the compounds?

solution:

In $S_2O_3^{2-}$, three oxygen atoms with an oxidation number of -2 give a -6 charge. The charge on the anion is -2, so the sulfur atoms must contribute a +4 charge. Each sulfur atom has an oxidation number of +2.

In Ag_2S, each silver atom has a +1 oxidation number, so sulfur has an oxidation number of -2.

In SO_4^{2-}, four oxygen atoms with oxidation numbers of -2 give a -8 charge. The charge on the anion is -2, so the sulfur atom must contribute a +6 charge.

One sulfur atom in thiosulfate loses four electrons as it is oxidized to an oxidation state of +6; the other sulfur atom gains four electrons as it is reduced to an oxidation state of -2.

Spectral sensitization:

Silver halide crystals absorb blue and ultraviolet light. Color films form images from light absorbed primarily by spectral sensitizers. Spectral sensitizers are organic dyes that adsorb to the surface of the grain and tune it to the incident wavelength of light. Spectral sensitization occurs by electron transfer from a dye to the silver halide grain. If the conduction band of a dye is at the right energy level in relation to the conduction band of the silver halide crystal, it can inject an electron into the conduction band.

A dye must adsorb to the silver halide surface correctly to be an effective spectral sensitizer. Some dyes may be effective on a 100 (cubic) surface, but ineffective on a 111 (octahedral) surface. Many dye molecules can aggregate on the surface which changes their adsorption wavelength.

If not all the sensitizing dye is removed during processing, dye stain results. Organic chemists work to change dye structures so that the dye molecules diffuse readily from the emulsion during processing, or so the dye molecules react with an ingredient in the processing solution to become decolorized.

DEVELOPMENT AND FIX:

A photon ($h\upsilon$) of the right energy generates a mobile photoelectron in a silver halide crystal which reduces interstitial Ag^+ ions in the crystal to Ag^0. A cluster of a minimum of three silver atoms forms a "latent image". On development, the latent image catalyzes reduction of the rest of the grain to Ag^0. Development of a silver halide grain is a kinetic process: given enough time in developer, even the unexposed grains will eventually be reduced.

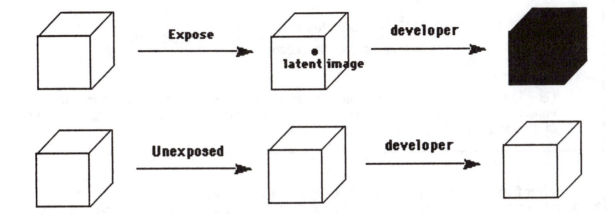

Amplification on development:

If a cubic grain, volume = 0.125 μ^3, contains a latent image of three silver atoms, what is the amplification factor on development (reduction) of the grain?

<u>solution</u>:

- The measured unit cell dimension for silver bromide is 5.77 Å (earlier exercise), so the volume of unit cell is: $(5.77 \text{ Å})^3 = 192 \text{ Å}^3$

- The volume of the grain in Å^3 is:

$$\text{volume} = \left(0.125 \ \mu^3\right) \times \left(\frac{10^4 \ \text{Å}}{1 \ \mu}\right)^3 = 1.25 \times 10^{11} \text{ Å}^3$$

- The number of unit cells in the cubic grain is therefore:

$$1.25 \times 10^{11} \text{ Å}^3 \times \frac{\text{unit cell}}{192 \ \text{Å}^3} = 6.51 \times 10^8 \text{ unit cells}$$

- Each unit cell contains 4 silver atoms:

$$\left(8 \text{ corners} \times \frac{1}{8}\right) + \left(6 \text{ faces} \times \frac{1}{2}\right) = 4 \text{ silver atoms/unit cell,}$$

so the number of silver atoms in the grain is:

$$6.51 \times 10^8 \text{ unit cells} \times \frac{4 \text{ Ag atoms}}{\text{unit cell}} = 2.60 \times 10^9 \text{ Ag atoms}$$

- The amplification factor on development is:

$$\frac{2.60 \times 10^9 \text{ Ag atoms after development}}{3 \text{ Ag atoms before development}} = 8.68 \times 10^8$$

(This is why the "relatively old" technology of silver halide photographic film remains important today. No other storage media (magnetic tape, disks, etc.) possesses this level of data amplification and data storage density.)

After film is developed, it is "fixed" to wash off undeveloped silver halide. A silver halide solvent such as thiosulfate solubilizes the silver halide allowing it to wash off the film.

$$AgBr(s) + 2S_2O_3{}^{2-}(aq) \rightarrow Ag(S_2O_3)_2{}^{3-}(aq) + Br^-(aq)$$

Now the customer has his photographic image. With a little luck, the image is clear, without defects, and the film will age well (won't lose its image over time, or turn black), and the customer is satisfied. If not, well, you'll probably hear about it in the form of a complaint - and then it's back to the lab to work out the "kinks". And so goes the challenge -and fun - of research.